KU-345-648

ANDERSONIAN LIBRARY
WITHDRAWN
FROM
LIBRARY
STOCK
UNIVERSITY OF STRATHCLYDE

THERMAL METHODS OF OIL RECOVERY

EXXON MONOGRAPHS

BIMETALLIC CATALYSTS: DISCOVERIES, CONCEPTS, AND APPLICATIONS
Sinfelt

A MODEL-MANAGEMENT FRAMEWORK FOR MATHEMATICAL PROGRAMMING
Palmer

STATISTICAL ANALYSIS OF MEASUREMENT ERRORS
Jaech

THERMODYNAMIC AND TRANSPORT PROPERTIES OF COAL LIQUIDS
Tsonopoulos, Heidman, and Hwang

PHYSICS OF COMPLEX AND SUPERMOLECULAR FLUIDS
Safran and Clark

THERMAL METHODS OF OIL RECOVERY
Boberg

WELL TESTING IN HETEROGENEOUS FORMATIONS
Streltsova

THERMAL METHODS OF OIL RECOVERY

THOMAS C. BOBERG

An Exxon Monograph

JOHN WILEY & SONS
New York • Chichester • Brisbane • Toronto • Singapore

***Library of Congress Cataloging in Publication Data*:**

Boberg, Thomas C.
Thermal methods of oil recovery/Thomas C. Boberg.
p. cm.

"An Exxon monograph."
Includes indexes.
ISBN 0-471-63300-3
1. Thermal oil recovery. I. Title.
TN871.B518 1987
622'.3382--dc19

87-22936
CIP

Printed in the United States of America

10 9 8 7 6 5 4 3 2 1

To my co-workers in the area of thermal oil recovery, both within and outside of Exxon

FOREWORD

A large fraction of the free world's oil reserves is recoverable only by the application of thermal enhanced recovery methods. Exxon Production Research Company and other Exxon affiliates, particularly Esso Resources Canada Limited, have devoted considerable effort over the past thirty years to development of practical and economic thermal recovery processes. One of the industry's earliest steamfloods was conducted by the Carter Oil Company, an Exxon affiliate, in the 1950s at Deerfield, Missouri.

Exxon was a pioneer in the development of cyclic steam stimulation and first tested this process in Quiriquire, Venezuela, in early 1961. The Quiriquire project was a forerunner to the successful application of steam stimulation in the highly productive Bolivar Coastal Fields of Western Venezuela. A wellbore insulation technique developed by Exxon Production Research Company in the late 1960s, and discussed in this book, made steam stimulation of these deep, high-pressure wells feasible.

When the large, highly viscous oil deposit at Cold Lake was discovered in Canada in the early 1960s, a major Exxon research effort was directed toward exploitation of this resource. In 1987, about twenty two years after the first successful steam stimulation of a pilot well at Cold Lake, commercial oil production from this resource has reached approximately 80,000 BPD.

The author of this book has been directly involved in the research and development activities for thermal recovery methods over a significant portion of the thirty plus years that these methods have been studied by Exxon. This book reflects that experience and consequently will be of interest to a broad spectrum of people concerned with the application of thermal recovery methods. It will be of particular value to the engineer who is faced with having to select the most suitable thermal process for a specific reservoir.

P. H. KELLY
VICE PRESIDENT—PRODUCTION
EXXON PRODUCTION RESEARCH COMPANY
NOVEMBER 1987

ABOUT THE AUTHOR

Thomas C. Boberg is a research advisor with Exxon Production Research Co. He joined Exxon in 1959 and has spent fifteen of his twenty-eight years with Exxon in the thermal recovery area. He holds a bachelor's degree in physics from Willam and Mary, SB and SM degrees in chemical engineering from MIT, and a Ph.D. in Chemical Engineering from the University of Michigan. He has been involved with the application of thermal recovery methods to most of Exxon's heavy oil reservoirs. His research led to Exxon's first test of cyclic steam stimulation in Quiriquire field in Venezuela in 1961. Boberg was awarded a patent (US3,292,702) on the process in 1966.

In 1964 Exxon's thermal recovery research effort became intensively involved in the development of a thermal recovery process for Esso Resources Canada's giant Cold Lake deposit. Boberg spent the year 1967 with Imperial Oil Ltd. working in Edmonton on that project. For the year prior to and the two years following that assignment he was supervisor of the Heavy Oil Recovery Research section for Exxon Production Research Co. From 1985 to the present he has again been directly involved with this resource in the area of thermal reservoir simulation.

His remaining 13 years with Exxon were spent in the area of conventional reservoir simulation for Aramco and Exxon's North Sea (U.K.) reservoirs, including a study of miscible displacement in the Brent field. During this nonthermal period he conducted studies of numerous light oil reservoirs. From 1980 to 1983 he directed studies of the application of steam injection methods, primarily cyclic, to several large Venezuelan reservoirs.

PREFACE

This book is intended for petroleum engineers interested in applying the fundamentals of thermal oil recovery. It stresses the practical aspects of this subject. Although this book includes a variety of both theoretical and field results, it is not intended to be a comprehensive research treatise or an all-encompassing review of the voluminous literature amassed over the last 40 years. Furthermore, the reader is presumed to have some background in conventional reservoir engineering technology as well as conductive heat transfer and thermodynamics.

This monograph has been developed in part from a set of lecture notes for a short course on thermal recovery methods which has been given for the Society of Petroleum Engineers of AIME since 1972. Concurrently with the preparation of these lecture notes, a *Thermal Recovery Methods Manual* was prepared for Exxon in-house courses. The latter has been used for company schools since 1973. This monograph is essentially an updated version of both of these predecessors. Its emphasis is on the reservoir engineering aspects of steamflooding, cyclic steam injection processes, and forward in situ combustion methods. Little used methods such as reverse combustion, cyclic combustion, and hot waterflooding are only briefly mentioned. Three chapters discuss actual field experiences with steam stimulation, steamflooding, and in situ combustion. Both dry and wet combustion are discussed. In addition, there is a chapter discussing wellbore heat losses and thermal well design.

Details of thermal project surface facilities, compressors, steam generators, and so on are not discussed in this monograph. Other publications such as the SPE monograph *Thermal Recovery Methods* by M. Prats cover these subjects in some detail.

The book contains numerous analytical equations and methods engineers can use without the aid of mainframe computers to make first estimates of the performance of steam stimulation, steamflooding, and in situ combustion operations. There is a heavy emphasis on process selection. If used judiciously, the calculation methods that are presented can give reasonable answers as well as enhance understanding of the physics of the processes involved. Recently, following a lecture on this subject, I was asked if I favored the use of analytic methods over numerical simulators for thermal recovery predictions. I replied, "Definitely not! Obviously, a comprehensive numerical model of a process can give the best and most complete answer that modern technology can provide. But before you undertake such an expensive calculation, you might want to make a simple calculation first to see if what you are simulating is worthwhile."

Exxon has been a strong supporter of the application of numerical methods for predicting thermal recovery behavior, and one of the longest chapters in this book discusses this subject. The numerical model examples discussed in that chapter highlight Exxon's experience. In more recent years, other companies have also published many fine papers involving the use of thermal simulators, but this chapter highlights the Exxon simulators and studies with which I am most familiar.

Because I have represented Exxon on the engineering committees for joint-interest in situ combustion projects conducted in Canada and in Oklahoma, my interest in the use of combustion is of long standing, but has never been as high as my interest in the application of steam. Industry interest in in situ combustion has also remained at a significantly lower level compared to that for steam. The *Oil and Gas Journal* reported in its April 14, 1986 issue that oil production resulting from steam applications had risen to about 469,000 STB/d in the United States compared to only 10,000 STB/d attributable to in situ combustion. The high cost of air compression and the toll of wells lost to corrosion and erosion by hot fluids has tempered interest in combustion processes for the time being. Yet they remain the only way in prospect to produce thin, deep, heavy-oil sands, so they are potentially still important and are discussed in some detail in this book.

The author wishes to acknowledge the helpful comments and suggestions of C. C. Mattax, S. A. Pursley, R. N. Healy, and M. B. Rotter of Exxon Production Research Company and of H. Y. Lo and E. S. Denbina of Esso Resources Canada.

THOMAS C. BOBERG

Houston, Texas
November 1987

CONTENTS

THERMAL METHODS OF OIL RECOVERY

Chapter 1

INTRODUCTION TO THERMAL RECOVERY METHODS

Although the readers of this monograph are likely to be familiar with the fundamentals of conventional oil recovery processes, the mechanisms and terminology for these processes are briefly reviewed in this section for completeness.

The three natural oil recovery mechanisms in oil reservoirs are solution gas drive, gas cap expansion drive, and water drive. Water or gas injection drives are known as secondary recovery methods. Enhanced oil recovery (EOR) usually involves tertiary recovery—the third method for recovery of additional oil not economically recoverable by secondary methods. All processes are strongly affected by reservoir description. Recovery process effectiveness depends so heavily on reservoir fluid flow and phase behavior that thorough knowledge of these factors is essential. Accurate reservoir description is particularly critical in determining the applicability of any enhanced recovery process. Information on reservoir properties comes from a combination of sources that include geologic studies, core analyses, logging, and pressure transient testing, as well as analysis of reservoir behavior and production history.

Ultimate oil recovery for conventional methods can vary from less than 10% of the oil originally in place with solution gas drive to well above 50% with water drive or water injection. Conventional recoveries might be only 5% or less for highly viscous oils and essentially zero in the case of extremely viscous bitumens. The average oil recovery from conventional oil fields produced by conventional methods worldwide is around 35%.

There are three major reasons why over half the oil in place in the average reservoir is unrecoverable by conventional means. First, only a portion of any reservoir can be contacted by the displacing fluid. Second, not all of the oil can be displaced from the rock that is contacted by the displacing fluid. And third, heavier, low-gravity oils are frequently too viscous to move to the production wells at rates sufficient to support an economic operation. The objective of an EOR process is to overcome one or more of these limitations at least partially.

OVERVIEW OF ENHANCED OIL RECOVERY METHODS[1-4]

Polymer flooding is used to increase the volume of reservoir contacted by the displacing fluid. Addition of a low concentration of polymer makes the floodwater more viscous, decreases its tendency to finger through and bypass oil, and leads to increased sweep of the reservoir. Polymer flooding does not recover residual oil trapped by capillary forces. Therefore, it can be viewed as an improved secondary recovery process rather than a tertiary process.

Trapped oil in swept zones can be mobilized by either surfactant processes or by miscible flooding. Surfactant flooding involves use of a surface-active chemical to substantially reduce the interfacial tension between oil and water. Capillary forces are reduced, and trapped oil can be forced through capillary restrictions and eventually produced. Miscible flooding eliminates the interfacial tension entirely by dissolving the oil in the injected solvent. Thus, it mobilizes residual oil in a way similar to surfactant flooding by eliminating interfacial forces. In general, because of limitations in contacting efficiency and oil mobility, surfactant and miscible flooding are practical only for low- to moderate-viscosity oils.

Thermal processes are used to recover viscous oils and tars by introducing heat into the reservoir, usually by steam injection or in situ combustion, to lower their viscosities.

Most successful commercial applications of EOR up to the present have involved thermal processes. As of 1986, about 78% of all EOR oil production in the United States has resulted from thermal projects. There are a number of reasons why the nonthermal methods have contributed a lesser amount.

Before we begin our discussions of thermal recovery methods, let us briefly review the various nonthermal techniques. The following sections discuss the principal EOR methods: polymer flooding, surfactant flooding, miscible flooding, and thermal methods, in a general way. The reasons for the subordinate role of nonthermal methods to thermal methods will become clearer through this discussion.

Polymer Flooding

In general, incremental recoveries for polymer flooding will be low, as this process cannot mobilize trapped residual oil. A major technical problem is that polymers are susceptible to thermal, chemical, or biological degradation. With currently available polymers, thermal degradation will occur in reservoirs with temperatures greater than 150–160°F.

Because the viscosity of a polymer solution is higher than that of water, polymer injection can result in reduced injectivity, increased project lives, and delayed production. In addition, the injected polymer may plug the sandface at the wellbore if the rock permeability is too low, particularly if an improper polymer is used or if it is not properly dissolved.

Many past applications of polymer flooding in the United States were encouraged by the economic incentives provided by tax laws, such as the Windfall Profits Tax. Much of the current field activity involves permeability profile modification treatments rather than true polymer flooding.[1] Small, near-wellbore treatments employing cross-linked or gelled polymers are being used to improve the vertical distribution of injected water.

The success of polymer floods has been difficult to evaluate. One must identify a small incremental recovery over the recovery expected from a conventional waterflood. In general, polymer flooding has relatively low ultimate potential because of the low incremental recoveries, the limited applicability of the process to oil viscosities below 100 cp and competition from other EOR processes.

Surfactant Flooding

Surfactant flooding can recover trapped residual oil. In most applications, a surfactant solution is injected to reduce the interfacial tension between the water and oil and mobilizes the trapped oil to form a flowing oil bank. Polymers are used in a surfactant flood to promote good reservoir sweep by increasing the viscosity of the water used to drive the surfactant bank through the reservoir. They are sometimes also used to increase the viscosity of the injected surfactant bank. Research has shown that high recovery levels of light oils can be obtained with well-designed surfactant floods in some reservoirs.

The high cost of injected chemicals is a major drawback of surfactant flooding and is frequently the most important factor in determining process economics. In addition, the high investment in chemical plant and facilities to manufacture and inject the surfactants and polymers causes heavy front-end loading of costs, which also detracts from favorable economics. Currently, there are no cost-effective surfactants for high-temperature, high-salinity conditions. In addition, there are no thermally stable polymers that are currently available. Both of these limitations represent major technical hurdles for high-temperature applications.

The application of surfactant flooding in a reservoir is a much more difficult technical challenge than for other processes. It is probably the most complex of all EOR processes. A significant amount of laboratory work is needed to tailor the surfactant system for each reservoir. In addition, field pilots are generally required before proceeding to commercial application. This usually necessitates longer lead times than for other EOR processes.

Surfactant flooding is currently in the field research stage. Although a large number of field pilot tests have been conducted, only one or two projects could be called commercial. Surfactant flooding will probably not contribute substantial commercial production on an industrywide basis before the late 1990s.

Substantial additional technical development is needed to improve the economics and solve other problems if surfactant flooding is to be broadly applied on a commercial scale. Low-cost surfactants are needed for reservoirs with high temperatures and salinities. Most surfactant processes will include the use of polymers, and stable polymers will be needed for higher temperature applications. For remote and/or offshore areas where logistics and storage of chemicals are difficult, it may be necessary to develop processes that utilize lower surfactant concentrations.

Miscible Flooding

Miscible flooding involves the injection of a fluid that is either miscible with the crude oil or generates a miscible solvent bank after contacting the oil. In most cases, the fluids normally injected—nitrogen, carbon dioxide, or light hydrocarbons—are gases at ambient conditions but have liquidlike densities at reservoir conditions.

Miscible flooding is intermediate in technical complexity. It is less complex than surfactant flooding but considerably more complex than polymer flooding. Miscible floods are capable of completely displacing oil from zones that have been swept with solvent. A major economic hurdle for miscible flooding in many reservoirs is the need for a low-cost source of injectants.

The major technical problem encountered in miscible flooding is poor volumetric sweep. The miscible gas is always less viscous and usually less dense than the reservoir crude oil. This causes fingering and gravity override of the injected solvent and can leave much of the oil uncontacted. This is a severe problem in thick, relatively homogeneous reservoirs of low dip as well as in highly heterogeneous reservoirs.

The water-alternating-with-gas (WAG) scheme is one attempt to overcome these limitations. This process uses cyclic injection of gas and water to lower the mobility of the injected fluid and improve the sweep.[5,6]

Gas fingering and override usually cause early gas breakthrough during a miscible flood. If a nonhydrocarbon gas (such as carbon dioxide or nitrogen) is used, the produced fluids usually have to be processed to separate out the solvent gas. This leads to higher capital investment and operating costs.

Despite these problems, miscible flooding is being applied on a commercial scale in a number of light-oil reservoirs. In the United States, field activity in carbon dioxide miscible flooding is increasing rapidly, particularly in West Texas. This is due to the availability of large quantities of natural carbon dioxide pipelined to West Texas from fields in nearby states. The major improvement needed in miscible flooding technology is better sweep efficiency. Several methods for achieving better mobility control—such as using "foams" or polymers to thicken injected gases—are being studied.

Thermal Recovery Methods

Most worldwide EOR activity to date has been concentrated in the United States. Of the 512 projects active in 1986, 201, or 39%, were thermal projects. However, these thermal projects are responsible, as we have said, for 78% of the total domestic EOR production. This constitutes about 7% of total oil production in the United States.[1-3] So, the vast majority of industry EOR production is obtained by thermal methods.

Basically, there are two classes of thermal recovery processes: displacement, or drive, processes and stimulation processes.

A drive process involves the displacement of reservoir fluids to adjacent wells, as in a waterflood. A thermal drive process involves the application of heat to reduce oil viscosity, thus improving the displacement efficiency of injected water and/or gas. Normally, a drive process involves propagation of heat within the reservoir over the entire distance between injection and producing wells. The injected fluid may carry heat generated at the surface, as with steam or hot-water injection, or the heat may be generated within the reservoir, as with in situ combustion, which results when air is injected.

Thermal stimulation involves the heating of a limited region around a producing well to reduce oil viscosity and to improve near-wellbore permeability by removal of fines and asphaltic deposits. The radius of the region heated might be only inches, as in the case of wellbore heaters used to heat the formation by conduction, or it may be many feet (10–100), as in cyclic steam stimulation. In all cases, the principal benefit is to improve well productivity.

Steam stimulation involves periodic injection of steam into a producing well. It has some aspects of a drive process, but unless wells are closely spaced or the treatments quite large, the effect of displacement is slight. Cyclic steam stimulation is also referred to by many other names, such as a steam soak, huff-and-puff, and steam-and-pump. Other fluids can be used in place of steam to heat the reservoir near the wellbore, including air, hot water, hot oil, or hot gas. These other fluids have not been widely used, however, the reasons for which will be discussed later.

THERMAL DRIVE PROCESSES

Hot-Fluid Injection

1. Hot-Water Injection. This is perhaps the simplest of all thermal drive processes and is the closest to a conventional waterflood in its ease of operation. It is not widely used, however, because of the limitations of the mechanisms aiding oil displacement. As temperature increases, oil mobility improves relative to that for water because viscosity decreases more rapidly with increasing temperature for a heavy oil than for water. Also, there is some thermal expansion of oil.

Ultimate recovery of oil by hot-water injection will exceed recovery by a cold-water flood only by the amount of this thermal expansion for such a case, provided residual oil saturation is not significantly altered by increasing temperature. On the other hand, for a given throughput of injected water, improved mobility means that oil recovery will be greater for hot-water than for cold-water injection.

A major disadvantage of hot-water injection, compared to steam injection, is that maximum energy injection rates are usually lower because saturated steam vapor has an energy content more than three times that of hot water below 423°F.[7] Heat losses both from the injection wellbore and from the oil sand will reduce the volume of oil sand that can be heated effectively. This effect is most severe at low injection rates and in thin sands.

A major commercal application of hot-water flooding has been conducted in the Schoonebeek field, Holland, where the sand thickness is about 80 ft and the oil viscosity is 180 cp.[8]

2. Steam Injection. This is the most widely used thermal drive process involving hot-fluid injection. Its major advantages over hot-water injection are that much higher recoveries and energy input rates to the formation are possible.[8] Oil recoveries are high because of the enormous volumes of gaseous steam that can pass through the oil sand to the condensation front, where the maximum penetration of steam has occurred.[9] The effects of the condensing gas drive of the oil within the steam-invaded region, together with oil vaporization, are combined with the effects of the hot water (steam condensate) moving ahead of the steam.

At elevated temperatures (above 500°F), thermal cracking can also increase oil recovery. Viscous fingering leads to poor sweep efficiency in the case of displacement by a low-viscosity fluid such as a light solvent. This effect is not as severe in the case of steamflooding, where, as a finger forms, heat transferred away from the finger causes condensation and retards its growth. Nevertheless, gravity effects, high oil viscosity, and reservoir heterogeneities can cause overriding of the steam and poor volumetric sweep efficiency under certain conditions.

Figure 1.1 is a schematic drawing of the distribution of temperature and fluid saturation for an idealized one-dimensional steamflood. Figure 1.2 compares oil recovery as a function of fluid injected (with steam expressed on a condensed basis) for steamflooding, hot waterflooding, and cold waterflooding as calculated by a simplified model.[8] For the case shown, which compares the injection of steam and hot water at 340°F, one pore volume of injected steam contains about three times the energy of one pore volume of hot water. Also shown for comparison are the results of two waterfloods conducted isothermally at 60°F (the original reservoir temperature) and 340°F. Oil recovery is much higher for the 340°F isothermal case than for the nonisothermal case because cap and base rock as well as oil sand are assumed to be at 340°F, and there is no energy expended in heating rock from 60°F, as in the nonisothermal case.

At higher injection pressures and temperatures, the difference in injected energy between steam and hot water per unit mass or fluid becomes less. At 3206 psia (705°F), the critical pressure of steam, there would be no difference between steam and hot water. Equipment available today limits maximum steam injection pressures to about 2500 psia (668°F). Where higher injection pressures are required, hot-water injection, in situ combustion, or some other process must be employed.

As for hot-water injection, heat loss considerations are important for the design of a steamflood. Ideally, high injection rates, thick sands, and close

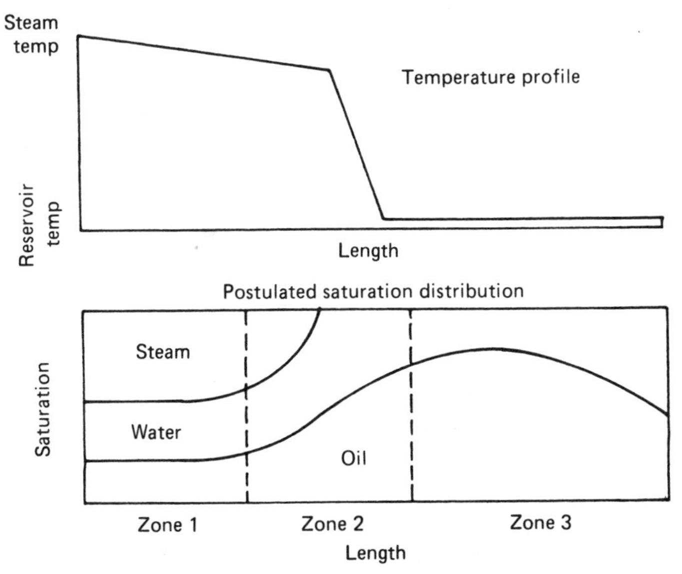

Figure 1.1 Concept of steam drive.[8] © 1965 *Oil & Gas Journal.*

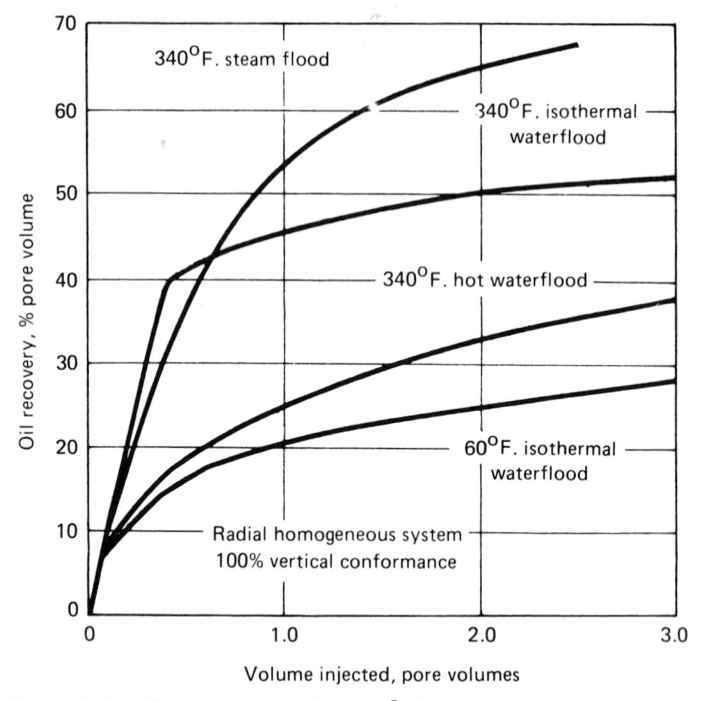

Figure 1.2 Oil recovery predictions.[8] © 1965 *Oil & Gas Journal.*

well spacings minimize these heat losses. In thick sands, however, if wells are too close and injection rates too high, channelling will result.[10] Normally, this occurs becasue of gravity effects and the tendency of steam to override the oil column. Thus, with close well spacing, early steam breakthrough can occur at producing wells, and unless injection rates are reduced, a large fraction of the injected heat will be produced.

Optimization of well spacing and injection rate is essential for favorable flooding performance and good economics. A major portion of the following material will be devoted to descriptions of prediction methods for steamflooding to aid in the selection of favorable steam drive candidates, as well as to estimate the effects of varying pattern size and well rates.

3. Hot-Gas Injection. This technique is almost never used because of the very low heat-carrying capacity of gas and the enormous compression required to achieve heating of a large volume of oil sand. Heat loss detracts more from hot-gas injection than for any of the three hot-fluid injection methods already discussed. This approach will not be discussed further.

Hot-gas injection has received limited attention as a method for in situ oil shale retorting because of the high gas temperatures that can be employed.

Any of the hot-fluid injection processes (hot water, steam, or hot gas) may employ a period of cold-water injection in the later period of the flood to drive the thermal bank through the remaining unheated oil sand. This approach appears especially advantageous to reduce the injection costs of hot waterflooding. A steam bank, however, cannot be propagated this way. Injection of cold water quickly condenses the steam in the oil sand, causing the process to revert to a hot waterflood.[9] When this happens, the hot condensing gas drive benefit associated with steamflooding is lost. In a steamflood, however, normally it is economically beneficial to switch to cold-water injection at an advanced stage in the flood. Further discussion of this will be found in later chapters.

In Situ Combustion Drive

1. Forward Dry Combustion. This process involves injection of oxygen-containing gas (normally air), ignition within the oil sand, and propagation of a combustion front through the reservoir.[8] Oil is displaced by hot flue gas (nitrogen and carbon dioxide) passing ahead of the combustion front and by the steam resulting both from the combustion and from the vaporization of connate water. The mechanism of displacement is, as in the case of steam, principally gas drive of the heated oil aided by oil vaporization and thermal cracking. Vaporized light ends condense farther ahead of the combustion front and provide some assistance to displacement by solvent dilution of the virgin crude. The rate of oil displacement is closely related to the rate of advance of the combustion front.

A schematic drawing of the temperature distribution and location of the various zones is shown in Figure 1.3 for an idealized one-dimensional case. Displacement of unheated oil by water and flue gas downstream of the combustion front is relatively inefficient for a viscous oil but has greater importance for a light oil. The combustion front's rate of advance is controlled by the rate at which residual hydrocarbon, or "fuel," is consumed by the injected oxygen. The process benefits from a high rate of air injection and a low fuel residuum deposited per unit volume of oil sand, which permits more rapid movement of the combustion front and hence a more rapid oil displacement. Since a major cost in the process is that for air compression, a low fuel concentration is generally desirable.

Forward dry combustion is not nearly as sensitive to heat losses as the hot-fluid injection processes. Therefore, combustion drive can be used with thin deep sands, where hot-fluid injection would be impractical. Nevertheless, heat losses become important if fuel concentrations and oxygen injection rates per unit area of combustion front (flux rates) are too low. In such cases, oxygen passing through the oil sand will be utilized only partially. Ultimately as the combustion front advances far enough radially from the

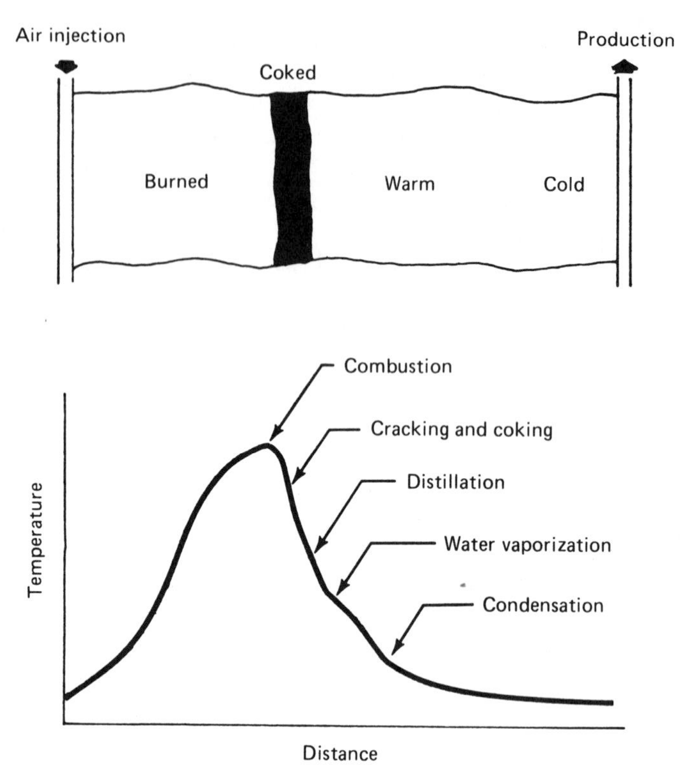

Figure 1.3 Underground combustion mechanism.[8] © 1965 *Oil & Gas Journal.*

injection well, the oxygen flux and combustion front temperatures will become so low that combustion will cease. As a way to improve combustion efficiency, interest is increasing in oxygen enrichment or nearly pure oxygen injection.[1]

In thin sands (<15 ft), experience is that the vertical coverage of the combustion front is better than might be expected, despite the tendency of air to override the oil column downstream of the combustion front. In thicker sands (>40 ft), however, a gravity override combustion front can be expected. Severe well completion damage can occur when attempting to produce a hot producing well. Heating at a producing well can occur soon after the process starts for the override type of flood if injection rates are too high and injection and production wells too closely spaced. In general, projects are designed to minimize the possibility of needing to produce a hot well in a combustion drive project.

Inefficiency of the forward dry combustion process stems from the low heat-carrying capacity of the injected air. Air moving through the hot oil sand behind the combustion front is a poor medium for moving the heat forward to a point downstream of the combustion front to aid in oil displacement. This inefficiency has created interest in the use of water injection simultaneously with air injection to reclaim the waste heat remaining behind the combustion front. This dual water and air injection creates a steam zone ahead of the combustion front and can be highly effective in aiding oil displacement.[22–24]

2. Forward Wet Combustion. This process is the one just described, involving simultaneous injection of air and water. Theoretically, it should be the most efficient of all the thermal drive processes. It has the features of both a steam drive and a dry-combustion drive process. It should be more efficient than either method because the energy is liberated in the reservoir and rapidly moved as a heated band through the oil sand so that the heating of the entire reservoir at any one time does not occur and heating is maintained only in the region where it aids oil displacement.

Unfortunately, gravitational forces cause separation of air and water, and in practice, the process is difficult to control. It is most applicable to thin sands. It may be useful for recovery of light- to medium-gravity oils left after waterflood, provided starting oil saturations and crude prices are sufficiently high to make it economically attractive. Because of the higher rate of heat propagation through the oil sand, air compression requirements to achieve a given level of oil recovery are lower than for dry combustion. In addition, fuel consumption is less than for dry combustion because of the greater volume of steam passing ahead of the combustion front. The steam drives the oil saturation down to a lower residual. At high water injection rates, incomplete fuel utilization occurs, and the process is termed *partially quenched*.[22]

Research laboratories continue to investigate optimum reservoir conditions and air–water ratios for this process. Major drawbacks compared to dry combustion are air–water segregation, reduced injectivity, and increased corrosion. Its practical application to date has been rather limited.

3. Reverse Combustion. This process might be likened to blowing through a cigarette, rather than inhaling, to propagate combustion. In reverse combustion, air flow is counter to the direction of movement of the combustion front. This process was originally conceived for preheating cold, highly viscous oils, as in the Athabasca tar sands, where there would be no movement of the highly viscous oil bank ahead of the combustion front in a forward drive.

The process consists of injecting air until there is communication with the producing wells, then using a downhole heater or other means to ignite the oil sand around the producing well. The combustion front burns back toward the injection well. Unfortunately, the reverse-combustion process is impractical for a number of reasons:

1. Only the lighter components rich in hydrogen burn, leaving a tarry residuum on the oil sand. This makes the ratio of injected air to oil recovered extremely high.
2. Damage to the producing well completion can occur on ignition.
3. Spontaneous ignition near the injection well will occur eventually, and the process will revert to forward combustion.[11] If the formation is over 100°F, this can occur in only a few days, even hours.

Because of these defects, reverse combustion is of little practical interest and will not be considered further in this book.

All thermal drive processes suffer in varying degrees from problems of reservoir contacting. Channelling and poor sweep are worse in flat reservoirs with extremely high viscosity oils. Well problems caused by premature heat breakthrough can be tolerated in the case of steam injection. They cannot, in general, be endured for in situ combustion. This problem, probably more than any other factor aside from high compression costs, is why the preponderance of thermal drives are steam drives. In 1986, about 480,000 STB/d† were produced by thermal methods in the United States, whereas only about 10,000 STB/d were produced by in situ combustion.

THERMAL STIMULATION PROCESSES

1. Wellbore Heating. Downhole electrical, gas, and steam heaters have been used to improve oil mobility in the near vicinity of the wellbore.[12] Electrical conduction heating, microwave heating, and other approaches have been tried experimentally to obtain greater penetration of heat into the formation than possible when using simple wellbore heaters that rely on heat conduction counter to the flow of fluids entering the well.

In most heavy-oil fields, steam stimulation, which can achieve much deeper heat penetration and, as a result, a much greater stimulation effect, has replaced these less effective techniques. Nevertheless, wellbore heating is sometimes used for low-production-rate wells (less than 30 bbl/d) where depth, excessive injection pressures, or swelling clays preclude the application of steam. Wellbore heating is used sometimes for low-productivity wells in which paraffin would deposit without heating.

2. Steam Stimulation. This technique[13–15] is by far the most popular

† STB = Stock tank barrels measured at standard conditions of 60°F, 14.7 psia.

thermal stimulation process and will be discussed in greater detail in subsequent chapters. The process consists of two steps:

1. a steam injection phase followed by a brief shut-in period and
2. an oil production phase that normally lasts much longer than the injection period.

As with steam drive, heat losses from the wellbore during injection and from the oil sand during production have a strong effect on performance. Normally, water–oil ratios will be quite high when the well is first returned to production, but the water cut quickly declines and the oil production rate will pass through a maximum, usually much higher than the original value. If permeability increases near the wellbore, as is frequently the case, there will be some permanent productivity improvement.

The major difficulty with steam stimulation has been the damage to casing and liners in wells that have been overheated during injection.[15–17] Because of this problem, industry practice is to avoid steam-stimulating deep, high-pressure wells. When steam injection pressures exceed 600–800 psi, tubing insulation or specially designed wells are usually required. Design considerations for steam injection wells will be discussed in Chapter 2.

3. Stimulation Using Other Hot Fluids. Hot water and gas are rarely used to stimulate wells because of their lower heat-carrying capacity compared to steam.[18] Hot-water stimulation has been used for high-pressure wells, where it is desired to heat to lower temperatures than possible with steam. Hot-gas injection for stimulation is practically unknown.

In a manner analogous to steam stimulation, air may be used to stimulate production.[19] In this process, reservoir oil is ignited (using a downhole heater or other means, if spontaneous ignition times are excessive), and air injection is continued until a heated region of desired size is created around the wellbore. The well is then returned to production. Prior to resuming production, however, flue gas (or other inert gas) or water is usually injected to cool and enlarge the heated region and to displace any air remaining through the combustion front so that no free oxygen moves back into the wellbore.

Severe liner damage can occur when stimulating wells by in situ combustion. In addition, there can be formation plugging from coke deposits and rust particles at the sand face.[18] These difficulties detract from the productivity benefits of the method, so it has not been widely used. There are no known commercial combustion stimulation projects active at the present time. This method will not be discussed further.

THE RESOURCE

The map in Figure 1.4 shows the location of the major heavy-oil-producing areas in the Western Hemisphere. In addition, it shows the Powder River basin in Montana and Wyoming, where in situ coal gasification may one day become commercially attractive, as well as the Piceance basin, where in situ methods of shale oil recovery may eventually be used. This book is devoted to methods useful for the areas hatched with horizontal lines, the heavy-oil

Figure 1.4 Selected major resources amenable to thermal recovery processes.[4] © 1978 SPE-AIME.

reservoirs, and tar sands. It does not discuss exotic or unusual thermal applications.

In the United States, most of the heavy oil is located in California, and the bulk of the thermal oil production in this country comes from that state. Approximately as much thermal oil is produced in Venezuela, mostly by cyclic steam injection in the conventional heavy-oil reservoirs of the Bolivar Coastal fields. A much larger deposit of highly viscous oil occurs in the Orinoco Oil Belt. Eventually it is anticipated that this area will become the major area for application of thermal recovery methods in Venezuela.

In Canada, in the province of Alberta, extremely large reservoirs containing highly viscous bitumen are found in the Cold Lake, Athabasca, and Peace River deposits. Commercial operations are expanding in Esso Resources Canada's cyclic steam injection operation at Cold Lake. Pilot operations are continuing in the Orinoco and in the other bitumen deposits of Alberta.[20]

Figure 1.5 shows the worldwide distribution of bitumen defined as oil having 10°API or lower gravity. In the Orinoco, many of the reservoirs have a high reservoir temperature so that the in situ oil viscosity allows the reservoirs to be produced to some extent without heating. In the Alberta tar sands, temperatures are lower and oil viscosities are so high that unheated production is impossible.

The worldwide total bitumen in place is estimated conservatively as 4.07×10^{12} bbl. A breakdown is given in Table 1.1. The numbers are speculative and probably low, however, because there is almost no reliable

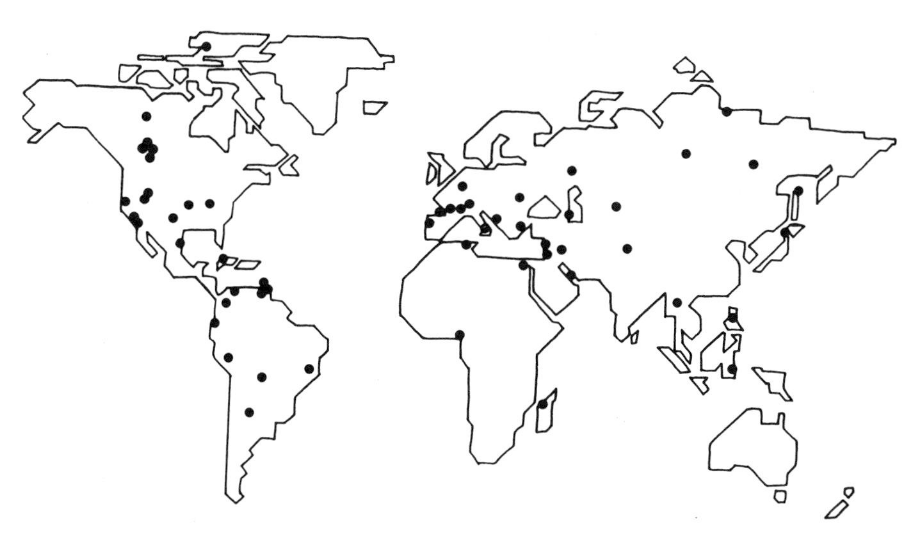

Figure 1.5 Worldwide distribution of tar sand deposits.[20] (After Meyer and Fulton[21].)

Table 1.1 World Bitumen Resources[a]

Country	Number of Deposits	Original Oil in Place (10^6 bbl)
Albania	1	371
Canada	7	2,432,600
Cuba	1	Undetermined
Italy	4	14,155
Madagascar	2	24,800
Mexico	3	Undetermined
Nigeria	1	Undetermined
Peru	1	66
Romania	1	25
Trinidad and Tobago	1	60
United States	53	34,390
USSR	10	557,024
Venezuela	4	1,000,012
West Germany	4	Undetermined
Zaire	1	1,925
Total	94	4,065,428

[a] After Meyer and Fulton.[21]

published quantitative information on this resource outside of Venezuela, Canada, and the United States. If the ultimately recoverable amount of oil were only one-quarter of this number, it would still be a quantity that is about double the entire world's proved reserves of about half a trillion barrels.[4] The energy resource that thermal methods can unlock is enormous indeed.

REFERENCES

1. EOR Survey, *Oil Gas J.* (April, 14, 1986), p. 71.
2. Winkler, R. O. "EOR Processes and their Potential Application in U.K. North Sea Reservoirs," Talk to British DOE, July 1984.
3. Welch, L. W. "Status of EOR Technology," SPE/DOE Enhanced Oil Recovery Symposium, Tulsa, OK, April 22, 1980.
4. Prats, M. "A Current Appraisal of Thermal Recovery," *J. Pet. Tech.* (August 1978), p. 1129.
5. Christian, L. D., Shirer, J. A., Kimbel, E. L., and Blackwell, R. J. "Planning a Tertiary Oil-Recovery Project for Jay/LEC Fields Unit," *J. Pet. Tech.* (August 1981), p. 1535.

6. Kane, A. V. "Performance Review of a Large-Scale CO_2-WAG Enhanced Recovery Project, SACROC Unit—Kelly-Snyder Field," *J. Pet. Tech.* (February 1979), p. 217.
7. Keenan, J. H., and Keyes, F. G. *Thermodynamic Properties of Steam*, 1st ed., Wiley, New York, 1956.
8. Boberg, T. C. "What's the Score on Thermal Recovery and Stimulation," *Oil Gas J.* (August 23, 1965), pp. 78–83.
9. Shutler, N. D. "Numerical Three Phase Model of the Linear Steam Drive Process," *Soc. Pet. Eng. J.*, (June 1969), pp. 232–246.
10. Shutler, N. D. "Numerical Three Phase Model of the Two Dimensional Steamflood Process," *Soc. Pet. Eng. J.* (December 1970), pp. 405–417.
11. Dietz, D. N., and Weijdema, J. "Reverse Combustion Seldom Feasible," *Prod. Monthly* (May 1968), p. 10.
12. Schild, A. "A Theory for the Effect of Heating Oil Producing Wells," *Trans. AIME*, **210**, 1 (1957).
13. Boberg, T. C. "Thermal Well Stimulation Method," U.S. Patent 3,292,702, December 20, 1966.
14. Boberg, T. C., and Lantz, R. B. "Calculation of the Production Rate of a Thermally Stimulated Well," *Trans. AIME*, **237**, I-1613 (1966).
15. Burns, J. A. "A Review of Steam Soak Operation in California," *J. Pet. Tech.* (January 1969), p. 25.
16. Willhite, G. P., and Dietrich, W. K. "Design Criteria for Completion of Steam Injection Wells," *J. Pet. Tech.* (January 1967), p. 15.
17. Leutwyler, K., and Bigelow, H. L. "Temperature Effect on Subsurface Equipment in Steam Injection Systems," *Trans. AIME*, **234**, I-93 (1965).
18. Dietz, D. N. "Thermal Recovery Techniques," Society of Petroleum Engineers of AIME, Reprint Series No. 10, 1972.
19. White, P. D., and Moss, J. T. "High-Temperature Thermal Techniques for Stimulating Oil Recovery," *J. Pet. Tech.* (September 1965), p. 1007.
20. Carrigy, M. A. "Thermal Recovery from Tar Sands," *J. Pet. Tech.*, (December 1983), p. 2149.
21. Meyer, R. F., and Fulton, P. A. "Toward an Estimate of World Heavy Crude Oil and Tar Sands Resources," *Proceedings of the Second International Conference on Heavy Crude and Tar Sands*, United Nations Institute for Training and Research, Caracas, 1982.
22. Dietz, D. N., and Weijdema, J. "Wet and Partially Quenched Combustion," *Trans. AIME*, **243**, I-411 (1968).
23. Parrish, D. R., and Craig, F. F., Jr. "Laboratory Study of a Combination of Forward Combustion and Waterflooding—The COFCAW Process," *Trans. AIME*, **246**, I-753 (1969).
24. Dietz, D. N. "Wet Underground Combustion, State of the Art," *Trans. AIME*, **249**, 1-605 (1970).

Chapter 2

PREDICTION OF WELLBORE HEAT LOSSES AND CASING TEMPERATURES

ABSTRACT

Hot-fluid injection processes such as steam or hot-water injection require, as a first consideration, whether heat losses from the injection wellbore are acceptable and how they can be minimized. This chapter describes the Ramey–Willhite[1,2] method for calculating how much of the steam injected at the wellhead will remain uncondensed at target formation depth. Nomographs presented by Huygen and Huitt are also presented for estimating heat loss rates from the wellbore. Both methods also permit estimating average casing temperatures, which are important for determining whether allowable casing stresses will be exceeded. In addition, Ramey's equations[1] for estimating bottom-hole injection temperature during hot-water injection are presented. The effect of various wellbore design parameters and types of tubing insulation on wellbore heat losses are discussed, as are computer-calculated pressure drops during steam injection. Practical design considerations for hot-fluid injection wells are covered. Finally, there is a discussion of a novel silicate foam insolation technique developed by Exxon and used successfully in Venezuela.[3]

INTRODUCTION

The prediction of wellbore heat losses is important in the design of a steam stimulation or steamflooding project. It is particularly important for deeper wells and at lower rates of injection since, under these conditions, energy lost from the wellbore can be a substantial fraction of the total. In addition, the higher steam injection pressures and temperatures required for injection into deeper wells greatly increase the danger of well-casing failure because of thermal expansion. Even a moderate rise in temperature is dangerous for casing not cemented, yet not free to expand, because of the possibility of failure by buckling. Washouts of the formation around uncemented casing greatly increase the danger of casing failure by buckling. Such wells should be avoided for steam injection operations.

In general, it is not advisable to inject steam down the casing because heat losses and casing temperatures are much higher than when injecting down tubing. For low-pressure sands many operators inject without a packer. For higher pressure applications, a packer is frequently used to maintain a low-pressure annulus, thereby helping to reduce casing temperatures and eliminate refluxing of steam in the annulus (which might occur downhole). Unfortunately, maintaining a low annulus pressure often is not feasible. Packers usually fail after a time that can be very short when operating at high temperatures. To prevent this, an annular fluid normally is required to provide weight on the packer. Insulating gels and viscous oils have been used successfully.

RAMEY-WILLHITE METHOD

Calculations by Willhite[2] show the effect of annulus pressure on casing temperature as a function of injection time for a steam temperature of 650°F for the case of injection down tubing (Fig. 2.1). Also shown is the effect of reducing radiation losses by coating the tubing with aluminium paint and the

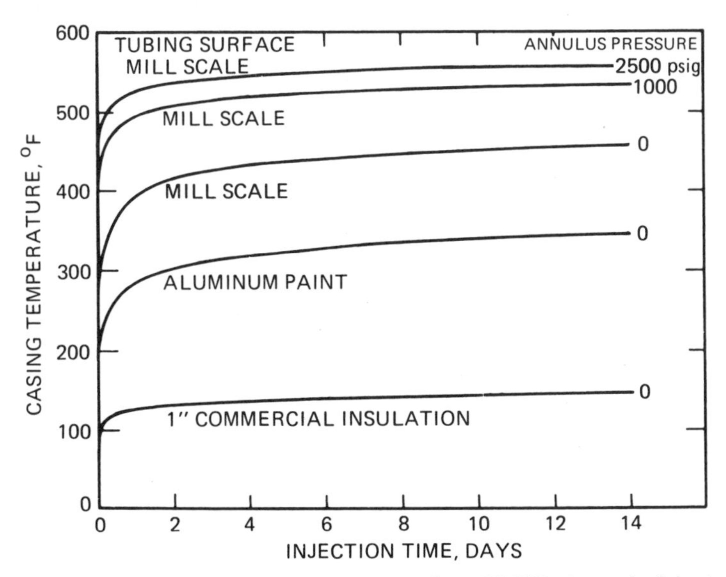

Figure 2.1 Calculated casing temperatures when 650°F steam is injected down tubing.[2] © 1967 SPE-AIME.

further reduction in losses that can be obtained by insulating the tubing. In practice, it is doubtful that coating with aluminum paint will be effective because it is likely that the outside wall of the tubing will be coated with rust and oil when it is run downhole. Insulation applied at the surface must be protected by suitable metal cladding lest it become oil and water saturated and lose effectiveness or become damaged when the tubing is run. Clad tubing is available but expensive. Industry research efforts are directed toward cheaper methods for effective tubing insulation.

One method insulates unclad tubing with a thick coating of foamed sodium silicate. The coating is formed when a solution of this material placed in the casing–tubing annulus is boiled off as steam is injected.[3] This approach is most effective when a dry, low-pressure annulus can be maintained. Because of the problem of packer failure, a practical compromise is to load the annulus with a viscous oil to protect the packer. This only partially decreases insulation effectiveness. More detail about the silicate insulation method is given later in this chapter.

Insulating cement has been proposed to reduce heat losses in wells completed expressly for thermal applications. Willhite has calculated (Fig. 2.2) the effect of reducing the cement thermal conductivity (K_{cem}) from 0.51 to 0.2 Btu/hr-ft-°F to enhance its insulating ability. This degree of change in thermal conductivity is believed to occur with ordinary cements that dry upon heating to elevated temperatures. Note that casing temperatures rise about 75°F when K_{cem} decreases from 0.51 to 0.2. A deliberate attempt to reduce the K_{cem} to reduce heat losses is not good engineering practice. It may cause the casing temperature, and thus thermally induced stresses, to exceed allowable values. For this reason, the use of an insulating cement is generally not recommended. The basic data used in Willhite's calculations are shown in Table 2.1.

In developing the method, Willhite considered the various resistances to heat transfer occurring from the fluid flowing in the tubing at T_f to the temperature T_h at the outside of the cement sheath in order to obtain an overall wellbore heat transfer coefficient, U_{to}. A schematic drawing of temperature as a function of radial distance from the center of the tubing is shown in Figure 2.3. At low annulus pressures and elevated tubing temperatures, the major contributor to heat transfer between the tubing and casing is radiation. However, at higher pressures, convection and conduction in the annulus become increasingly important. These factors influence the variation of U_{to} with tubing temperature and annulus pressure, as can be seen in Figure 2.4. These values apply rigorously only for the parameters given in Table 2.1 but can be used as reasonable estimates for other casing dimensions.

Using Ramey's approximation,[1] the average heat loss rate per foot of

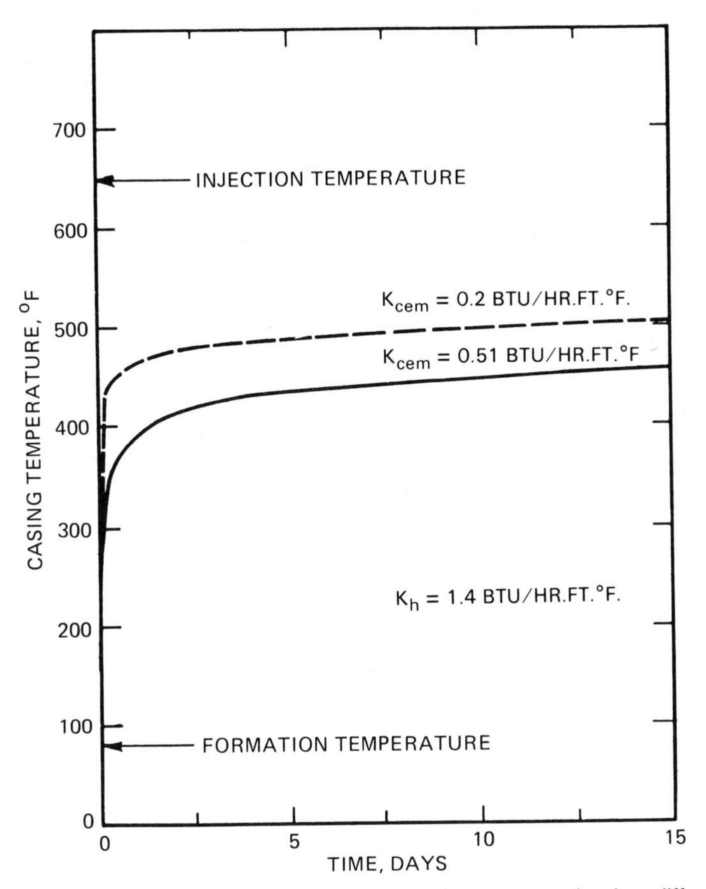

Figure 2.2 Comparison of casing temperatures for cements having different thermal conductivities.[2] © 1967 SPE-AIME.

depth, Q, is obtained:

$$Q = \frac{2\pi K_{\mathrm{h}}(T_{\mathrm{h}} - T_{\mathrm{e}})}{f(t)} \tag{2.1}$$

where K_h = average formation thermal conductivity (Btu/hr-ft-°F)

T_{e} = average undisturbed formation temperature $= \frac{1}{2}(T_{\mathrm{bh}} + T_{\mathrm{surf}})$ (°F)

T_{bh} = normal (original) bottom-hole temperature (°F)

T_{surf} = normal (geothermal) surface temperature (°F)

T_{h} = temperature at outside of cement sheath (°F)

Table 2.1 Parameters for Figure 2.4[a]

Hole Size	9.625 in.
Casing (7 in., 26 lb, J-55)	
OD	7.000 in.
ID	6.276 in.
Tubing	
($2\frac{7}{8}$ in., 6.4 lb, J-55)	
OD	2.875 in.
ID	2.441 in.
K_h	1.4 Btu/hr-ft-°F
K_{cem}	0.51 Btu/hr-ft-°F
K_{ins}	$0.0256 + (T\text{-}50)(3.67 \times 10^{-5})$ Btu/hr-ft-°F
α	0.04 ft^2/hr
$\epsilon_{to} = \epsilon_{ci}$ (mill scale)	0.9
ϵ_{to} (aluminum paint)	0.4
T_e	80°F

[a] From Ref. 2. Abbreviations: OD, outside diameter; ID, inside diameter. © 1967 SPE-AIME.

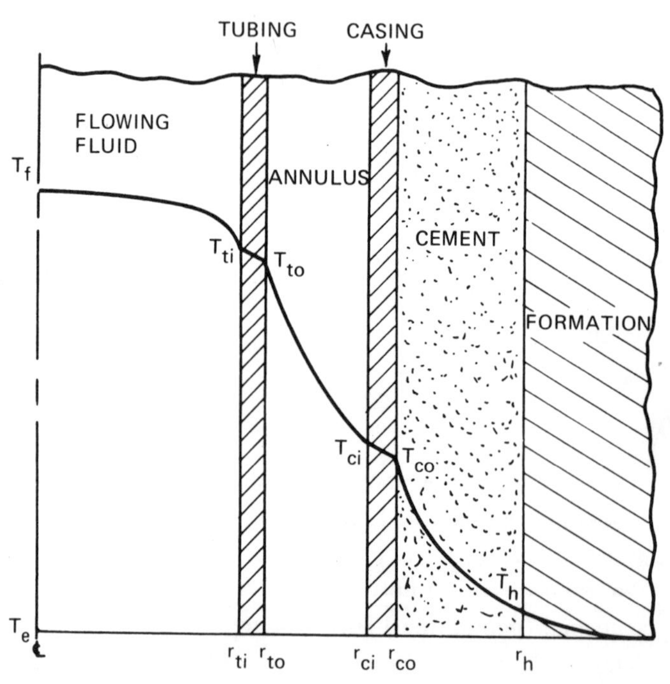

Figure 2.3 Temperature distribution in an annular completion.[2] © 1967 SPE-AIME.

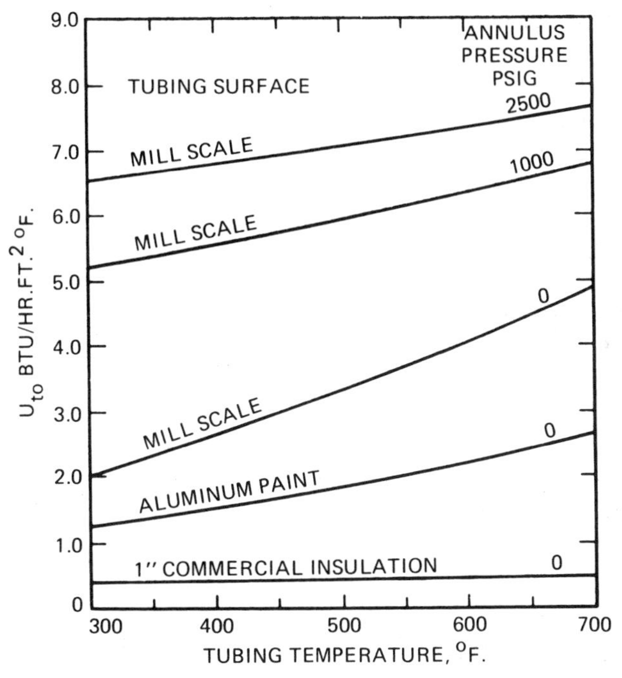

Figure 2.4 Variation of U_{to} with tubing temperature for parameters of Table 2.1. © 1967 SPE-AIME.

and

$$f(t) = \ln\left(\frac{2\sqrt{\alpha t}}{r_h}\right) - 0.29 \tag{2.2}$$

where α = average formation thermal diffusivity (ft^2/d)
r_h = radius of outside of cement sheath (ft)
t = time of steam injection (days)

Equation (2.2) for $f(t)$ is highly accurate for $t > 11$ days but can be used for t as small as 1 day with only about 11% error.[1]

T_h is obtained by

$$T_h = \frac{T_s f(t) + \gamma T_e}{f(t) + \gamma} \tag{2.3}$$

where T_s = average injected steam temperature (°F)
$\gamma = K_h / r_{to} U_{to}$
r_{to} = outside tubing radius (ft)

The steps in the computation are:

1. Estimate the value of U_{to} from Figure 2.4.
2. Calculate T_h from Eq. (2.3).
3. Calculate Q using Eqs. (2.1) and (2.2).

Using the equations for U_{to} (see Appendix), an improved estimate can be made for casing and tubing sizes other than 7 and $2\frac{7}{8}$ in., and the final result for Q is obtained by successive approximation.

The casing wall temperature is calculated by

$$T_{ci} = T_h + \frac{r_{to} U_{to} \ln (r_h / r_{co})}{K_{cem}} (T_s - T_e) \tag{2.4}$$

which neglects the heat transfer resistance of the inside tubing film, the tubing wall, and casing wall. In Equation (2.4) r_{co} is the outside casing radius in feet.

HUYGEN–HUITT METHOD

A useful set of graphs for estimating the heat loss rate Q and T_{ci} have been prepared by Huygen and Huitt.[4] The equations on which these plots are based are similar to those for the Ramey–Willhite method with the following exceptions.

1. Convection in the annulus is ignored. This introduces little error for the low-pressure annulus case since radiation and conduction are controlling.
2. The thermal conductivity of the cement sheath, K_{cem}, is taken as equal to that of the formation.

The conductivity of the formation, K_h, and the thermal diffusivity α have values equal to those given in Table 2.1. The average undisturbed formation temperature is assumed to be 80°F. For cases that deviate substantially from these conditions, the Ramey–Willhite method should be used.

The plots are given for a combination of casing and tubing sizes given in Figures 2.5–2.15. Table 2.2 indicates which plot to use.

Table 2.2 Casing and Tubing Combinations for Figures 2.5–2.15[a]

Tubing Size (in.)	Casing Size (in.)				
	$4\frac{1}{2}$	$5\frac{1}{2}$	7	$8\frac{5}{8}$	$10\frac{3}{4}$
2	Fig. 2.5	Fig. 2.6	—	—	—
$2\frac{1}{2}$	—	Fig. 2.7	Fig. 2.8	Fig. 2.10	Fig. 2.13
$3\frac{1}{2}$			Fig. 2.9	Fig. 2.11	Fig. 2.14
$4\frac{1}{2}$				Fig. 2.12	Fig. 2.15

[a] The sizes listed are the API numerical sizes.

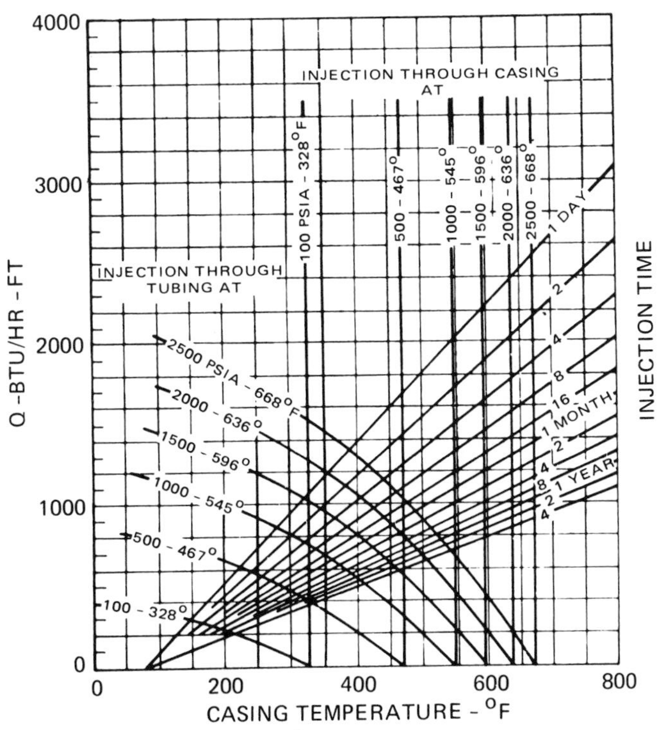

Figure 2.5 Heat lost in wellbore and casing temperature during steam injection: 2-in. tubing, $4\frac{1}{2}$-in. casing.[4] © 1966 *Producers Monthly*.

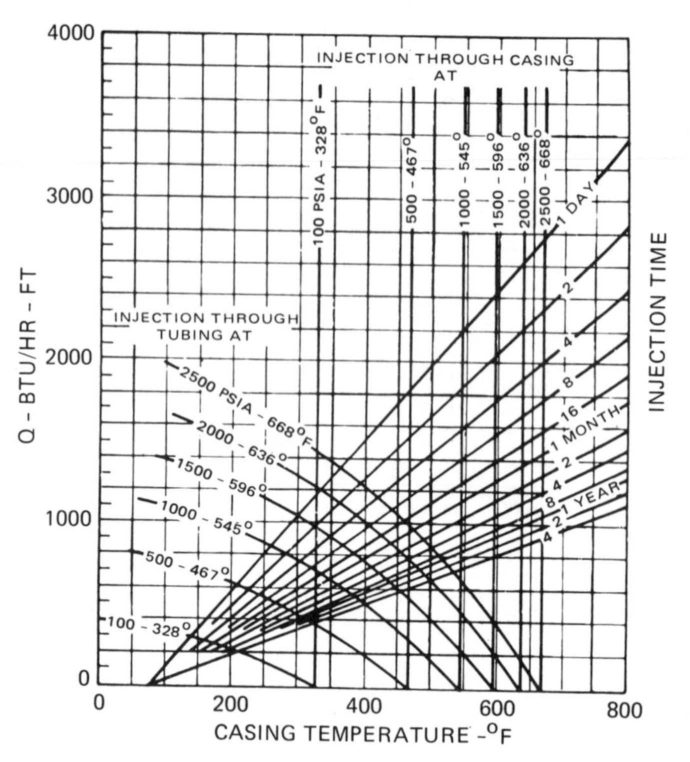

Figure 2.6 Heat loss in wellbore and casing temperatures during steam injection: 2-in. tubing, $5\frac{1}{2}$-in. casing.[4] © 1966 *Producers Monthly.*

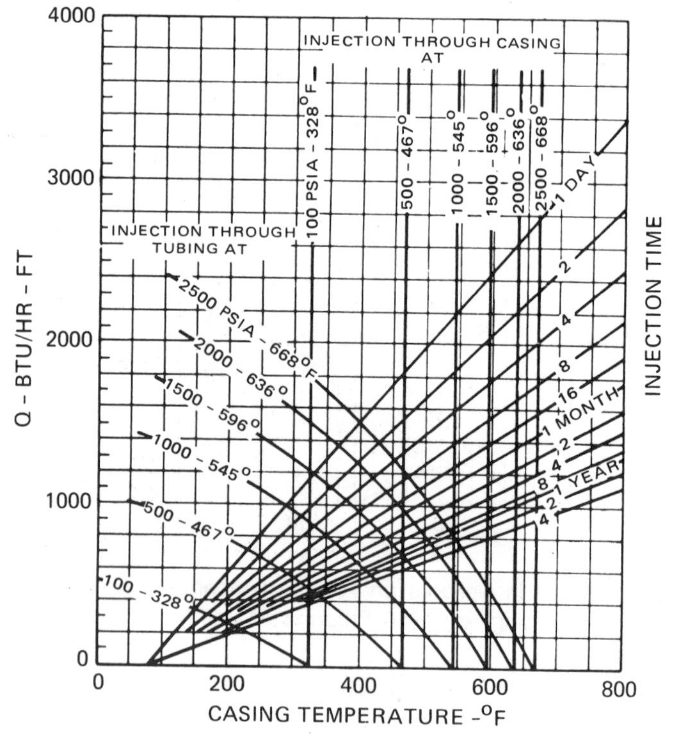

Figure 2.7 Heat loss in wellbore and casing temperature during steam injection: $2\frac{1}{2}$-in. tubing, $5\frac{1}{2}$-in. casing.[4] © 1966 *Producers Monthly.*

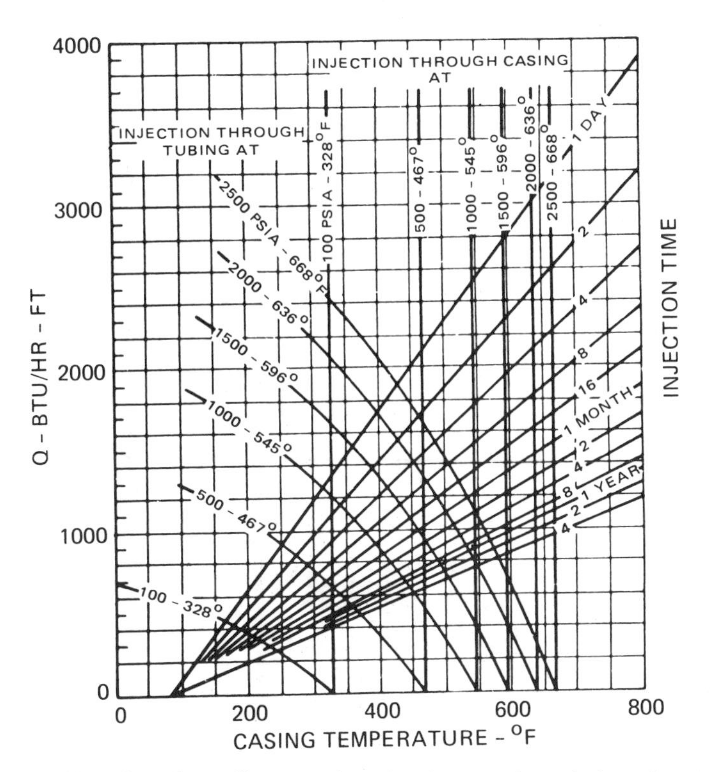

Figure 2.8 Heat loss in wellbore and casing temperature during steam injection: $2\frac{1}{2}$-in. tubing, 7-in. casing.[4] © 1966 *Producers Monthly*.

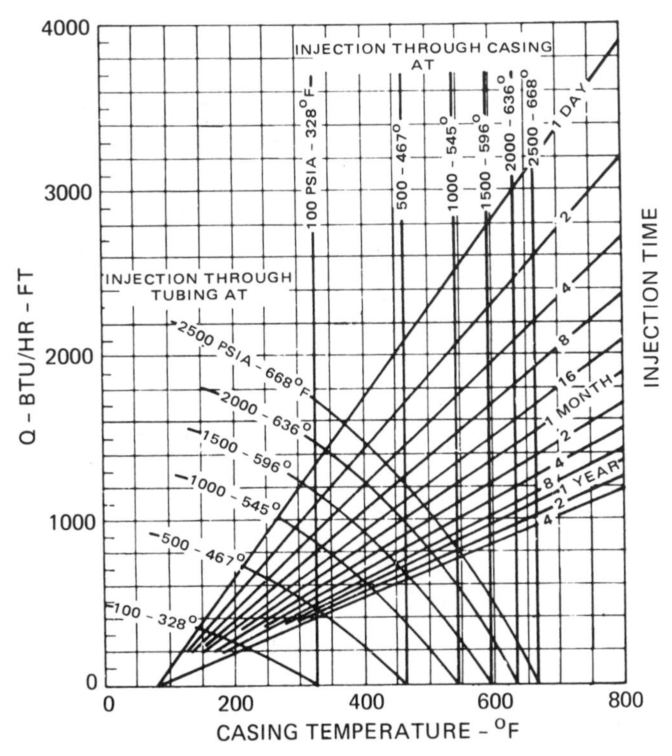

Figure 2.9 Heat loss in wellbore and casing temperature during steam injection: $3\frac{1}{2}$-in. tubing, 7-in. casing.[4] © 1966 *Producers Monthly*.

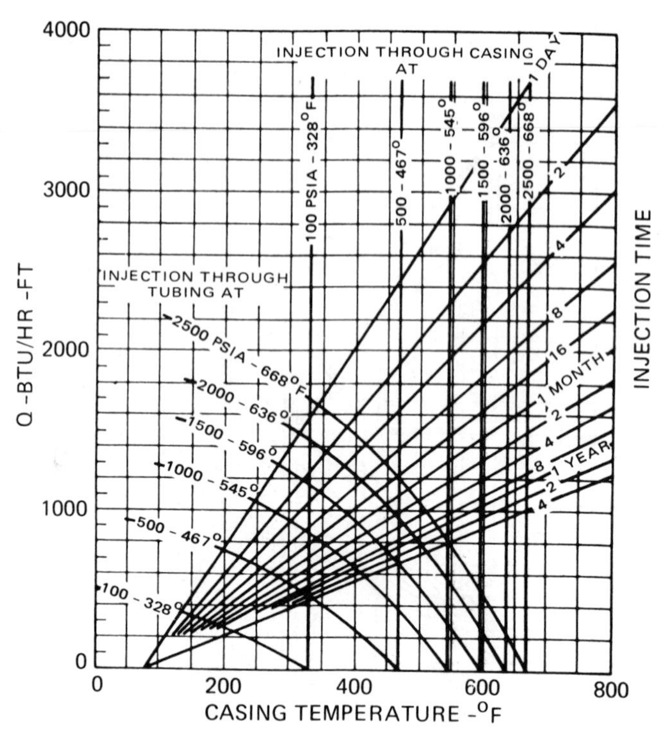

Figure 2.10 Heat loss in wellbore and casing temperature during steam injection: $2\frac{1}{2}$-in. tubing, $8\frac{5}{8}$-in. casing.[4] © 1966 *Producers Monthly*.

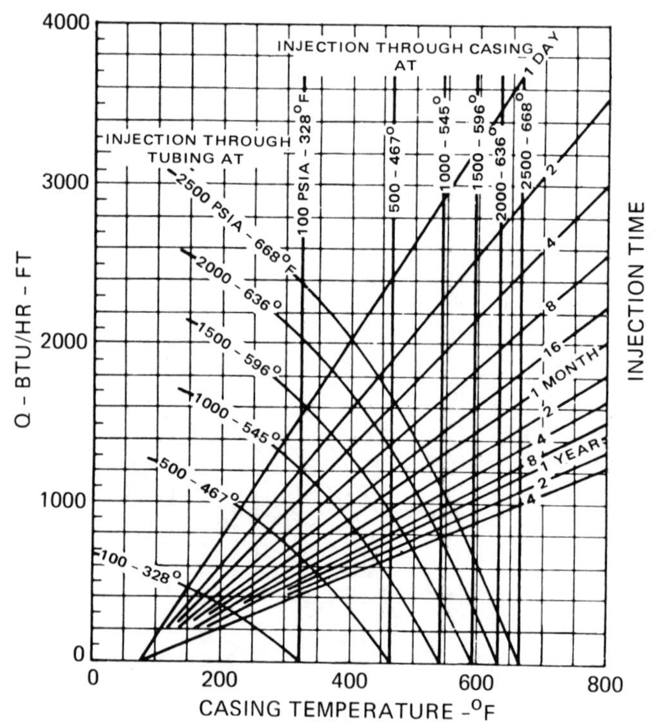

Figure 2.11 Heat loss in wellbore and casing temperature during steam injection: $3\frac{1}{2}$-in. tubing, $8\frac{5}{8}$-in. casing.[4] © 1966 *Producers Monthly*.

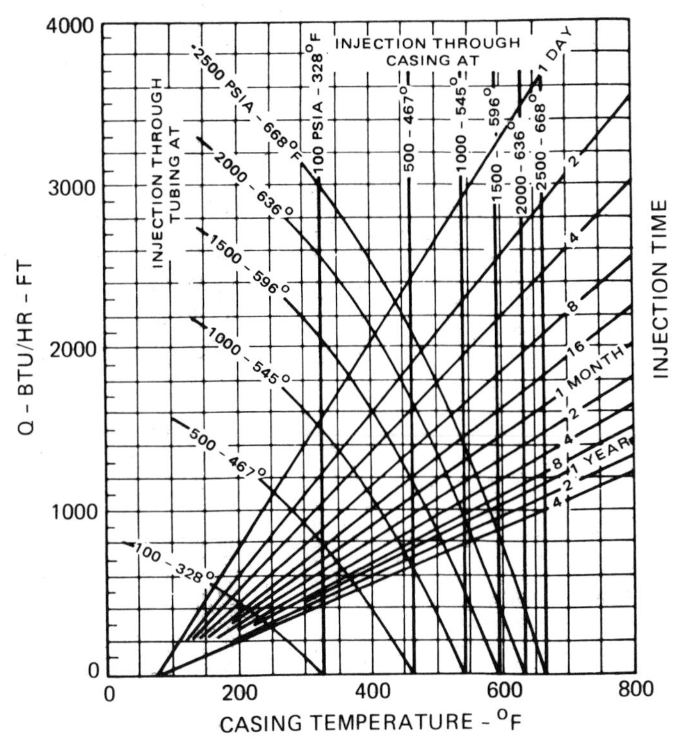

Figure 2.12 Heat loss in wellbore and casing temperature during steam injection: $4\frac{1}{2}$-in. tubing, $8\frac{5}{8}$-in. casing.[4] © 1966 *Producers Monthly.*

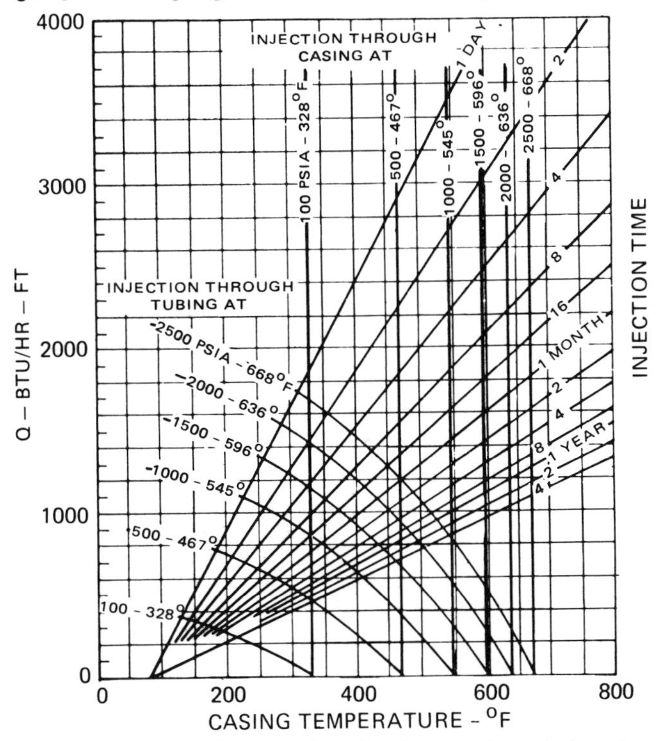

Figure 2.13 Heat loss in wellbore and casing temperature during steam injection: $2\frac{1}{2}$-in. tubing, $10\frac{3}{4}$-in. casing.[4] © 1966 *Producers Monthly.*

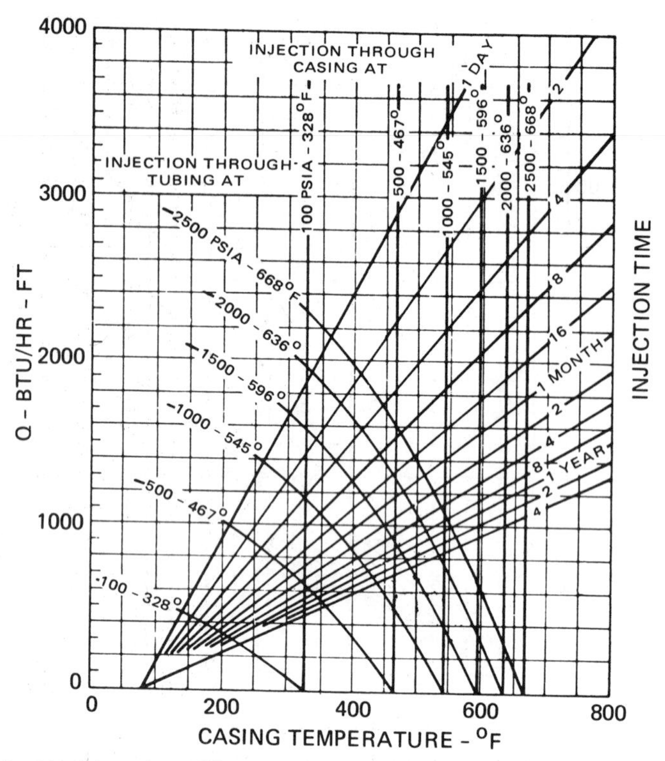

Figure 2.14 Heat loss in wellbore and casing temperature during steam injection: $3\frac{1}{2}$-in. tubing, $10\frac{3}{4}$-in. casing.[4] © 1966 *Producers Monthly*.

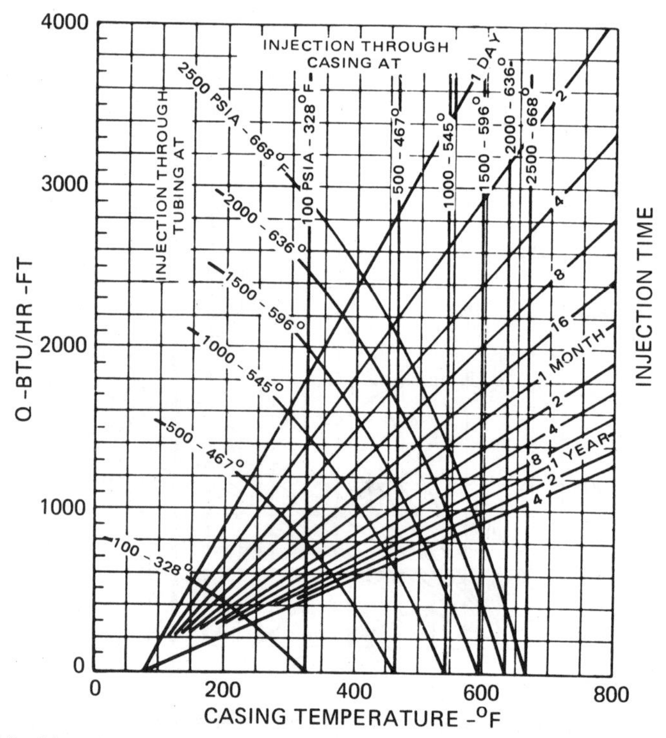

Figure 2.15 Heat loss in wellbore and casing temperature during steam injection: $4\frac{1}{2}$-in. tubing, $10\frac{3}{4}$-in. casing.[4] © 1966 *Producers Monthly*.

ESTIMATE OF DOWNHOLE STEAM QUALITY

The bottom-hole steam quality X_i is estimated by

$$X_i = X_{surf} - \frac{QD}{m_s H_{wv}} \tag{2.5}$$

where X_{surf} = surface-injected steam quality (lb vapor/lb vapor plus liquid)
Q = wellbore heat loss rate (Btu/hr-ft)
m_s = mass rate of steam injection (lb/hr)
H_{wv} = specific latent heat of vaporizaton at injection temperature and pressure (Btu/lb)
D = depth (ft)

DOWNHOLE TEMPERATURE DURING HOT-FLUID INJECTION

Hot-water injection, although of lesser commercial importance than steam, has application in some cases. Ramey[1] derived the following relation for hot-liquid injection down tubing: the $f(t)$ term is defined as before. The bottom-hole temperature of the injected liquid, $T_{f_{bh}}$, is given by

$$T_{f_{bh}} = aD + b - aA + (T_{f_{surf}} + aA - b)e^{-D/A} \tag{2.6}$$

where a = geothermal gradient (°F/ft)
b = average surface geothermal temperature (°F)
$T_{f_{surf}}$ = fluid temperature at surface (°F)

and

$$A = \frac{wc[K_h + r_{to} U_{to} f(t)]}{2\pi r_{to} U_{to} K_h} \tag{2.7}$$

where w = fluid injection rate (lb/hr)
c = specific heat of liquid (Btu/lb − °F)

The other quantities have already been defined. Note that Eqs. (2.6) and (2.7) can also be used to determine the bottom-hole temperature for the case where the steam condenses completely before entering the formation. To do this, we must first find the depth Z where the steam quality goes to zero. By setting $X_i = 0$ and transposing Eq. (2.5), the depth at which condensation is complete occurs at

$$Z = \frac{X_{surf} m_s H_{wv}}{Q} \tag{2.8}$$

To obtain the bottom-hole temperature of the steam condensate for this case, solve Eqs. (2.6) and (2.7) with

$$D \text{ replaced by } D - Z$$

$$b \text{ replaced by } b + aZ$$

$$T_{f_{surf}} = T_s$$

SPECIFIC EXAMPLES OF EFFECT OF WELLBORE HEAT LOSS PARAMETERS

Figures 2.16–2.18 give examples showing the effect of injection rate and depth on the percentage of injected heat lost as a function of time during steam injection. These are intended only as a qualitative guide and do not exempt the engineer from using the more rigorous calculation techniques presented here. The figures are taken from Satter's paper,[5] where he employed techniques similar to those of the Ramey–Willhite method. Figure 2.16 gives the percentage of input heat lost after one year of steam injection for the case of saturated steam at 500 psia as a function of depth and injection rate. It indicates that for injection rates in excess of 9000 lb/hr, less

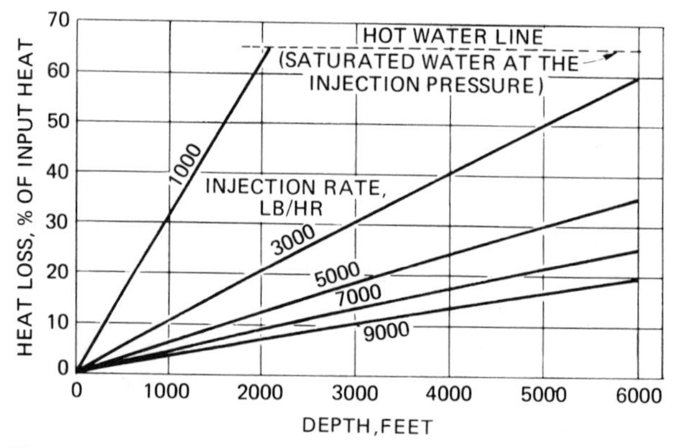

Figure 2.16 Heat loss as function of injection rate and depth. Wellhead condition of steam, saturated steam at 500 psia and 467°F; injection time, 1 year.[5] © 1965 SPE-AIME.

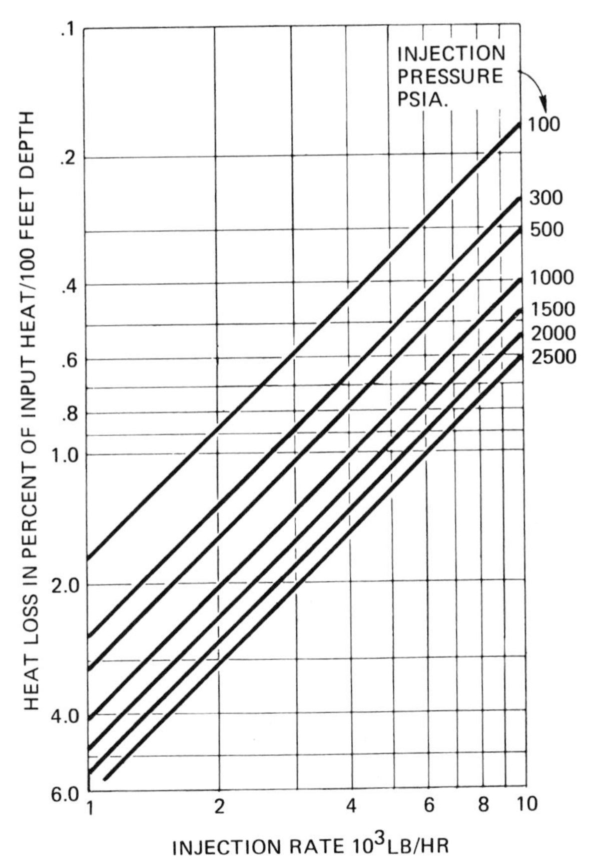

Figure 2.17 Heat loss as a function of injection rate and pressure in case of injecting saturated steam for 1 year.[5] © 1965 SPE-AIME.

than 20% of the injected heat will be lost for wells shallower than 6000 ft. The effects of injection pressure and rate are shown in Figure 2.17. Heat losses are greater, as this figure indicates, for higher injection pressures because the condensing steam temperature is higher than at lower pressures.

Figure 2.18 gives the percentage heat lost as a function of injection time in years for the case of using a packer versus not using a packer. For the conditions assumed (5000 lb/hr, 500 psia, 3000 ft), not maintaining a dry, low-pressure annulus results in about 30% greater heat loss. Notice that after the first year the heat loss percentage decreases very slowly with time. Figure 2.19 gives the location of the point in the well where all the steam is condensed to hot water as a function of injection pressure and rate.

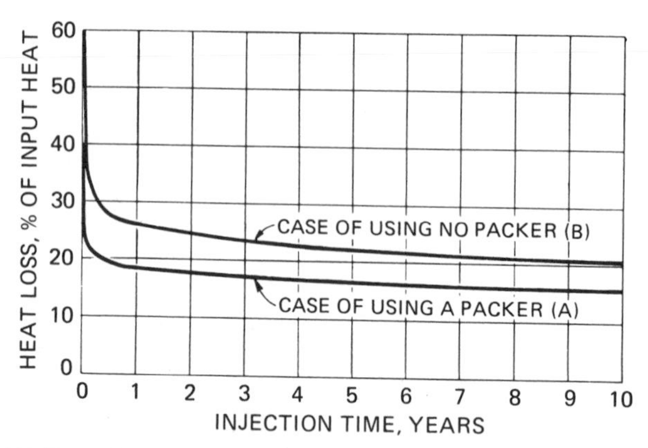

Figure 2.18 Wellbore heat losses with and without packer. Injection rate, 5000 lb/hr of saturated steam; injection pressure, 500 psia; saturated temperature, 467°F; depth, 3000 ft.[5] © 1965 SPE-AIME.

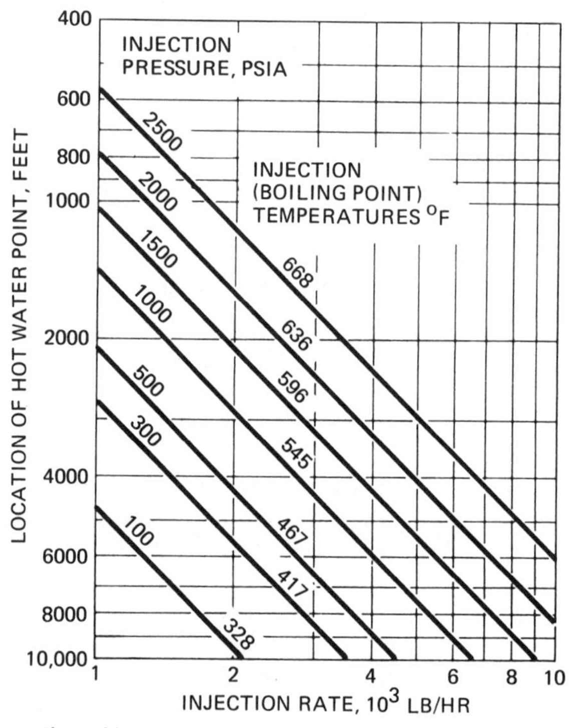

Figure 2.19 Location of hot-water point as function of injection rate and pressure in case of injecting saturated steam for 1 year.[5] © 1965 SPE-AIME.

EFFECT OF TUBING PRESSURE DROP

None of the above techniques account for pressure drop in the tubing as a function of fluid head or friction. During injection, a properly designed steam injection well should not have a greatly different tubing pressure at the wellhead from that at the bottom of the hole since the pressure drop resulting from friction will be offset by the gravity head. However, if the tubing is undersized for the desired injection rate, a significant pressure drop can occur. Equations for estimating tubing pressure at bottom hole are given in the literature.[6] It is suggested that T_s be estimated at the average tubing pressure when using Eqs. (2.3) and (2.4) and for evaluation of H_{wv} in Eq. (2.5).

Results of calculations of steam pressure drop are given in Figures 2.20 and 2.21. A computer program similar to that developed by Earlougher[6] was used. These figures plot the tubing pressure drop as a function of depth in feet. Figure 2.20 shows that at a low surface injection pressure, the pressure drop can be substantial because friction losses will exceed the gain in pressure owing to the head of the fluid. The head is much greater at higher surface pressures.

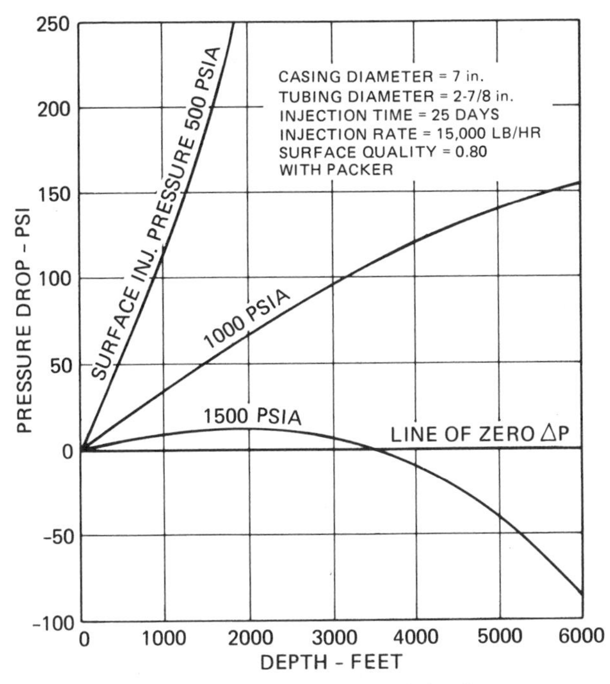

Figure 2.20 Effect of injection pressure and depth on pressure drop.[7]

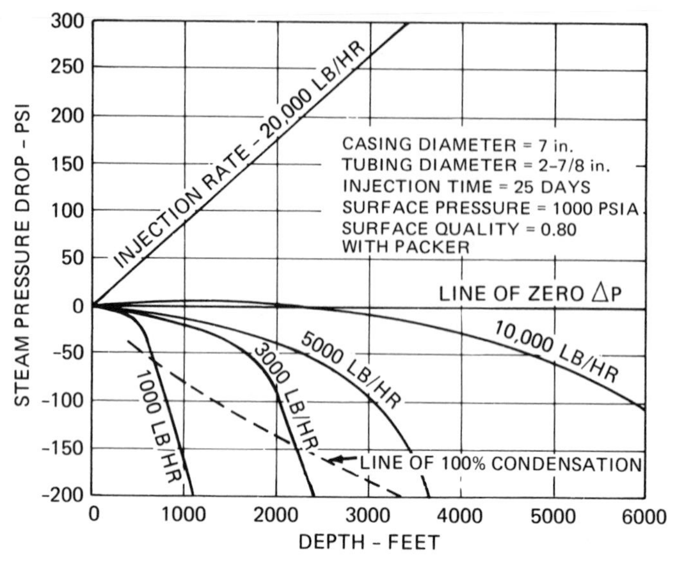

Figure 2.21 Effect of injection rate and depth on steam pressure drop.[7]

For the injection rate shown (15,000 lb/hr), frictional losses become greatly reduced at greater depths where steam qualities are low and vapor velocities are reduced. For the 1500 psia surface pressure, there is an increase of about 40 psia at a depth of 5000 ft for the case shown. Figure 2.21 shows the effect of injection rate. At higher rates, the pressure loss is positive, indicating a lower than surface pressure at bottom hole. For low injection rates, the gravity head dominates, and the pressure will be higher at reservoir depth than at the surface.

THERMAL WELL COMPLETIONS

Wells completed for thermal operations generally employ the following principles to minimize casing stresses.

Maintain Freedom for Casing to Expand

Over the uncemented portion of the casing, it is not unusual for the casing to be free to expand linearly and rise from several inches to several feet. This will depend on the freedom of the casing, its increase in temperature above the formation temperature, and the length of the uncemented portion. To permit this expansion, it is advisable to unbolt the casing head from the surface casing. Buckling or other failure could occur if this is not done. Some thermal wells especially designed for steam injection service have employed a tarry coating on the outside of the casing.[8] This permits movement of the pipe within the cement sheath. This technique is recommended only for steam temperature service below 500°F. At higher temperatures, the tar can coke, binding the pipe and preventing contraction when it cools.

The thermal elongation ΔL of the casing over a length L where it is free to expand is given by

$$\Delta L = \alpha_l (\Delta T) L$$

where α_l = coefficient of thermal expansion (7.0×10^{-6} in./in.-°F)

ΔT = increase in temperature over average formation temperature (°F)

ΔL = free length of tubing or casing (ft)

Prestress Casing

Casing that is cemented can be prestressed in tension by pulling on the casing while cementing in stages. The compressive stress S_c developed upon heating is given by

$$S_c = \alpha_l (\Delta T) E - S_t \tag{2.8}$$

where E = modulus of elasticity (29×10^6 psi)

S_t = tensile stress developed during prestressing (psi)

Note that the compressive stress S_c is approximately $S_c = 200\,\Delta T - S_t$. If the casing has not been prestressed, $S_t = 0$.

It is recommended that S_c not exceed the yield strength (Fig. 2.22). This is plotted as a function of temperature for various casing grades. The ultimate tensile strength is plotted in Figure 2.23.

The type of casing joint used is important. Threaded joints with short threads and collars or long threads and collars will have lower yield strengths than the pipe body. The joint yield stress becomes controlling in these cases. It is recommended that buttress threads (or the equivalent), which have the highest joint strength, be used in new thermal wells.

Ordinary neat cement has a substantial compressive strength loss at temperatures above 230°F. Compressive strength may drop from 4000 to 300 psi in 30 days at high temperatures. Temperature stability can be obtained by adding silica quartz or silica flour to the cement (add 30–60%

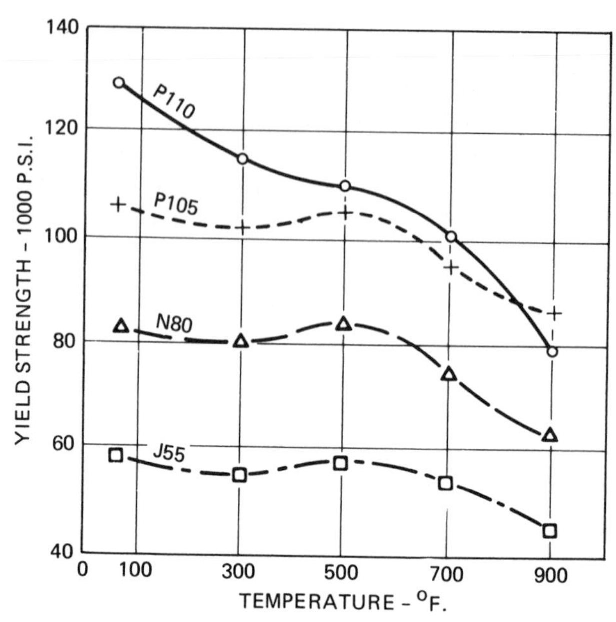

Figure 2.22 Yield strength of casing versus temperature.[9] © 1967 The World Petroleum Congress, Elsevier Publishing Co. Ltd.

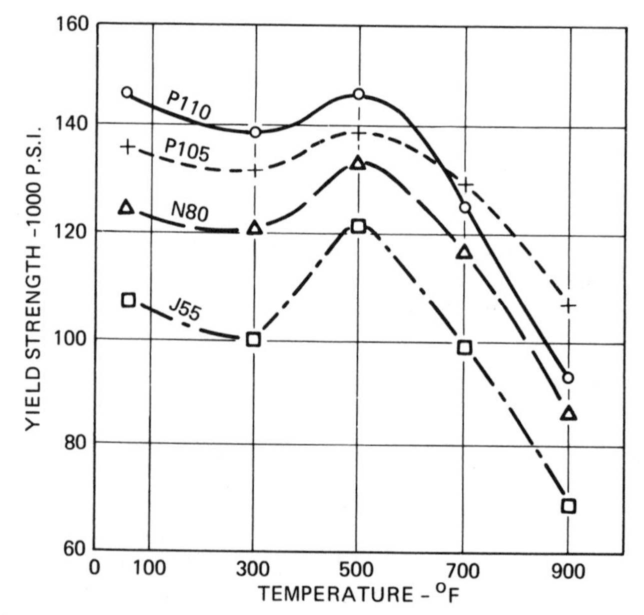

Figure 2.23 Tensile strength of casing versus temperature.[9] © 1967 The World Petroleum Congress, Elsevier Publishing Co. Ltd.

Table 2.3 Compressive Strengths of Cementing Compositions[a]

Silica Flour (%)	Gel (%)	ft³ Perlite	Strength (psi) 100°F 8 hr	100°F 1 day	400°F 1 day	400°F 7 days
		API Class G Cement				
30	0	0	585	2,515	4,915	6,640
40	0	0	330	1,490	4,890	6,500
30	0	1	315	925	3,915	3,565
30	2	1	230	710	3,075	1,990
40	0	1	355	775	3,825	3,130
40	2	1	205	400	2,560	1,905
30	0	2	120	420	2,145	1,415
30	2	2	60	325	1,595	1,480
40	0	2	145	375	1,850	2,025
40	2	2	30	265	1,440	1,220
30	0	3	60	270	1,080	1,220
30	2	3	20	180	840	880
		50–50 Class G—Pozzolan Cement				
30	0	0	130	690	4,050	4,100
30	2	0	100	480	2,925	3,280
40	0	0	135	670	3,930	3,850
40	2	0	95	460	1,505	2,945
		Calcium Aluminate Cement				
40	0	0	5	965	900	620

[a] From Ref. 10. © 1966 SPE-AIME.

Table 2.4 Pressure Ratings (psi) at Various Temperatures of Metallic Parts of API Wellheads, Valves, and Flanges.[a,b]

Metal Temperature (°F) −20–250°F	300	350	400	450	500	550	600	650
2000	1995	1905	1860	1810	1753	1635	1540	1430
3000	2930	2860	2785	2715	2605	2455	2310	2145
5000	4880	4765	4645	4525	4340	4090	3850	3575

[a] Does not apply to 5000 psi 6BX API flanges.
[b] From Ref. 13. © 1986 API.

silica by weight). Compressive strengths are given in Table 2.3. Further details on well completion design can be found in the literature.[8–12]

Elevated temperatures and pressures may require replacement of metallic wellhead parts used in conventional operations with equipment designed for thermal operations. Table 2.4 gives pressure–temperature ratings of metallic parts of API wellheads, valves, and flanges.

SILICATE FOAM INSULATION

A novel technique was developed by Exxon for use in high-pressure steam injection wells in the Venezuelan Bolivar coastal fields.[3,14–16] The technique has been used successfully for a number of years by Lagoven. It involves coating the tubing with a layer of foamed sodium silicate that is formed by boiling a sodium silicate solution. The foam is an excellent insulator, having a thermal conductivity of approximately 0.017 Btu/hr-ft-°F.[17,18]

In a field operation, a solution of sodium silicate is placed in a packed-off annulus, and then steam is injected down the tubing. The hot tubing causes the silicate solution to boil, leaving a coating of insulating foam, usually about $\frac{1}{4}$–$\frac{1}{2}$ in. thick, on the hot-tubing surface. Since the foam immediately becomes an effective insulation, none is deposited on the inner wall of the casing. Silicate solution that remains in the annulus after steaming for several hours is removed from the annulus by displacing it with water (if the solution is not removed, it may solidify in the annulus). The water is removed by gas lifting or by swabbing.[19] Figure 2.24 is a schematic showing the steps of the insulation process.

Once the insulation is formed, heat loss is reduced, and lower casing temperatures and higher sandface steam qualities result. A comparison showing the foam's effectiveness is presented in Figure 2.25, which shows the calculated maximum casing temperature in a well with packed-off tubing. The three cases show the relationship between casing temperature and steam injection time for uninsulated tubing, commercially available insulated tubing, and tubing with a $\frac{1}{4}$-in.-thick coating of silicate foam. Lowest casing temperatures are calculated using the silicate foam insulation.

The silicate foam insulation process has been used routinely in the field since 1970 in Lagoven's Lake Maracaibo steam injection operations. To date, it has been used in wells whose depths have varied from 2600 to 4500 ft, with steam injection pressures that are normally 1250–1650 psi. This technique appears to be as effective as and less expensive than conventionally used methods of reducing wellbore heat loss and preventing casing failures.

To reduce or prevent venting of steam past leaking packers and subsequent dissolving of silicate foam during protracted injection, it is found

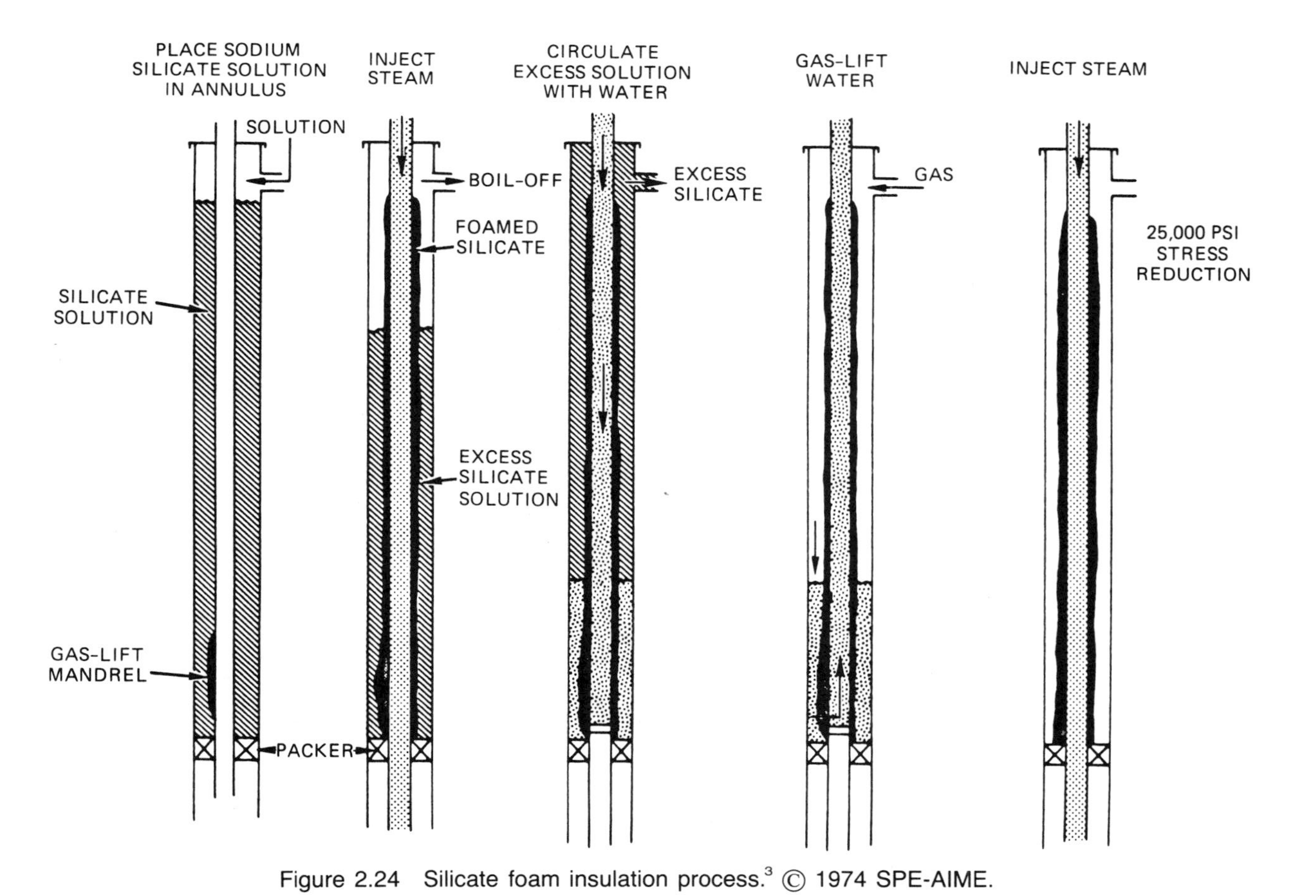

Figure 2.24 Silicate foam insulation process.[3] © 1974 SPE-AIME.

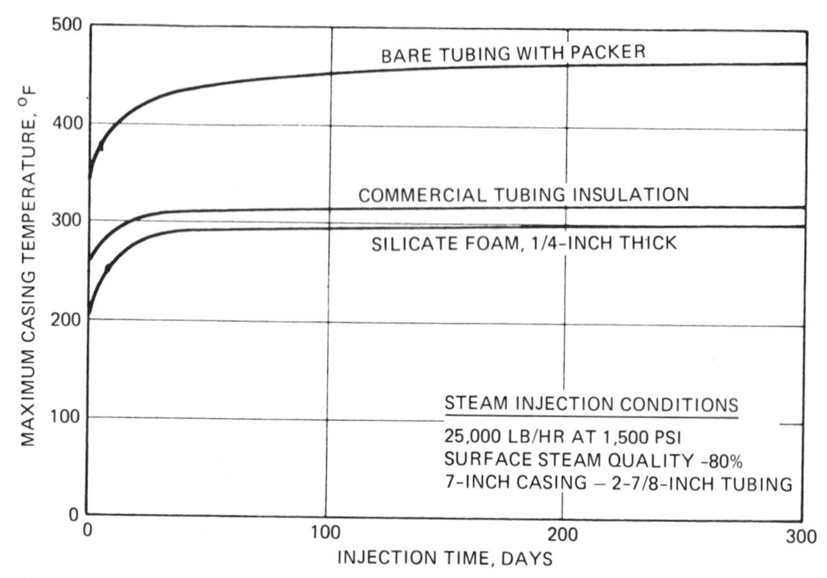

Figure 2.25 Casing temperature as function of time.[3] © 1974 SPE-AIME.

effective to load the annulus with Bunker-C or heavy oil. This does not greatly reduce insulation effectiveness during steam injection. Tubing or packer leaks must be avoided when using this insulating technique because steam entering the annulus will quickly dissolve the silicate foam and could cause a casing failure. Maintaining an oil-filled annulus appears to be the best approach to avoid this problem.

OTHER WELLBORE INSULATION METHODS

Various other wellbore insulation methods have been tested as a part of the U.S. Department of Energy Program, Project Deep Steam.[20] A comparison of the insulating effectiveness of various approaches is reported in a cooperative study by Sandia National Laboratories and Husky Oil.[21]

APPENDIX

Equations for U_{to}

Willhite[2] gives the following equations, which may be used to obtain improved estimates of U_{to} for casing and tubing sizes other than 7 and

$2\frac{7}{8}$ in., respectively:

$$\frac{1}{U_{to}} = \frac{1}{h_c + h_r} + \frac{r_{to} \ln (r_h/r_{co})}{K_{cem}} \tag{2A.1}$$

The convection and conduction in the annulus is accounted for by h_c:

$$h_c = \frac{0.049(\text{GrPr})^{0.333}\text{Pr}^{0.074}K_{ha}}{r_{to} \ln (r_{ci}/r_{to})} \tag{2A.2}$$

where

$$Gr = \frac{(r_{ci} - r_{to})^3 g\rho_{an}^2 \beta_{an}(T_{to} - T_{ci})}{\mu_{an}^2} \tag{2A.3}$$

$$\text{Pr} = \frac{c_{an}\mu_{an}}{K_{ha}}$$

K_{ha} = thermal conductivity of gas in annulus at average temperature and pressure of annulus (Btu/hr-ft-°F)
g = gravitational acceleration (4.17×10^8 ft/hr^2)

and ρ_{an}, β_{an}, μ_{an}, and c_{an} are the density, thermal coefficient of expansion, viscosity, and specific heat of the annulus fluid, respectively.

Radiation in the annulus is accounted for by h_r:

$$h_r = \sigma F_{t_{ci}}(T_{to}^{*2} + T_{ci}^{*2})(T_{to}^* + T_{ci}^*) \tag{2A.4}$$

where σ is the Stefan–Boltzmann constant, $=1.713 \times 10^{-9}$ (ft^2-hr-°R^4)$^{-1}$.

The asterisks indicate that the designated temperatures are absolute temperature (°R):

$$F_{t_{ci}} = \frac{1}{\epsilon_{to}} + \frac{r_{to}}{r_{ci}}\left(\frac{1}{\epsilon_{ci}} - 1\right) \tag{2A.5}$$

where ϵ is the emissivity (dimensionless) and the subscripts to and ci indicate outside of tubing and inside of casing, respectively.

REFERENCES

1. Ramey, H. J. "Wellbore Heat Transmission," *J. Pet. Tech.* (April 1962), p. 427.
2. Willhite, G. P. "Overall Heat Transfer Coefficients in Steam and Hot Water Injection Wells," *J. Pet. Tech.* (May 1967), p. 607.

3. Penberthy, W. L., Jr., and Bayless, J. H. "Silicate Foam Wellbore Insulation," *J. Pet. Tech.* (June 1974), pp. 583–588.
4. Huygen, H. A., and Huitt, J. L. "Wellbore Heat Losses and Casing Temperatures During Steam Injection," *Producers Monthly* (August 1966), p. 2.
5. Satter, A. "Heat Losses During Flow of Steam Down a Wellbore," *J. Pet. Tech.* (July 1965), p. 845.
6. Earlougher, R. C., Jr. "Some Practical Considerations in the Design of Steam Injection Wells," *J. Pet. Tech.* (January 1965), p. 79.
7. Hagedorn, A. R. Exxon Production Research Co., unpublished data, May 1966.
8. Greer, F. C., and Shryock, S. H. "New Technique Improves Steam Stimulation Completion," *J. Pet. Tech.* (June 1968), p. 691.
9. Gates, C. F., and Holmes, B. G. "Thermal Well Completions and Operations," *The World Petroleum Congress Proceedings*, Vol. 3, Elsevier, New York, 1967, pp. 419–429.
10. Cain, J. E., and Shryock, S. H. "Cementing Steam Injection Wells in California," *J. Pet. Tech.* (April 1966), p. 431.
11. Willhite, G. P., and Dietrich, W. K. "Design Criteria for Completion of Steam Injection Wells," *J. Pet. Tech.* (January 1967), p. 15.
12. Leutwyler, K., and Bigelow, H. L. "Temperature Effects on Subsurface Equipment in Steam Injection Systems," *J. Pet. Tech.* (January 1965), p. 93.
13. API Specification for Wellhead and Christmas Tree Equipment, Spec. 6A, 15th ed. (April 1986), p. 149.
14. Bayless, J. H., and Penberthy, W. L., Jr. "Technique for Insulating a Wellbore with Silicate Foam," U.S. Patent 3,525,399, August 25, 1970.
15. Bayless, J. H., and Penberthy, W. L., Jr. "Technique for Insulating a Wellbore with Silicate Foam," U.S. Patent 3,718,184, February 27, 1973.
16. Penberthy, W. L., Jr., West, R. C., and Bayless, J. H. "Method for Retarding the Solution of Sodium Silicate Foam," U.S. Patent 3,664,424, May 23, 1972.
17. Baker, E. J., Jr. "New Applications for Sodium Silicate," Southwest Research Institute paper presented at the Twelfth National SAMPE Symposium, 1968.
18. Weldes, H. H., and Lange, K. R. "Properties of Soluble Silicates," *Ind. and Eng. Chem.* (April 1969), pp. 29–44.
19. Penberthy, W. L., Jr., Bayless, J. H., and Ayers, R. C. "Method of Removing Excess Sodium Silicate Solution from Insulated Well Annulus," U.S. Patent 3,664,425, May 23, 1972.
20. Johnson, D. R., Ed. "Project Deep Steam Quarterly Report, July 1–September 30, 1980," Sandia National Laboratories, SAND80-2873, March 1981.
21. Aeschliman, D. P., Noble, N. J., and Meldau, R. F. "Thermal Efficiency of Steam Injection Test Well with Insulated Tubing," Paper 83-34-37, Presented at the Thirty-Fourth Annual Technical Meeting of the Petroleum Society of CIM in Banff, May 1983.

Chapter 3

HEAT TRANSMISSION BY HOT-FLUID INJECTION

ABSTRACT

When one has defined the average conditions of steam quality or temperature at formation depth in the injection wellbore, simplified methods described in this chapter can be used to approximate temperatures that will be established in the formation itself for various quantities of total hot-fluid input. Use of the Lauwerier equation for hot-liquid injection and the Marx–Langenheim equation for steam injection are discussed in this chapter. These equations account for vertical heat losses occurring by conduction to unproductive strata above and below the oil sand as well as the heating of the invaded oil sand itself. The equations represent a considerable oversimplification of the heating process in that reservoir heterogeneities, gravity effects, and areal fluid flow effects are not considered. Nevertheless, they have been used with remarkable success as the heat transfer basis for numerous simplified methods (to be discussed in later chapters) that account for oil displacement with various degrees of rigor. Results of other more sophisticated methods, including numerical simulator results, for the simplified case of one-dimensional hot-fluid flow through an oil sand are also presented.

INTRODUCTION

Predictions of the behavior of thermal stimulation and drive processes involving hot-fluid injection require an accurate accounting of the energy injected into the reservoir. In this chapter, we will discuss some of the simplified analytical models that describe reservoir heating and temperature change as a hot fluid such as hot water or steam is injected. In all cases discussed, uniform vertical invasion of the oil sand by the injected fluid is assumed. Flow is one dimensional (radial or linear) with heat losses by conduction to overburden and underburden taken into account. Conduction is allowed only vertically in the simplest methods. In the real world, three-dimensional flow of both heat and fluids occurs, so that solutions presented here are only a crude approximation of an actual reservoir situation. They more closely approximate behavior in a thin homogeneous sand where gravity and stratification effects are minimal. In all cases, a constant rate of injection is assumed. Other assumptions are as follows:

1. Flow velocities are sufficiently low that thermal equilibrium between injected and reservoir fluids and reservoir sand is instantaneous.
2. Heat transfer results from the physical movement of the injected fluids and thermal conduction from warmer to cooler oil sand and to impermeable rock above and below the oil sand. The simpler methods neglect conduction in the direction of flow and assume that temperatures do not vary with depth within the oil sand.

HOT-LIQUID INJECTION

The general differential equation for three-dimensional heat transfer during hot-liquid injection is

$$\nabla \cdot K_h \nabla T - \nabla \cdot (\rho_f c_f \bar{V} T) = \frac{\partial[(\rho c)_{R+F} T]}{\partial t} \tag{3.1}$$

where K_h = formation thermal conductivity (Btu/hr-ft-°F)

T = temperature (°F)

t = time (hr)

$\rho_f c_f \bar{V} = \rho_o c_o \bar{V}_o + \rho_w c_w \bar{V}_w$

$\rho c \bar{V}$ = product of density, specific heat, and average flow velocity of water or oil (Btu/ft^2-hr-°F)

and

$(\rho c)_{R+F}$ = product of density and specific heat of rock including interstitial fluids (Btu/ft^3-°F)

$$= \phi(\rho_o c_o S_o + \rho_w c_w S_w) + (1 - \phi)\rho_r c_r \tag{3.1a}$$

where ϕ = porosity (dimensionless)

S = fluid saturation (dimensionless)

The subscripts f, o, w, and r refer to fluid, oil, water, and rock, respectively.

Equation (3.1) also applies to the impermeable strata surrounding the oil reservoir where $V = \phi = 0$. Here Eq. (3.1) reduces to

$$\nabla \cdot K_h \nabla T = \frac{\partial(\rho_r c_r T)}{\partial t} \tag{3.2}$$

LAUWERIER SOLUTION

Perhaps the most widely known analytical solution to Eqs. (3.1) and (3.2) is that of Lauwerier.[1] This model assumes the following:

(a) Linear, one-dimensional, incompressible flow in a homogeneous sand.
(b) Constant physical properties and fluid saturations (the effect of this assumption is further discussed in Chapter 12).
(c) Temperature uniform with vertical position within the oil sand at any given horizontal position ($K_h = \infty$ is the vertical direction in the oil sand).
(d) $K_h = 0$ in the horizontal direction in both sand and surrounding impermeable rock.
(e) Constant injection temperature and rate.

Temperatures in the oil sand and impermeable rock are given by

$$\bar{T} = \operatorname{erfc}\left[\frac{\xi + |\eta| - 1}{2\sqrt{\theta(\tau - \xi)}}\right] U(\tau - \xi) \tag{3.3}$$

where

$$\bar{T} = \frac{T - T_r}{T_I - T_r} \tag{3.3a}$$

and T = temperature at any point in oil sand or impermeable rock (°F)
T_r = original reservoir temperature (°F)
T_I = injection temperature at sand face (°F)

For a linear system,

$$\xi = \frac{4 K_{hob} x}{h^2 \rho_f c_f \bar{V}} \tag{3.3b}$$

where K_{hob} = thermal conductivity of overburden (Btu/day-ft-°F)
x = distance from injection well (ft)
h = gross oil sand thickness (ft)

For a radial system,[2]

$$\xi = \frac{4K_{\text{hob}}\pi r^2}{h\rho_f c_f Q} \tag{3.3c}$$

where r = radial distance from injection well (ft)
Q = volumetric flow rate (ft^3/d)

For either system,

$$\eta = \begin{cases} 1 & \text{for temperatures in oil sand} \\ 2z/h & \text{for temperatures in impermeable rock } (z \geqq h/2) \end{cases}$$

where z is the vertical distance from the midplane of the sand in feet,

$$\tau = \frac{4K_{\text{hob}}t}{h^2(\rho c)_{\text{R+F}}} \quad \text{(dimensionless time)} \tag{3.3d}$$

where t is the time since the start of injection in days,

$$\theta = (\rho c)_{\text{R+F}}/(\rho c)_{\text{ob}} \tag{3.3e}$$

where $(\rho c)_{\text{ob}}$ is the product of specific heat and density for the impermeable rock (overburden) in Btu per square foot per degree Fahrenheit, and

$$U(\tau - \xi) = \begin{cases} 0 & \text{when } (\tau - \xi) \leqq 0 \\ 1 & \text{when } (\tau - \xi) > 0 \end{cases}$$

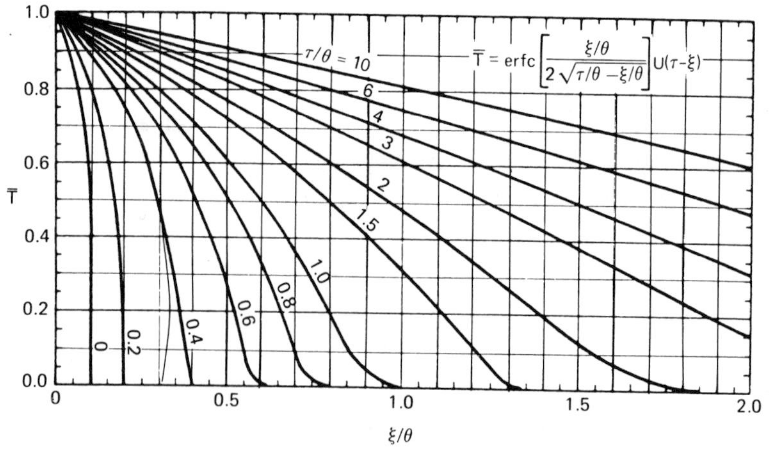

Figure 3.1 Dimensionless temperature distribution for Lauwerier solution for hot waterflooding.[1] © 1955 *Appl. Sci. Research.*

Equation (3.3) for the case of T within the oil sand can be presented in graphical form as plots of $\bar{T}$ as a function of ξ/θ and τ/θ. This is given in Figure 3.1.

OTHER SOLUTIONS FOR HOT-LIQUID INJECTION

Numerous workers have developed other analytical and numerical solutions to Eqs. (3.1) and (3.2). The solutions involve less restrictive assumptions than the Lauwerier solution but are more complex. Spillette[3] has summarized these methods. Some of the methods permit calculation of the temperature in the oil sand. Others give equations for the fraction of injected energy not lost to the impermeable rock, commonly referred to as the thermal efficiency. Table 3.1 summarizes the various techniques.

Spillette compared the results of his numerical solution to the results of Avdonin and Lauwerier for the linear and radial cases. The comparisons are shown in Figures 3.2 and 3.3, which plot dimensionless temperature T against linear or radial distance from the injection well. The thermal properties used for these calculations are given in Table 3.2. For the linear case, it is seen that neglecting conduction in the direction of flow causes the Lauwerier solution to be conservative in that it does not predict as great a heat advance into the formation as do the other methods. This is also true for the radial case but to a lesser degree. Note that temperatures predicted by the Lauwerier method are higher near the wellbore than those predicted by the more exact methods.

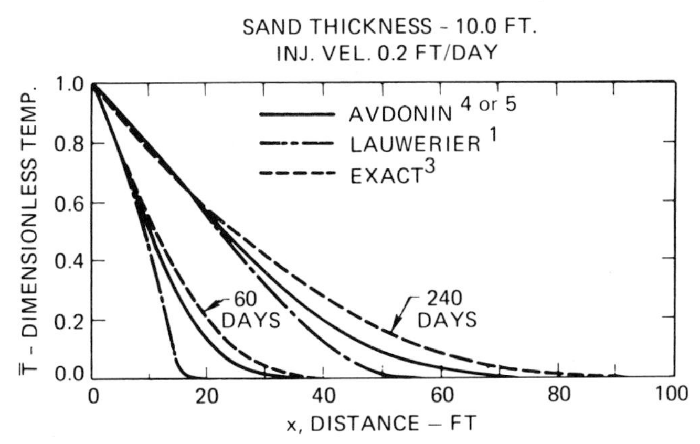

Figure 3.2 Sand temperature distribution, linear flow.[3] © 1965 *J. Can. Pet. Tech.*, *CIM*.

Table 3.1 Summary of One-Dimensional Hot-Liquid Injection Solutions

			Thermal Conductivity				Solution	
			Sand		Impermeable Rock			
Author	Ref.	Flow Type	Horizontal	Vertical	Horizontal	Vertical	Temperature Distribution	Thermal Efficiency
Lauwerier	1, 2	Linear, radial	0	∞	0	Finite	Yes	Yes
Avdonin	4	Linear, radial	Finite	∞	0	Finite	Yes	No
Avdonin	5	Linear, radial	0	Finite	0	Finite	Yes	No
Rubinshtein	6	Radial	Finite	Finite	Finite	Finite	No	Yes
Rubinshtein	7, 8	Radial	Finite	∞	Finite	Finite	Yes	No
Spillette[a]	3	Linear, radial	Finite	Finite	Finite	Finite	Yes	Yes

[a] Numerical Solution: Other methods are analytic.

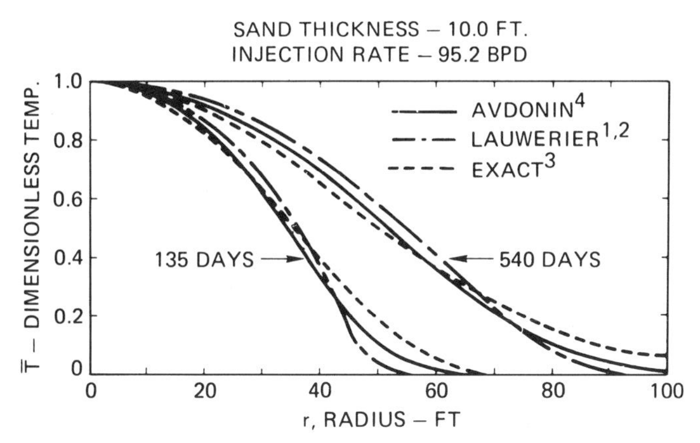

Figure 3.3 Sand temperature distribution, radial flow.[3] © 1965 *J. Can. Pet. Tech.*, *CIM.*

Table 3.2 Thermal Properties Used for Calculations of Figures 3.2 and 3.3[a]

Volumetric heat capacity of injected fluid, $\rho_f c_f$	62.4 Btu/ft-°F
Volumetric heat capacity of porous media including interstitial fluids, $(\rho c)_{R+F}$	42.45 Btu/ft-°F
Volumetric heat capacity of rock and of overburden, $\rho_r c_r = (\rho c)_{ob}$	36.3 Btu/ft-°F
Thermal conductivity of oil, formation and overburden, K_h	1.4 Btu/hr-ft-°F

[a] From Ref. 3. © 1965 *J. Can. Pet. Tech.*, CIM.

STEAM INJECTION

Equation (3.1) must be modified for the case of steam injection to account for the latent heat of vaporization of steam. To faciliate this, we shall use the specific enthalpies (H_o, H_w, H_g) of the phases rather than the heat capacities:

$$\nabla \cdot K_h \nabla T - \nabla \cdot (\bar{V}\rho_f H_f) = \frac{\partial[(\rho c)_{R+F} T]}{\partial t} \tag{3.4}$$

where

$$\bar{V}\rho_f H_f = \bar{V}_o \rho_o H_o + \bar{V}_w \rho_w H_w + \bar{V}_g \rho_g H_g$$

and

$$(\rho c)_{R+F} = \frac{\phi(S_o \rho_o H_o + S_w \rho_w H_w + S_g \rho_g H_g)}{T - T_r} + (1 - \phi)(\rho_r c_r)$$

Note that this equation for $(\rho c)_{R+F}$ is more general than the previous one in that it accounts for the enthalpies of all three phases. For a system containing no free gas, $H_g = H_s$, the enthalpy of steam. Whenever both steam and condensed water exist together in equilibrium, we say that the steam is saturated. At saturated conditions, only one steam temperature will exist for each pressure. In general, there will always be some condensed water present in the reservoir during steam injection, and we assume conventionally that steam and water are at equilibrium at saturated conditions. For the case of no free gas and saturated conditions,

$$H_g = H_s = H_{wv} + H_{ws} - H_{wr}$$

where H_{wv} = latent heat of vaporization at P_s, T_s (Btu/lb)
H_{ws} = enthalpy of liquid water at P_s, T_s (Btu/lb)
H_{wr} = enthalpy of liquid water at P_r, T_r (Btu/lb)

Enthalpy data must be referred to a base temperature and pressure. The H_{wr} term must be included to reference enthalpy to original reservoir conditions, which is the base condition usually assumed in this text. In the standard water enthalpy tables (steam tables), all enthalpies are referenced to liquid water at 32°F and pressure at saturated conditions.

Note that Eq. (3.2) remains the same as for hot-liquid injection to describe heat transport in the impermeable rock surrounding the oil sand.

MARX–LANGENHEIM SOLUTION

Probably the most popular approximate analytical solution to Eqs. (3.4) and (3.2) is that of Marx and Langenheim.[9] The assumptions are similar to those employed in the Lauwerier solution:

(a) One-dimensional flow in a homogeneous sand; area swept can be of any arbitrary shape.

(b) Constant physical properties and fluid saturations.

(c) $K_h = \infty$ in the vertical direction in the oil sand.

(d) $K_h = 0$ in the horizontal direction in both the oil sand and surrounding rock.

(e) Oil sand is assumed to be either at T_I, the injected steam temperature, or T_r, the original reservoir temperature. (This assumption neglects the gradual drop-off in temperature in the hot-water bank ahead of the steam that occurs in reality. It becomes an increasingly poorer approximation as injection rates decrease, but it is reasonably accurate at high injection rates. A comparison between the idealized and actual temperature distribution is shown in Fig. 3.4.)

(f) Constant injection temperature and rate.

Some laboratory results giving temperature as a function of distance from the inlet end of a core are given by Willman[10] and are shown in Figure 3.5. For the conditions of the experiment shown in Figure 3.5, the Marx–Langenheim assumption (e) would be a rather crude approximation. Other authors[11,12] have improved the realism of their calculations by taking this hot-water movement into account.

Under the Marx–Langenheim assumptions, Eq. (3.4) is integrated to obtain

$$m_s H_t = 2 \int_0^t \frac{K_{\text{hob}}\, \Delta T}{\sqrt{\pi a(t-y)}} \frac{dA}{dy}\, dy + (\rho c)_{R+F} h\, \Delta T \frac{dA}{dt} \tag{3.5}$$

Briefly, Eq. (3.5) states

$$\begin{matrix}\text{rate} \\ \text{of energy} \\ \text{injected}\end{matrix} = \begin{matrix}\text{rate of} \\ \text{energy loss} \\ \text{to cap and} \\ \text{base rock}\end{matrix} + \begin{matrix}\text{rate of} \\ \text{energy accumu-} \\ \text{lation in} \\ \text{heated oil sand}\end{matrix}$$

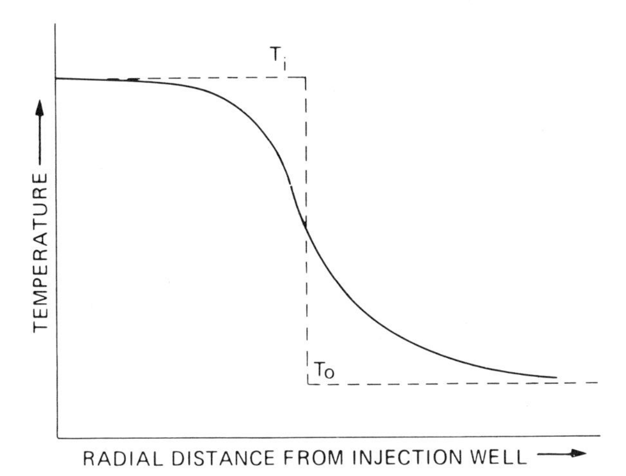

Figure 3.4 Qualitative true temperature distribution (solid line) vs. Marx–Langenheim step approximation (dashed line)[9] © 1959 SPE-AIME.

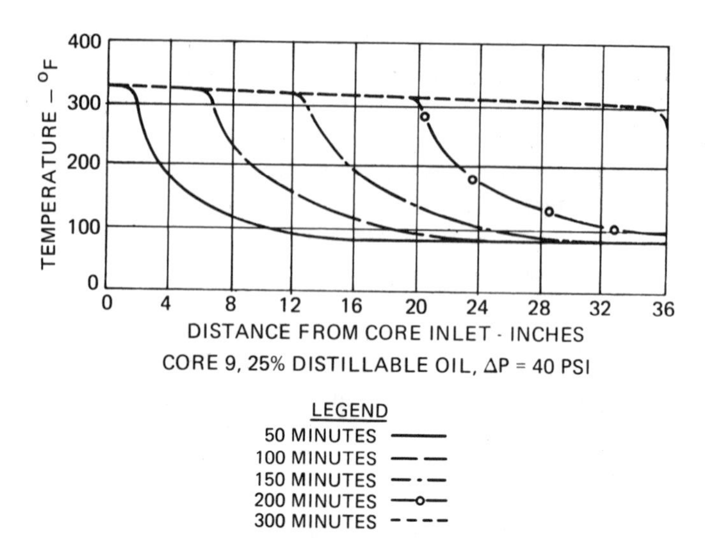

Figure 3.5 Temperature profiles during steam injection in laboratory sand-packed tube.[10] © SPE-AIME.

where m_s = mass rate of steam injection (lb/hr)

H_t = specific enthaphy of steam at P_I, T_I referenced to T_r, P_r, $= X_i H_{wv} + H_{ws} - H_{wr}$ (Btu/lb)

X_i = injected steam quality at sand face (lb water vapor/lb total fluid injected)

$\Delta T = T_I - T_r$ (°F)

α = overburden thermal diffusivity, $= K_{hob}/(\rho c)_{ob}$ (ft²/d)

A = area heated to T_I at any time t (ft²)

Equation (3.5) is analogous to the equation describing fluid flow into a growing hydraulic fracture bounded by permeable surfaces. Laplace transformation is applied to obtain the heated area:

$$A = \frac{m_s H_t h(\rho c)_{R+F} \alpha}{4K_{hob}^2 \Delta T} \frac{\xi_s}{\theta} \tag{3.6}$$

where

$$\frac{\xi_s}{\theta} = e^{\tau/\theta} \operatorname{erfc}\left(\sqrt{\frac{\tau}{\theta}}\right) + \frac{2\sqrt{\tau/\theta}}{\sqrt{\pi}} - 1 \tag{3.7}$$

Here, τ and θ are defined as in Eq. (3.3). A table giving values of ξ_s/θ as a function of $\sqrt{\tau/\theta}$ is found in Appendix A. Figure 3.6 is a plot of ξ_s/θ versus τ/θ for easy reference when making steam stimulation or flooding calculations.

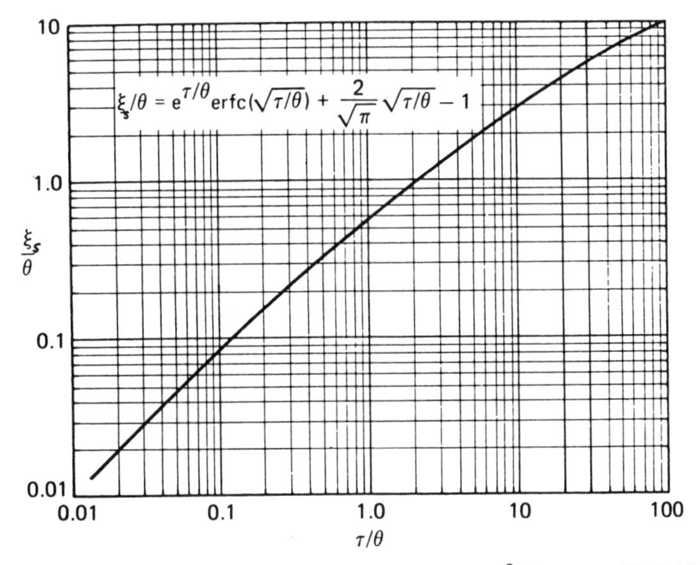

Figure 3.6 Dimensionless position of steam front.[9] © 1959 SPE-AIME.

Note that Eq. (3.6) may also be written as

$$A = \frac{m_s H_t h(\rho c)_{R+F}}{4K_{hob}\,\Delta T(\rho c)_{ob}}\,\frac{\xi_s}{\theta} \tag{3.6a}$$

For a linear system,

$$A = xw$$

where w is the system width in feet.

For a radial system,

$$A = \pi r_h^2$$

where r_h is the radius of the region in the oil sand heated to T_I in feet.

Equation (3.6a) or slightly modified forms of it will be used in later chapters concerning calculations of steam stimulation and steamflooding performance. A more general form of Eq. (3.6a) for the case of variable steam injection rate is given in Appendix C.

MANDL–VOLEK SOLUTION

An extension of the model of Marx and Langenheim was developed by Mandl and Volek to account for the fall-off of temperature in the region of the hot-water bank.[11] This extension does not account for horizontal conduction, but it does account for heat transport by the flow of hot water beyond the leading edge of the steam zone. Their solution for the area of the steam zone, A, which is valid only for $t > t_c$, is

$$A = \frac{m_s H_t (\rho c)_{R+F} h}{4K_{hob}(\rho c)_{ob}\,\Delta T}\left[\frac{\xi_s}{\theta} - \sqrt{\frac{\epsilon}{\pi}}\left(\beta + \frac{\epsilon e^{\tau/\theta}}{3}\operatorname{erfc}\sqrt{\frac{\tau}{\theta}} - \frac{\epsilon}{3\sqrt{\pi\tau/\theta}}\right)\right] \tag{3.7}$$

where $\beta = \left(1 + \dfrac{H_{wv} X_i}{c_w\,\Delta T}\right)^{-1}$

$\epsilon = (\tau/\theta)(1 - t_c/t)$

t_c = critical time, defined by root of Eq. (3.8)

and

$$e^s \operatorname{erfc}\sqrt{s} = \beta \tag{3.8}$$

where

$$s = \left(\frac{\tau}{\theta}\right)\left(\frac{t_c}{t}\right) \tag{3.9}$$

The function $e^s \operatorname{erfc}\sqrt{s}$ is tabulated as a function of $\sqrt{s}$ in Appendix A. The Myhill–Stegemeier method described in Chapter 10 employs the Mandl–Volek refinement.

WILLMAN SOLUTION

Willman and co-workers[10] obtained a solution specific to the radial case, which comes from equating the cumulative injected heat with that stored in the impermeable rock and oil sand (the same restrictive assumptions apply as in the case of the Marx–Langenheim model):

$$r_h^2 = \frac{1.46 m_s H_t}{K_{hob}\,\Delta T}\sqrt{\frac{\alpha}{\pi}}\left[\frac{\sqrt{t}}{2} - \frac{h\theta}{8}\sqrt{\frac{\pi}{\alpha}}\ln\left(\frac{4\theta}{h}\sqrt{\frac{\alpha}{\pi}}\sqrt{t} + 1\right)\right] \tag{3.10}$$

The relation has utility since tables or graphs of the special functions required for the Marx and Langenheim solution are not needed. It is not as general as the Marx–Langenheim solution and is less frequently used.

NUMERICAL SOLUTION

Shutler[12] has solved Eq. (3.4) numerically for the case of one-dimensional flow in combination with the appropriate differential equations for three-

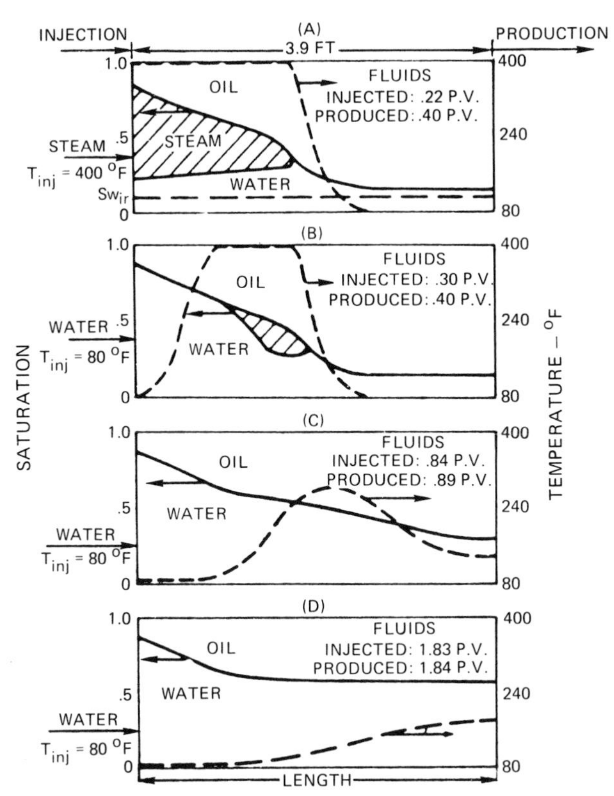

Figure 3.7 Calculated steam bank saturation and temperature profiles.[12] © 1969 SPE-AIME.

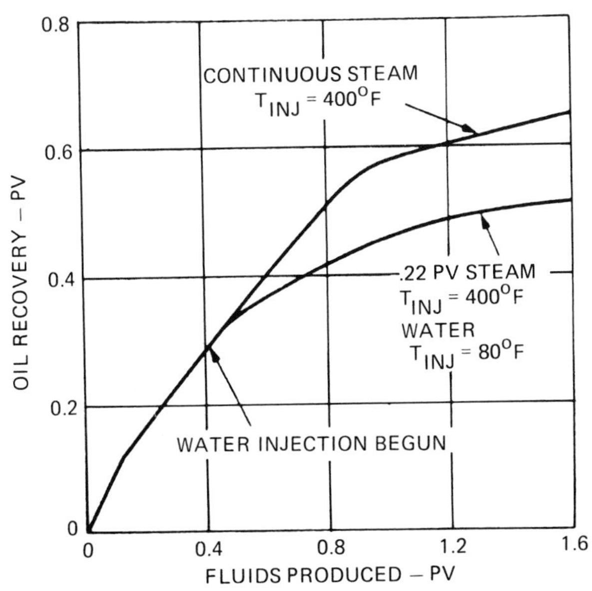

Figure 3.8 Comparison of calculated steamflood and steam bank recovery curves.[12] © 1969 SPE-AIME.

phase flow of oil, steam, and water. A more complete discussion of the equations solved may be found in Chapter 11. Figure 3.7 shows the calculated temperature and fluid saturation distribution computed for four different stages during a flood of a laboratory model. Note that after 0.22 pore volumes (PV) of steam are injected, water injection is commenced. The steam zone has been mostly invaded by the injected water in stage B. In stages C and D, the thermal bank continues to dissipate, and the process has become a hot waterflood. Figure 3.8 compares the calculated oil recovery from the thermal bank flood of Figure 3.7 with a continuous steamflood of the same model. Appendix B gives data for the laboratory model.

APPENDIX A

Functions Used in Marx–Langenheim Equation[a]

$\sqrt{\tau/\theta}$, x	e^{x^2} erfc x	ξ_s/θ, e^{x^2} erfc $x + \frac{2x}{\sqrt{\pi}} - 1$	$\sqrt{\tau/\theta}$, x	e^{x^2} erfc x	ξ_s/θ, e^{x^2} erfc $x + \frac{2x}{\sqrt{\pi}} - 1$
0.00	1.00000	0.00000	0.50	0.61569	0.17988
0.02	0.97783	0.00039	0.52	0.60588	0.19234
0.04	0.95642	0.00155	0.54	0.59574	0.20507
0.06	0.93574	0.00344	0.56	0.58618	0.21807
0.08	0.91576	0.00603	0.58	0.57687	0.23133
0.10	0.89646	0.00929	0.60	0.56780	0.24483
0.12	0.87779	0.01320	0.62	0.55898	0.25858
0.14	0.85974	0.01771	0.64	0.55039	0.27256
0.16	0.84228	0.02282	0.66	0.54203	0.28676
0.18	0.82538	0.02849	0.68	0.53387	0.30117
0.20	0.80902	0.03470	0.70	0.52593	0.31580
0.22	0.79318	0.04142	0.72	0.51819	0.33062
0.24	0.77784	0.04865	0.74	0.51064	0.34564
0.26	0.76297	0.05635	0.76	0.50328	0.36085
0.28	0.74857	0.06451	0.78	0.49610	0.37624
0.30	0.73460	0.07311	0.80	0.48910	0.39180
0.32	0.72106	0.08214	0.82	0.48227	0.40754
0.34	0.70792	0.09157	0.84	0.47560	0.42344
0.36	0.69517	0.10139	0.86	0.46909	0.43950
0.38	0.68280	0.11158	0.88	0.46274	0.45571
0.40	0.67079	0.12214	0.90	0.45653	0.47207
0.42	0.65912	0.13304	0.92	0.45047	0.48858
0.44	0.64779	0.14428	0.94	0.44455	0.50523
0.46	0.63679	0.15584	0.96	0.43876	0.52201
0.48	0.62609	0.16771	0.98	0.43311	0.53892

Functions Used in Marx–Langenheim Equation[a] (*continued*)

$\sqrt{\tau/\theta}$, x	e^{x^2} erfc x	ξ_s/θ, e^{x^2} erfc $x + \frac{2x}{\sqrt{\pi}} - 1$	$\sqrt{\tau/\theta}$, x	e^{x^2} erfc x	ξ_s/θ, e^{x^2} erfc $x + \frac{2x}{\sqrt{\pi}} - 1$
1.00	0.42758	0.55596	3.00	0.17900	2.56414
1.05	0.41430	0.59910	3.10	0.17372	2.67169
1.10	0.40173	0.64295	3.20	0.16873	2.77954
1.15	0.38983	0.68746	3.30	0.16401	2.88766
1.20	0.37854	0.73259	3.40	0.15954	2.99602
1.25	0.36782	0.77830	3.50	0.15529	3.10462
1.30	0.35764	0.82454	3.60	0.15127	3.21343
1.35	0.34796	0.87127	3.70	0.14743	3.32244
1.40	0.33874	0.91847	3.80	0.14379	3.43163
1.45	0.32996	0.96611	3.90	0.14031	3.54099
1.50	0.32159	1.01415	4.00	0.13700	3.65052
1.55	0.31359	1.06258	4.10	0.13383	3.76019
1.60	0.30595	1.11136	4.20	0.13081	3.87000
1.65	0.29865	1.16048	4.30	0.12791	3.97994
1.70	0.29166	1.20991	4.40	0.12514	4.09001
1.75	0.28497	1.25964	4.50	0.12248	4.20019
1.80	0.27856	1.30964	4.60	0.11994	4.31048
1.85	0.27241	1.35991	4.70	0.11749	4.42087
1.90	0.26651	1.41043	4.80	0.11514	4.53136
1.95	0.26084	1.46118	4.90	0.11288	4.64194
2.00	0.25540	1.51215	5.00	0.11070	4.75260
2.05	0.25016	1.56334	5.20	0.10659	4.97417
2.10	0.24512	1.61472	5.40	0.10277	5.19602
2.15	0.24027	1.66628	5.60	0.09921	5.41814
2.20	0.23559	1.71803	5.80	0.09589	5.64049
2.25	0.23109	1.76994	6.00	0.09278	5.86305
2.30	0.22674	1.82201	6.20	0.08986	6.08581
2.35	0.22255	1.87424	6.40	0.08712	6.30874
2.40	0.21850	1.92661	6.60	0.08453	6.53184
2.45	0.21459	1.97912	6.80	0.08210	6.75508
2.50	0.21081	2.03175	7.00	0.07980	6.97845
2.60	0.20361	2.13740	7.20	0.07762	7.20195
2.70	0.19687	2.24350	7.40	0.07556	7.42557
2.80	0.19055	2.35001	7.60	0.07361	7.64929
2.90	0.18460	2.45690	7.80	0.07175	7.87311

Functions Used in Marx–Langenheim Equation[a] (*continued*)

$\sqrt{\tau/\theta}$, x	$e^{x^2}\operatorname{erfc} x$	ξ_s/θ, $e^{x^2}\operatorname{erfc} x + \frac{2x}{\sqrt{\pi}} - 1$	$\sqrt{\tau/\theta}$, x	$e^{x^2}\operatorname{erfc} x$	ξ_s/θ, $e^{x^2}\operatorname{erfc} x + \frac{2x}{\sqrt{\pi}} - 1$
8.00	0.06999	8.09702	9.00	0.06231	9.21772
8.20	0.06830	8.32101	9.20	0.06097	9.44206
8.40	0.06670	8.54508	9.40	0.05969	9.66645
8.60	0.06517	8.76923	9.60	0.05846	9.89090
8.80	0.06371	8.99344	9.80	0.05727	10.11539
			10.00	0.05614	10.33993

[a] From Ref. 9. © 1959 SPE-AIME.

APPENDIX B: LABORATORY MODEL DATA FOR FIGURES 3.7 AND 3.8

Parameter	Value
k, darcies	132
ϕ	0.372
L, ft	3.9
Sand thickness, ft	0.83
$K_{h,R+F}$, Btu/ft-day-°F	24[a]
K_{hob}, Btu/ft-day-°F	12
$(\rho c)_{R+F}$, Btu/ft^3-°F	36
$(\rho c)_{ob}$, Btu/ft^3-°F	30
T_r, °F	80
S_{wi}	0.1
S_{gi}	0
T_I, °F	400
P_{inj}, psi	260
$P_{prod, psi}$	$\geqq 190$[b]

T, °F	$\mu(cp)$
50	5000
150	126
250	20
350	7.1
450	4.0

[a] Conductivity is infinite in vertical direction.
[b] $q_{prod} \approx 1$ bbl/d.

APPENDIX C: MARX–LANGENHEIM SOLUTION FOR VARIABLE-RATE CASE

For the case of a variable steam injection rate, a superposition solution of Eq. (3.5) may be used[13]:

$$A(t) = \frac{H_t h(\rho c)_{R+F}}{4K_{hob}\,\Delta T(\rho c)_{ob}}\left[m_{s,n}\left(\frac{\xi_s}{\theta}\right)_n + \sum_{m=1}^{m=n-1}(m_{s,m} - m_{s,m+1})\left(\frac{\xi_s}{\theta}\right)_m\right] \tag{3C.1}$$

where $m_{s,n}$ is the steam rate during the nth time period in pounds per hour.

In general, for most engineering calculations, the use of an average m_s is sufficient.

REFERENCES

1. Lauwerier, H. A. "The Transport of Heat in an Oil Layer Caused by the Injection of Hot Fluid," *Appl. Sci. Res., Sec. A.*, **5**, 145 (1955).
2. Malofeev, G. E. "Calculation of the Temperature Distribution in a Formation when Pumping Hot Fluid into a Well," *Neft'i Gaz*, **3**(7), 59 (1960).
3. Spillette, A. G. "Heat Transfer During Hot Fluid Injection into an Oil Reservoir," *J. Cdn. Pet. Tech.* (October–December 1965), p. 213.
4. Avdonin, N. A. "Some Formulas for Calculating the Temperature Field of a Stratum Subject to Thermal Injection," *Neft'i Gaz*, **7**(3), 37 (1964).
5. Avdonin, N. A. "On the Different Methods of Calculating the Temperature Fields of a Stratum During Thermal Injection," *Neft'i Gaz*, **7**(8), 39 (1964).
6. Rubinshtein, L. I. "The Total Heat Losses in Injection of a Hot Liquid Into a Stratum," *Neft'i Gaz*, **2**(9), 41 (1959).
7. Rubinshtein, L. I. "A Contact Thermal Conduction Problem," *Dan SSSR*, **135**(4), 805 (1960).
8. Rubinshtein, L. I. "An Asymptotic Solution of an Axially Symmetric Contact Problem in Thermal Convection for High Values of the Convection Parameter," *Dan SSSR*, **146**(5), 1043 (1962).
9. Marx, J. W., and Langenheim, R. N. "Reservoir Heating by Hot Fluid Injection," *Trans. AIME*, **216**, 312 (1959).
10. Willman, B. T., Valleroy, V. V., Runberg, G. W., Cornelius, A. J., and Powers, L. W. "Laboratory Studies of Oil Recovery by Steam Injection," *J. Pet. Tech.* (July 1961), p. 681.
11. Mandl, G., and Volek, C. W. "Heat and Mass Transport in Steam Drive Processes," *Soc. Pet. Eng. J.* (March 1969), p. 59.
12. Shutler, N. D. "Numerical Three-Phase Model of the Linear Steam Drive Process," *Soc. Pet. Eng. J.* (June 1969), p. 232.
13. Ramey, H. J. "Discussion of Reservoir Heating by Hot Fluid Injection by J. W. Marx and R. N. Langenheim," *Trans. AIME*, **216**, 364 (1959).

Chapter 4

INTRODUCTION TO STEAM STIMULATION PRINCIPLES

ABSTRACT

This chapter provides introductory material about cyclic steam stimulation, which is probably the lowest cost and most successful application of any thermal method. In those instances where it is feasible, it is the thermal method that should receive first consideration. This chapter covers basic process mechanics and presents a simplified steady-state analysis that can be used to predict the level of oil rate increase in the case of an oil that is mobile under cold conditions. Equations for calculating the stimulated rate assume a known average temperature and radius of the region heated around the stimulated well. The effects of permeability damage near the wellbore and the removal of such damage during steaming are discussed. The results of simplified calculations illustrating the effect of various operating and reservoir parameters on cyclic steam stimulation production behavior are presented.

INTRODUCTION

Steam injection is used for thermal stimulation of a producing well by providing localized heating in the vicinity of the wellbore. Steam stimulation is sometimes referred to as "steam soaking" or the "huff-and-puff" steam injection process.

Stimulation of heavy-oil-producing wells by cyclic steam injection has received intense interest by heavy-oil producers since the early 1960s. Steam stimulation is being applied on a commercial scale, particularly in California, Venezuela, and Canada.

This chapter will present a physical description of the steam stimulation process, provide a method to estimate the recovery of viscous crude oil by steam stimulation, and give guidelines for recognition of prospective steam stimulation candidates.

BASIC CONCEPTS

The steam stimulation process consists of injecting wet steam (a mixture of steam and water at steam temperature) into a viscous oil-bearing formation for a specified period of time, allowing the heat to "soak" (or penetrate) into the reservoir rock and fluids for several days, and subsequently placing the well back on production. Following flow-back of some of the injected steam and condensate, oil production commences at an elevated rate that eventually declines as heat is dissipated. Then the well is restimulated. A typical cyclic stimulated history is illustrated in Figure 4.1.

The success of steam stimulation stems from the large effect of temperature on the viscosity of heavy crude oils. As temperature increases, the viscosity of crude oil decreases markedly. As shown in Figure 4.2, the degree of reduction is much more pronounced for low-gravity crude oils. This explains why thermal stimulation is presently being applied primarily in reservoirs producing low-gravity crudes. Since the productivity of a well is inversely proportional to oil viscosity, any reduction in viscosity will result in an increase in the well's production rate. The primary objective in a thermal stimulation process, therefore, is to get thermal energy into the formation and allow the rock to act as a heat exchanger for storage of the injected heat. This heat may then be used effectively to lower the viscosity of the oil flowing through the heated region.

Besides the benefit obtained from reduced viscosity, an additional amount of stimulation is often produced because of the removal of certain

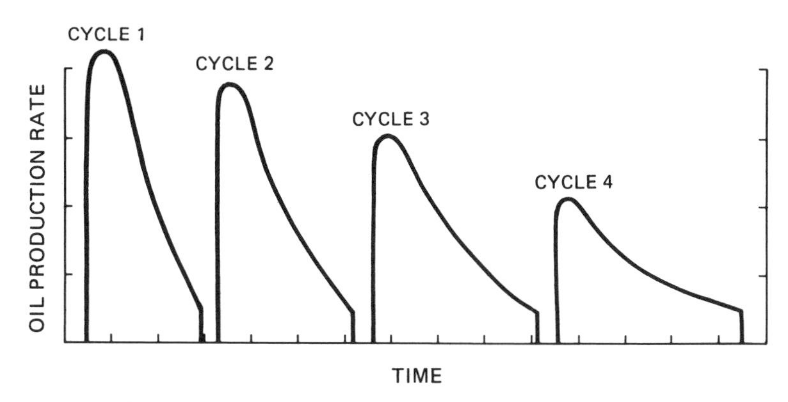

Figure 4.1 Typical steam stimulation performance.

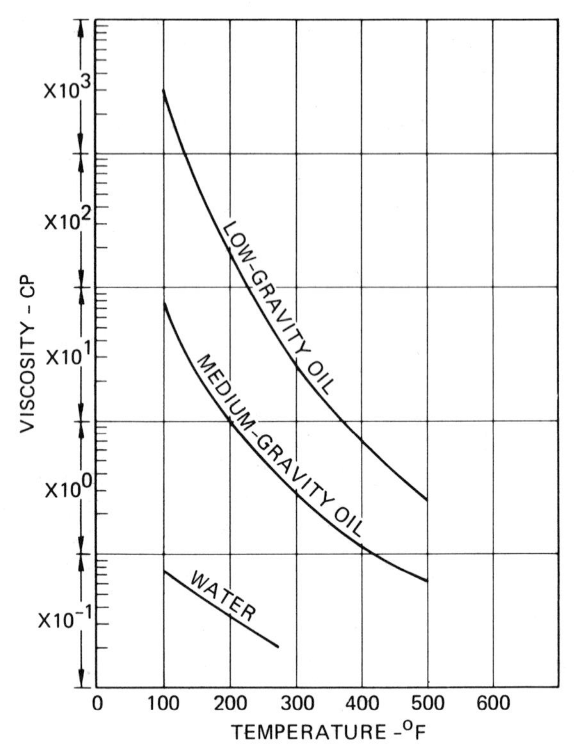

Figure 4.2 Effect of temperature on liquid viscosity.

types of near-wellbore damage, such as fine solids, asphaltic deposits, and paraffinic deposits. As with other stimulation techniques, the removal of this damage often results in productivity increases higher than those attainable by the reduction in oil viscosity alone.

Thermal Stimulation and Damage Removal

We can illustrate the separate and combined effects of thermal stimulation and damage removal with the help of the simplified radial flow model shown in Figure 4.3.

The model contains a zone of damage (radius r_d, permeability k_d) near the wellbore and an unaltered zone (radius r_e, permeability k) extending to the drainage radius; in addition, it contains a third zone (radius r_h) that has been heated to some uniform temperature higher than the remainder of the reservoir. In the heated region, the oil is assumed to be at a uniform, reduced oil viscosity, μ_{oh}. The permeability in the heated region has the

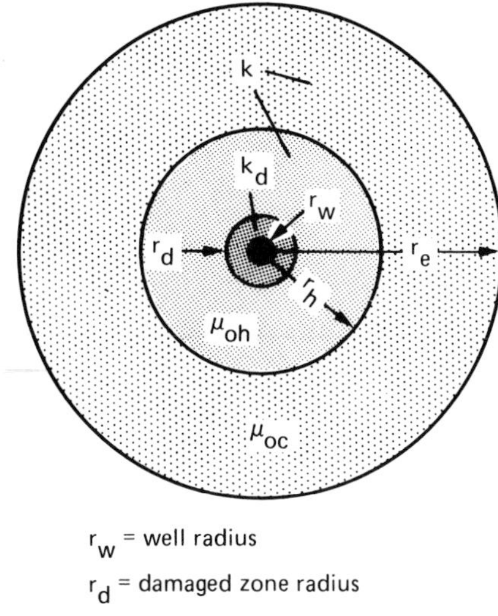

Figure 4.3 Simplified radial flow model for thermal stimulation calculations.

value k_d for $r < r_d$ and a value k for $r_d < r < r_h$. For cases where damage is removed, it is assumed that k_d is restored to k (the original undamaged value). Using this simplified model, the amount of stimulation possible for various combinations of damage removal and thermal stimulation can be calculated.

The effectiveness of steam stimulation can be seen by examining the steady-state radial flow equation. Under unstimulated conditions, this equation is, for a homogeneous system:

$$q_{oc} = \frac{2\pi hk\,\Delta P}{B_o \mu_{oc} \ln r_e / r_w} \tag{4.1}$$

where q_{oc} = cold, or unstimulated, oil production rate (STB/d)
h = net oil sand thickness (ft)
k = formation permeability (darcies)
ΔP = difference in pressure between that at drainage radius r_e and that at wellbore radius r_w, $= P_e - P_w$ (psi)
B_0 = formation volume factor (RB/STB)
μ_{oc} = original, or cold, oil viscosity (cp)
STB = stock tank barrels (at 60°F and 14.7 psia)
RB = reservoir conditions barrels

The unstimulated productivity index J_c is defined as

$$J_c = q_{oc}/\Delta P \tag{4.2}$$

For the general case where permeability varies with increasing radial distance from the wellbore,

$$J_c = 2\pi h \Big/ \int_{r_w}^{r_e} (B_o \mu_o / k) d \ln r = 2\pi h \Big/ B_o \mu_o \left[\frac{1}{k_d} \ln \frac{r_d}{r_w} + \frac{1}{k} \ln \frac{r_e}{r_d} \right] \tag{4.3}$$

where r_d is the radius of the reduced permeability or damaged zone near the wellbore in feet, and k_d is the permeability of the damaged zone in darcies.

Defining the skin factor as

$$S_c = \left(\frac{k}{k_d} - 1 \right) \ln \frac{r_d}{r_w} \tag{4.4}$$

and rearranging;

$$J_c = 2\pi h \Big/ \frac{B_o \mu_{oc}}{k} \left[\frac{k}{k_d} \ln \frac{r_d}{r_w} + \ln \frac{r_e}{r_d} \right]$$

$$= 2\pi h \Big/ \frac{B_o \mu_{oc}}{k} \left[\frac{k}{k_d} \ln \frac{r_d}{r_w} - \ln \frac{r_d}{r_w} + \ln \frac{r_d}{r_w} + \ln \frac{r_e}{r_d} \right]$$

$$= 2\pi h \Big/ \frac{B_o \mu_{oc}}{k} \left[S_c + \ln \frac{r_e}{r_w} \right] \tag{4.5}$$

After steam stimulation has occurred, a similar integration from r_w to r_e gives the stimulated productivity index J_h:

$$J_h = 2\pi h \Big/ \int_{r_w}^{r_e} (B_o \mu_o / k) d \ln r$$

$$= 2\pi h \Big/ \left[B_o \mu_{oh} \left(\frac{1}{k_d} \ln \frac{r_d}{r_w} + \frac{1}{k} \ln \frac{r_h}{r_d} \right) + B_{oc} \mu_{oc} \left(\frac{1}{k} \ln \frac{r_e}{r_h} \right) \right] \tag{4.6}$$

Neglecting the effect of heating on the formation volume factor yields

$$J_h = 2\pi h \Big/ \frac{B_o}{k} \left[\mu_{oh} \left(\frac{k}{k_d} \ln \frac{r_d}{r_w} + \ln \frac{r_h}{r_d} \right) + \mu_{oc} \ln \frac{r_e}{r_h} \right]$$

and similarly defining the skin factor under heated conditions as

$$S_h = \left(\frac{1}{k_d} - 1\right) \ln \frac{r_d}{r_w}$$

then

$$J_h = 2\pi h \Big/ \frac{B_o \mu_{oc}}{k} \left[\frac{\mu_{oh}}{\mu_{oc}} \left(S_h + \ln \frac{r_h}{r_w} \right) + \ln \frac{r_e}{r_h} \right] \tag{4.7}$$

Dividing Eq. (4.7) by Eq. (4.5) gives Eq. (4.8), the ratio of the heated to the cold productivity indices:

$$\frac{J_h}{J_c} = \left(S_c + \ln \frac{r_e}{r_w} \right) \Big/ \left[\frac{\mu_{oh}}{\mu_{oc}} \left(S_h + \ln \frac{r_h}{r_w} \right) + \ln \frac{r_e}{r_h} \right] \tag{4.8}$$

Using this equation, the effect of near-wellbore oil viscosity reduction for various conditions of wellbore damage can be calculated.

Assume as an illustration that the heat reduces the oil viscosity 100-fold in the heated zone. Results of calculations are shown in Table 4.1 for the indicated assumed zone radii. Table 4.1 gives the ratio of the heated to the cold productivity indices (J_h/J_c) for an undamaged, moderately damaged, and a severely damaged case. The ratios assume that pressure transients have substantially passed and that radial steady-state flow is occurring. Studying the numbers in this table leads us to conclude the following:

1. Damaged wells will attain much greater improvement ratios than undamaged wells, whether the damage is removed or not, when the producing zone is heated.
2. Removal of damage has principal benefit after the formation has cooled substantially.

While use of the steady-state equation calculates only a threefold improvement ratio for the undamaged case, it should be recognized that pressure transients will cause improvement ratios for a brief period of time to be greatly in excess of this immediately after returning a well to production following steam stimulation. Often this effect is academic, however, if the capacity of available fluid-lifting equipment is exceeded, which is often the case at the beginning of the production cycle. The steady-state equation has been found to be adequate for most practical cases, and its use will be illustrated in the next chapter.

Table 4.1 Calculated stimulation ratios for flow model shown in Figure 4.3[a]

	Amount of Damage		Stimulation Ratio, Heated ($\mu_{oh} = 0.01\mu_{oc}$)			Stimulation Ratio, Unheated		
	Permeability Damage Ratio k_d/k	Skin Factor S_c	Damage Not Removed	50% Removal[b]	100% Removal[b]	Damage Not Removed	50% Removal[b]	100% Removal[b]
Undamaged	1	0	2.94	2.94	2.94	1	1	1
Moderately damaged	0.1	21	10.8	11.3	11.8	1	1.6	4.0
Extremely damaged	0.01	228	50.3	67.4	100.0	1	1.94	34.0

[a] Well radius $r_w = 0.3$ ft, damaged radius $r_d = 3.0$ ft, heated radius $r_h = 30$ ft, drainage radius $r_e = 300$ ft, S_c = skin prior to heating, S_h = skin after heating;

$$\frac{J_h}{J_c} = \left(S_c + \ln\frac{r_e}{r_w}\right) \Big/ \left[\frac{\mu_{oh}}{\mu_{oc}}\left(S_h + \ln\frac{r_h}{r_w}\right) + \ln\frac{r_e}{r_h}\right]$$

[b] For 50% removal, $S_h = 0.5S_c$; for 100% removal, $S_h = 0$. *Note:*

$$S_c = \left(\frac{k}{k_d} - 1\right)\ln\frac{r_d}{r_w}$$

RESERVOIR FACTORS AFFECTING STEAM STIMULATION BEHAVIOR

Many factors should be considered when selecting a thermal stimulation candidate. Factors such as fuel costs, water treatment, the market price of oil, and equipment requirements (steam generators, treating equipment, flow lines, produced water disposal, etc.) lend themselves readily to economic analysis. Many other reservoir parameters, which are not so readily evaluated, have been studied using the calculation method of Boberg and Lantz[2] (see Chapter 7).

Results of several reservoir parameter studies are presented in Figures 4.4–4.8. (Reservoir and injection data used for these figures are given in Tables 4.2 and 4.3.) The incremental oil–steam ratio was selected as the primary dependent variable for these studies since it can be directly related to the economics of the process. The incremental oil–steam ratio is defined as the ratio of the stimulated less primary oil production to the amount of steam injected expressed as barrels of condensate. Values of this ratio referred to in the following discussion are maximum values that are reached when the computed oil production rate has returned to its prestimulation value.

Effect of Skin Damage

As already illustrated in Table 4.1, it is apparent that the amount of skin damage present in a well prior to stimulation can have a tremendous effect on the production response of the well when it is steam stimulated. This is

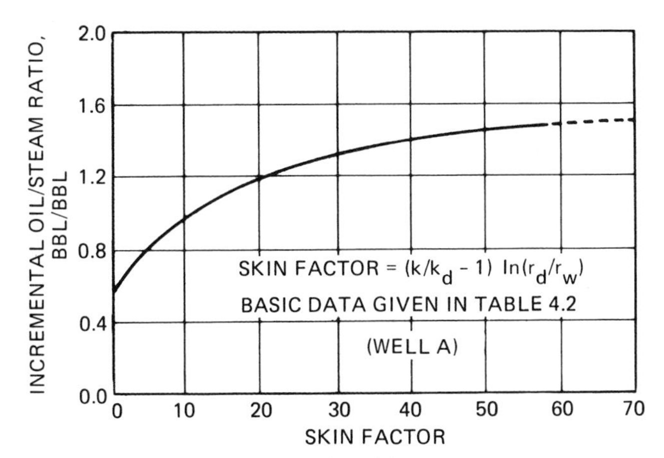

Figure 4.4 Effect of skin factor on calculated incremental oil–steam ratio.[2] © 1966 SPE-AIME.

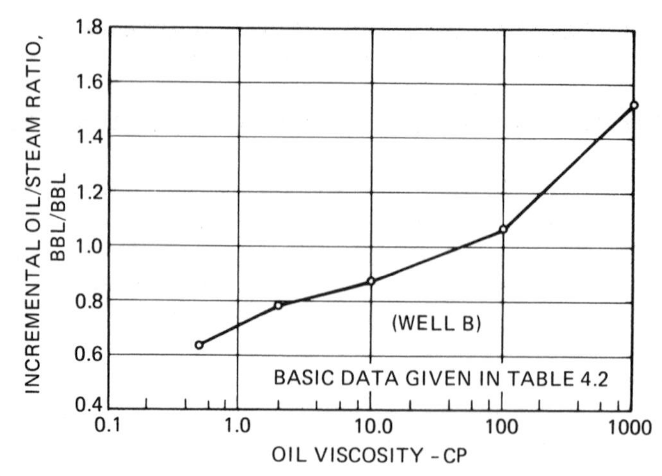

Figure 4.5 Effect of initial viscosity on incremental oil–steam ratio.[2] © 1966 SPE-AIME.

Table 4.2 Steam Stimulation Test and Calculation Data.[a]

	Well A	Well B	Well C
Reservoir Characteristics			
Depth, ft	1250	3000	3740
Section thickness, ft	400	200	1088
Net sand thickness, ft	400	67	234
Number of sands	1	6	18
Reservoir temperature, °F	100	100	120
Oil viscosity, cp			
At T_r	2000	0.5 –1000	133
At 300°F	9	0.08–6	8
Skin factor	0–100	0–5	0
Effective well radius,[b] ft	0.181	0.00176	0.00176
Prestimulation			
Oil production rate, bbl/d	10	300	99
Oil productivity index, bpd/psi	0.100	1.0	0.30
Water–oil ratio, bbl/bbl	0.029	1.0	0.57
Gas–oil ratio, scf/bbl	63	1000	600
Stimulation			
Steam injected, 10^6 lb	3.0	40.0	16.6
Wellhead injection surface conditions			
Pressure, psig	290	780	770
Temperature, °F	420	520	5.8
Steam quality, dimensionless	0.8	0.95	0.95
Injection time, days	7	80	55
Shut-in time, days	3	4	5

[a] From Ref. 2. © 1966 SPE-AIME.
[b] Includes effect of well completions, perforations, etc., if any.

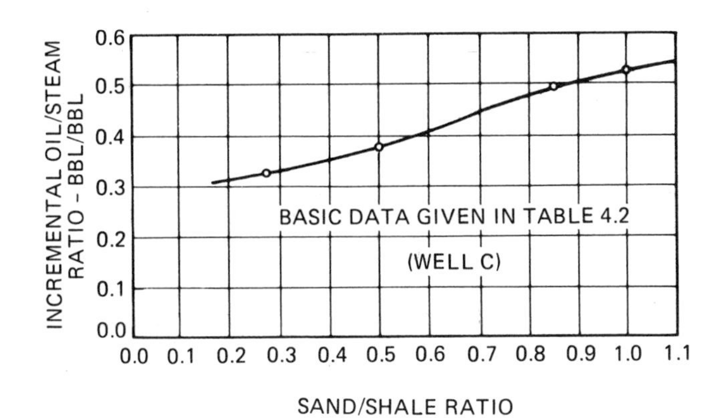

Figure 4.6 Calculated effect of sand–shale ratio on incremental oil–steam ratio.[2] © 1966 SPE-AIME.

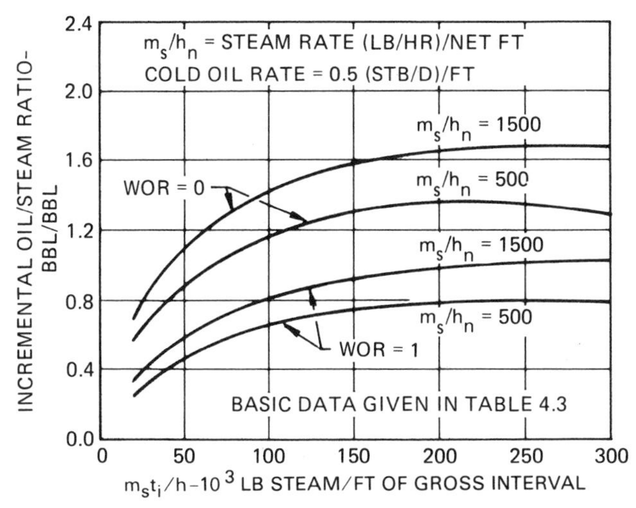

Figure 4.7 Theoretical prediction of incremental oil–steam ratio versus steam injected.[2] © 1966 SPE-AIME.

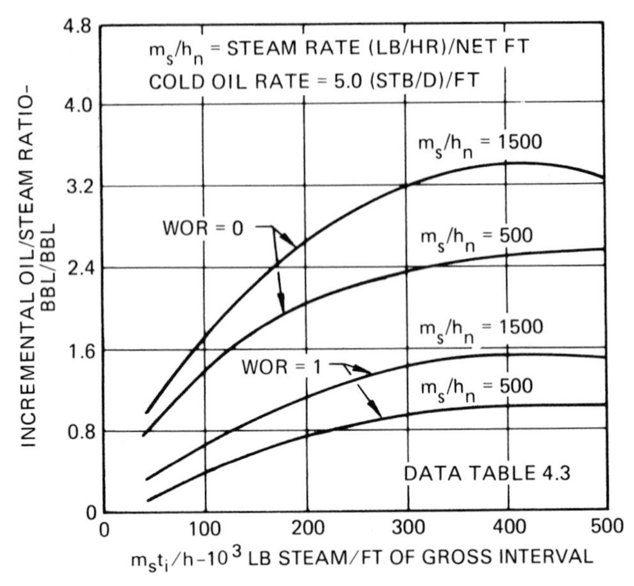

Figure 4.8 Theoretical prediction of incremental oil–steam ratio versus steam injected.[2] © 1966 SPE-AIME.

Table 4.3 Reservoir and Fluid Data for Figures 4.6 and 4.7[a]

Original reservoir temperature, °F	125
Original reservoir oil viscosity, cp	70
Oil gravity, °API	16.5
Oil specific heat, Btu/lb-°F	0.482
Formation thermal diffusivity, ft^2/d	0.632
Formation thermal conductivity, Btu/d-ft-°F	24.0
Sand–shale ratio	0.5
Average individual sand thickness, ft	13
Formation depth–section thickness ratio	14.6
Effective well radius, ft	0.25
Effective drainage radius, ft	1000
Normal producing bottom-hole pressure, psia	300
Static formation pressure, psia	1000
Producing gas–oil ratio, scf/bbl	500
Shut-in time following injection, days	3

[a] From Ref. 2. © 1966 SPE-AIME.

true even if no wellbore cleanup (i.e., damage removal) is obtained, although the stimulation benefits are greater when the damage is removed.

Figure 4.4 shows the effect of skin damage on the incremental oil–steam ratio for a stimulation cycle for well A of Table 4.2. These results are based on the assumption that no damage was removed by the steam. Calculations were made assuming this well had skin factors ranging from 0 to 60. Figure 4.4 shows that the calculated cycle-end incremental oil–steam ratio increases by a factor of about 2.5 as skin factor increases over this range.

Effect of Oil Viscosity

For a given temperature rise, the viscosity reduction of a low-viscosity oil is much less pronounced than for a high-viscosity crude. For this reason, productivity increase following steam injection will be smaller for lower original oil viscosities. On the other hand, it is not obvious how much less incremental low-viscosity oil will be produced over the entire cycle length since the lower heat removal rate with the produced fluids will cause stimulated production to last longer.

If there were no heat losses from the heated region of the oil sand, the increased cycle length for the lower viscosity oil should make up the difference in stimulated oil rate and give the same incremental oil recovery. Consequently, heat losses to unproductive rock are important in evaluating the effect of oil viscosity on incremental oil recovery.

The effect of initial oil viscosity (calculated for well B of Table 4.2) is shown in Figure 4.5. The results show a 50% increase in incremental oil recovered for a 1000-cp oil as compared with a 50-cp oil, all other factors the same. In this calculation, the cold production rate prior to stimulation was held constant as viscosity was increased. This situation assumes that the higher viscosity oils would have to be produced from reservoirs where $k_o \Delta P$ is proportionately higher. Had cold rate decreased as would be the case where μ_o increases and $k_o \Delta P$ is held constant, the result shown may not occur, and performance would normally be poorer for a more viscous oil.

Effect of Sand–Shale Ratio and Sand Thickness

The effect of the sand–shale ratio on the incremental oil–steam ratio for one cycle is shown in Figure 4.6 for well C of Table 4.2. The net sand thickness and number of sands were held constant in these calculations, and the gross section thickness was varied in order to vary the sand–shale ratio. The decline in the incremental oil–steam ratio as the sand–shale ratio decreases (increasing gross section thickness) is the result of increased heat losses to the interbedded shales.

The above calculations about the sand–shale ratio hold only as long as the average individual sand thickness is roughly constant. For example, a well that has two 100-ft-thick sands each separated by 800 ft of shale would probably respond to steam stimulation more favorably than a well that had twenty 10-ft-thick sands separated by shale 42 ft thick, even though both wells would have the same gross and net sand thicknesses and the same sand–shale ratios. The two sands in the first case would be so thick that they would cool much more slowly than those in the second case.

Effect of Reservoir Pressure, Producing Mechanism, and Oil Saturation

Reservoir pressure and the reservoir producing mechanism are important in steam stimulation. A reservoir with high formation pressure and a low rate of pressure decline is desirable. A depleted reservoir will give poorer response. A rapidly declining reservoir pressure may mean that stimulation will be economic for only one or two cycles. Wells open to depleted thief zones likely will not be good candidates as the steam will follow the path of least resistance in entering the formation. A workover prior to steam operations, if feasible, is desirable in such circumstances. In general, open intervals should have as high an oil saturation as possible. Reservoirs having higher oil saturations and porosities will generally be better steam stimulation candidates than those having lower ones.

A thick gravity drainage reservoir, such as the Potter Sand in the Midway-Sunset field in California, can give good stimulation response over many cycles despite its pressure depletion because of the substantial fluid head of the oil column. Richly saturated sands over 200 ft thick with good vertical permeability should be considered good candidates, even if pressure is depleted, because of the gravity drainage potential.

In any case, for a moderately viscous oil, steam stimulation mainly accelerates the primary recovery, although some additional oil recovery will come from the region heated around the well. In a highly viscous oil reservoir with no significant primary production, some oil that is heated will be produced as long as there is some mechanism to displace it from the heated region (as by fluid expansion, formation compaction, or solution gas drive).

Effect of Cold Oil Rate, Water–Oil Ratio, and Gas–Oil Ratio

One important factor in evaluating a potential steam stimulation prospect is the cold production rate prior to stimulation or, more particularly, the cold productivity index. This variable can have a large effect on the the incremental oil–steam ratio. As an example, compare a steam stimulation prospect that has a cold oil production rate of 1 bbl/d with one having a cold rate of 4 bbl/d. Assume that the same amount of steam is injected into each well and that the poststimulation rates are four times the cold rate. The first producing day after stimulation, the high-rate well will produce 12 barrels of oil more than the lower rate well. The same will be the case on succeeding days. Hence, up to a point, everything else being equal, the well with the higher rate will produce the most incremental oil. But for high-initial-rate wells (over about 200 bbl/d), the rate of heat removal from the formation becomes important, and a point of diminishing returns is reached.

The effect of the cold rate can be seen by comparing the incremental oil–steam ratios calculated in Figures 4.7 and 4.8 for the well of Table 4.3. These figures show the incremental oil–steam ratio as a function of steam injected per foot of gross section, with the water–oil ratio and injection rate per foot of sand as parameters. The incremental oil–steam ratios in Figure 4.8 (cold rate 5 bbl/d-ft net sand) are roughly double those of Figure 4.7 (cold rate 0.5 bbl/d-ft net sand). These figures apply for undamaged wells.

Damaged wells can have high incremental oil–steam ratios even if the cold rate is low. Rather than have a high cold rate, it would be better to have a high product of reservoir transmissibility and reservoir pressure. The latter criterion is more general than the cold rate since it would include damaged wells as favorable candidates even if their cold rates were low. Figure 4.9 shows the effect of cold-producing rate and treatment size on the incremental oil–steam ratio for a Venezuelan well on gas lift. Note that the maximum incremental oil–steam ratio occurs at a higher level of cold rate as treatment size is increased.

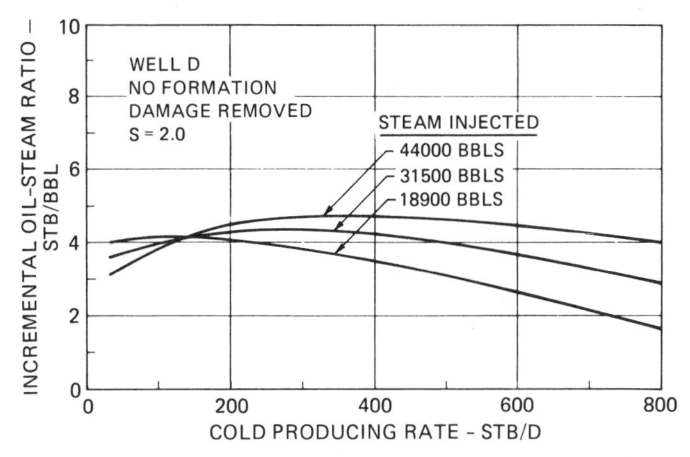

Figure 4.9 Effect of cold rate on oil–steam ratio, well B. Venezuelan Bolivar Coastal Field (BCF) well on Gas Lift.[3] © 1973 SPE-AIME.

The effect of a high water–oil ratio is detrimental, as Figures 4.7 and 4.8 indicate. The production of a large volume of water through the heated region will accelerate cooling, owing to the high heat capacity of water (which is approximately twice that of oil). Further, if the well is drawn down to too low a producing bottom-hole pressure early in the production cycle, some of the water will vaporize. This will greatly increase the rate of heat removal, owing to the high latent heat of vaporization of water. In this regard, a high gas–oil ratio can also be detrimental since gas production will reduce the partial pressure of water vapor in the heated zone, cause increased flashing of water to steam, and thus allow even more heat to be removed.

Formation Depth

Formation depth seriously limits the widespread application of steam stimulation because of excessive heat losses and casing failures experienced with deeper wells.

Steam injection technology is advancing so that it may become feasible to steam-stimulate wells deeper than 5000 ft using tubing insulation. Wellbore heat losses for deep wells can cause sufficient condensation that steam injected at the bottom hole will have low quality or be completely condensed to become hot water, depending on the depth and injection rate, unless insulation is used. Some manufacturers are currently making dual-wall insulated tubing that permit steam injection at relatively high bottom-hole steam qualities at considerable depth. A more economical insulating technique consists of boiling off the tubing–casing annulus filled with commerical grade (36°Be) sodium silicate solution to form a thick, porous, water-soluble insulation on the tubing wall.[4] Further discussion of this approach is found in Chapter 2.

In summary, the following reservoir factors are favorable for good steam stimulation performance:

1. High oil saturation and porosity.
2. High cold-oil productivity, but probably below 600 bbl/d.
3. High formation transmissibility.
4. High reservoir pressure, but below 1500 psi to permit injection of saturated steam.
5. Permeability damage near the wellbore.
6. Thick sands and high sand–shale ratio.
7. Higher initial oil viscosity for wells of comparable cold rates.
8. Low water–oil and gas–oil ratios.
9. Shallow depth.

OPERATING FACTORS AFFECTING STEAM STIMULATION BEHAVIOR

A number of factors influence the growth rate of the heated radius as steam enters the oil formation. The rate at which the steam is injected is important. High steam injection rates are desirable to minimize the fraction of the injected heat lost from the wellbore as well as minimize the time the well must be off production while steam is being injected.

During injection, an energy balance as the steam enters the formation states that the heat input at the bottom hole must equal the heat lost to overburden and underburden plus that stored within the oil sand at a specified time. Hence, the radius of the region heated is maximized if a high injection rate is used since, for a given volume injected, less energy will be lost to unproductive rock. The larger heated radius benefits the stimulation effect. In thick sands, there will be the tendency for steam to override as gravity segregation occurs. When steaming a thick section, it is common to hang the tubing near the bottom to force the steam to pass by the entire sandface even though it may tend to enter high in the sand.

During the soak portion of a steam stimulation cycle, the producing well is shut in to let the temperatures near the wellbore decrease somewhat and to allow the steam in the formation to condense. The length of the soak period is generally 1–4 days. Most studies indicate that the benefit of extending the soak period is offset by lost production while the well remains shut in, so extended soak periods are usually not recommended.

The production portion of the steam stimulation cycle consists of producing the well as in normal field operations.[1] Where possible, the well should be drawn down to a bottom-hole pressure that is slightly above the steam saturation pressure in order to prevent water from flashing to steam as it enters the wellbore. Steam flashing during production robs heat from the formation and causes it to cool too rapidly. During production, the temperature level gradually decreases throughout the heated volume as a result of heat losses by conduction radially, by conduction to overburden and underburden, and because some of the heat is produced with the reservoir fluids. As the temperature drops, the oil viscosity increases and the stimulation effect is diminished.

The stimulation process may be repeated to maintain the overall production rate at a higher level than its prestimulation value. For a given level of energy input, however, the stimulation benefits derived from succeeding treatments will generally decrease as a result of pressure and oil saturation depletion in the reservoir. As oil is produced, there will be an increase in the water saturation in the heated region. The resulting decrease in the relative permeability to oil will tend to make succeeding cycles less attractive. The maximum number of successfully applied treatments will depend on many reservoir and economic factors.

Probably the most important operating variable is the size of the steam treatment. Certainly, the stimulation cycle will last longer for a larger treatment. The effect of treatment size on the performance of a stimulation well will be discussed in a later section.

Effect of Treatment Size

The size of the steam treatment is the principal variable the engineer can use to optimize a steam stimulation operation. In general, small treatments can be used for wells where the major problem is wellbore damage. Figures 4.7 and 4.8 indicate that increasing treatment size will increase the incremental oil–steam ratio up to a point but that the oil–steam ratio will reach a maximum and eventually decline. This is also shown in Figure 4.9.

For damaged wells or formations in which steam channelling occurs, the maximum oil–steam ratio will occur for relatively small treatment sizes. Channelling prevents a substantial increase of the heated radius with increasing treatment size over a large fraction of the sand thickness. In practice, steam treatments usually range from 3000 to 100,000 bbl with 5000–40,000 bbl being more common.

For highly viscous oils in wells with little or no productivity prior to steaming, it is necessary to use more massive volumes of heat than for conventional heavy oil wells (since the only significant production will come from the region of the oil sand heated). For this type of situation, the cyclic injection of steam must become substantially more than a well stimulation method. It becomes, in fact, a combination of stimulation and thermal drive.

The most economical treatment size must be determined by an economic analysis for the total project under consideration. An individual well optimization will not suffice. One must consider the relative merits of continuing to steam well A versus moving off well A to stimulate well B. All wells must be considered as candidates for stimulation during each unit of time for such an investigation. The problem is complex and is still the subject of research.

Multiwell analyses to date have favored smaller treatments because for a given steam-generating capacity, they permit placing more stimulation wells on production sooner than when larger treatments are used.[1] In general, the treatments must be larger for wells with higher fixed stimulation costs (costs of well workover, making connections, etc.) and for more productive wells producing from thicker sections.

Other Operating Factors

In the case of gas lift wells, the wellhead pressure and gas lift gas injection rate are important variables. The Boberg–Lantz method[2] was combined

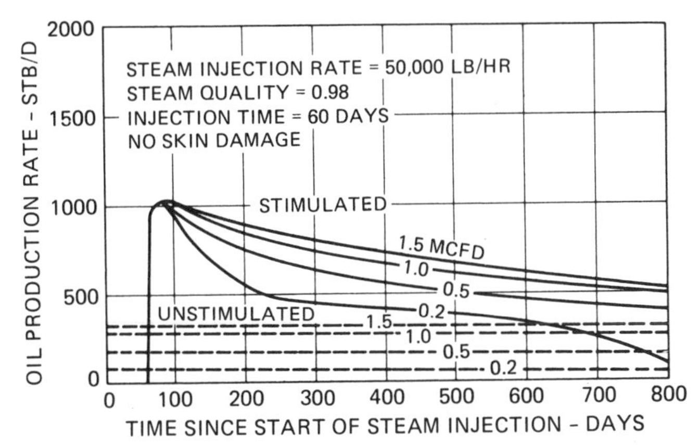

Figure 4.10 Effect of increase in gas lift injection rate, well A. Venezuelan (BCF) well.[3] © 1973 SPE-AIME.

with the two-phase vertical pipe flow correlations of Orkiszewski to develop Figures 4.10 and 4.11, which illustrate the effect of gas lift gas injection rate and wellhead pressure.[3] Increasing gas lift gas injection rate and reducing wellhead back pressure appear to be greatly beneficial over the range studied. Data for this well are found in Table 4.4.

In summary, the following operating factors affect steam stimulation performance:

1. *Treatment Size.* Use larger treatments for
   (a) high-productivity wells,
   (b) thick section,
   (c) undamaged wells,
   (d) wells having high fixed costs per stimulation job, and
   (e) high-viscosity oils with essentially no initial productivity.
2. *Injection Rate.* Use high injection rates to
   (a) minimize time well is off production,
   (b) maximize heated radius, and
   (c) minimize heat lost from wellbore.
3. *Soak Time.* Use a short soak time (1–4 days) unless field data indicate a distinct benefit to increasing the length of the shut-in period.
4. *Producing Bottom-Hole Pressure.* Maximize drawdown but keep bottom-hole pressure above the water vapor pressure at the bottom-hole temperature to minimize steam flashing. Optimal gas lift gas injection rates should be used along with maximum wellhead pressures consistent with the above condition in the case of gas lift wells.

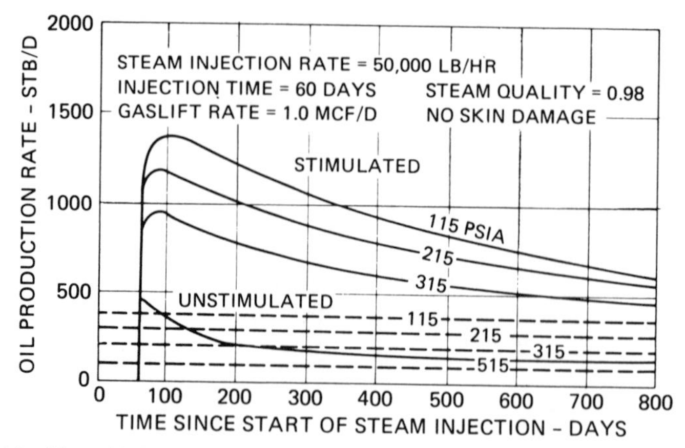

Figure 4.11 How high wellhead pressures caused by restricted flow line affect steam stimulation response, Venezuelan (BCF) well, well A.[3] © 1973 SPE-AIME.

Table 4.4 Data for Venezuelan (BCF) Steam-Stimulated Wells, Shown in Figures 4.9–4.11[a]

Reservoir	Well A	Well B
Depth, ft	2600	2650
Reservoir pressure,[b] psia	944	940
Original reservoir temperature, °F	121	125
Net thickness, ft	73	100
Gross thickness, ft	222	260
Number of individual sands	10	4
Drainage radius, ft	900	1000
Well radius, ft	0.25	0.25
Oil viscosity, cp (at original reservoir temperature)	312	300
Prestimulation		
Water–oil ratio, STB/STB	0.002	0.012
Gas–oil ratio, scf/STB	500	300
Oil productivity index, STB/d-psi	0.6209	—
Gas Lift		
Inside tubing diameter, in.	2.992	2.992
Gas lift entry depth, ft	2371	2350

[a] From Ref. 3.
[b] At start of injection.

REFERENCES

1. Niko, H., and Troost, P. J. P. M. "Experimental Investigation of the Steam-Soak Process in a Depletion-Type Reservoir," SPE 2978 Presented at Meeting of the SPE of AIME, October 4–7, 1970.
2. Boberg, T. C., and Lantz, R. B. "Calculation of the Production Rate of a Thermally Stimulated Well," *J. Pet. Tech.* (1966), p. 1613.
3. Boberg, T. C., Penberthy, W. L., Jr., and Hagedorn, A. R. "Calculating the Steam Stimulated Performance of Gas-Lifted and Flowing Heavy-Oil Wells," *J. Pet. Tech.* (1973), p. 1207.
4. Penberthy, W. L., and Bayless, J. H. "Technique for Insulating a Wellbore with Silicate Foam," U.S. Patent 3,525,399, August 25, 1970.

Chapter 5

STEAM STIMULATION FIELD OBSERVATIONS

ABSTRACT

Actual published field results for a variety of steam stimulation projects are summarized in this chapter.[1-22] From this information, certain observations can be made that generally support earlier conclusions about the kinds of reservoirs and operating parameters that favor successful application of this method. Specific information about certain large steam stimulation operations in the Tia Juana field in Venezuela and in Cold Lake, Alberta, Canada, are presented. In Tia Juana, oil–steam ratios were initially observed to be quite high owing to the high oil mobility at original reservoir temperature. At Cold Lake, oil mobility at original conditions is essentially zero, and fracturing must be employed for steam to enter the formation. Nevertheless, low, but quite acceptable, oil–steam ratios are obtained for over eight cycles of injection and production. The chapter concludes with a brief discussion of costs incurred during steam stimulation operations.

INTRODUCTION

Steam stimulation and steam drive have found widespread use in both California and Venezuela. Table 5.1 gives 1982 estimated incremental production by steamflooding and steam stimulation for the United States, Venezuela, and other areas of the world.

Table 5.1 shows that, while worldwide incremental production is predominantly from steam stimulation, domestic production by steam drive is now slightly greater. This is a relatively recent development. Domestic production by steam drive was only about 20,000 bbl/d in the late 1960s but has risen steadily throughout the seventies to its present level. This occurred because many steam stimulation projects were converted to steam drive as natural reservoir energy became depleted.

Table 5.1 Incremental Production by Steam Methods (1982)

	Steam Stimulation (STB/d)	Steam Drive (STB/d)	Total (STB/d)
United States	142,100	153,700	295,800
Venezuela	142,900	8,200	151,100
Others	39,000	11,000[a]	50,000
Total	324,000	172,900	496,900

[a] Includes oil production by hot-water drive (about 9000 bbl/d).

Expanded steam stimulation operations have offset these conversions to steam drive, and production by steam stimulation has remained relatively constant in recent years. Venezuelan production by steam processes has remained relatively constant over the last several years but is expected to rise in the future because of plans to increase production in the Orinoco Heavy Oil Belt. A more recent survey conducted in 1986 places combined oil production by steam stimulation and steam drive at about 470 kSTB/d in the United States.[2]

Table 5.2 displays reservoir characteristics for 19 mostly successful steam stimulation projects.[1] Several items in Table 5.2 are of interest. First, all of these projects are in sandstone reservoirs. The characteristic low porosities of carbonate reservoirs make them poorer steam stimulation candidates. Net sand thicknesses are high, averaging more than 150 ft. Depths range from just under 1000 to about 3000 ft. Porosities are high, 35% being common. Permeabilities are high, usually above 1 darcy. Most of these reservoirs have reasonably high reservoir pressure. Those with low reservoir pressures are usually thick and are probably produced by gravity drainage. Oil saturations are for the most part high, and water saturations are correspondingly low. Although it is not tabulated, these reservoirs probably had low cold-water cuts. Finally, oil viscosities are usually in the intermediate range, 1000–10,000 cp.

Table 5.3 presents operating data and results from the same 19 projects listed in Table 5.2. Several items from this table are noteworthy. First, treatment sizes varied considerably. On a net sand basis, though, the variations are smaller, and most treatments are between 50 and 150 bbl/ft of net sand. Injection periods generally require about 2 weeks, and stimulated production lasts anywhere from 6 months to a year or more. Soak periods of about 1 week are usually employed. Stimulation ratios (initial hot production rate/cold production rate) are generally between 5 and 15. Oil–steam ratios range from 0.1 to 5.0.

Tables 5.2 and 5.3 present a representative sampling of the types of reservoirs in which steam stimulation has been applied, and observations

Table 5.2 Steam Stimulation Field Experience: Reservoir Characteristics[a]

Project Number	Location	Year	Formation	Net Pay (ft)	Depth to Top (ft)	Porosity (%)	Permeability (md)	Reservoir Temperature (°F)	Reservoir Pressure (psi)	Oil Saturation[b] (%PV)	Water Saturation (%PV)	Oil Gravity (°API)	Oil Viscosity (cp[°F])	Primary Recovery (%OIP)
1	Guadalupe field, CA	1964	Unconsolidated sand	40	2700	—	—	—	—	—	—	10	—	—
2	Midway-Sunset field, CA	1964	Potter sand	~40	1140	—	—	110	75	—	—	15.4	150 (110)	—
3	Midway-Sunset field, CA	1964	Potter sand	75	1395	—	—	110	100	—	—	13	1700 (110)	—
4	Midway-Sunset field, CA	1964	Mya Tar sand	28	1225	—	—	102	75	—	—	13.5	4000 (102)	—
5	Wilmington field, CA	1965	Tar zone	125	2050	35–40	1000–8000	120	300–600	(2150)	18	12–15	150–400 (120)	12
6	Wilmington field, CA	1965	Ranger zone	150	2300	36	1500–2000	135	300–1000	(2050)	22	14–20	20–100 (135)	12
7	Cat Canyon field, CA	1965	Sandstone	150	3200	35	293	130	1400	70 (1780)	30	12.2	1000 (130)	—
8	Midway-Sunset Section 5A, CA	1964	Potter sand	271	800	36.5	1133	100	60	(1822)	25	14	1100 (100)	11
9	Midway-Sunset Section 23A, CA	1964	Tulare sand	240	890	25	956	105	95	(1075)	36	14	1450 (100)	9
10	West Newport, CA	1960?	B sand	250	1400	37	1000	105	85	69 (1940)	28	15	700 (105)	10–15
11	Huntington Beach, CA	1964	Upper Tar sand	~100	2000	38	2300	110	600	(1800)	35	13	1800 (80)	—
12	Santa Barbara, Venezuela	1964	Sacacual sand	40–250	~700?	35?	1000–2000	85	—	—	—	16–19	120–510 (85)	6.9
13	Emlichheim, West Germany	1966	Valendis sand	79	2625	35	800–900	97	1250	86?	14	24.5	175 (97)	11.5
14	Tia Juana, Venezuela	1964	Lr. Lagunillas sand	130	1500–2000	40	2000	110	780	80	20	11–15	700–7000 (110)	5–11
15	Oxnard field, CA	1965	Vaca Tar sand	300–400	1870	34.3	5548	100	930	(1962)	23	5	38,000 (140)	—
16	Oxnard field, CA	1965	Vaca Tar sand	300–400	1870	34.3	5548	100	930	(1962)	23	5	38,000 (140)	—
17	Quiriquire field, Venezuela	1960	Sandstone	216–422	1300–4700	—	—	—	—	(<1000)	—	14.4	—	—
18	Huntington Beach, CA	1964	TM sand	50	2000	35	400	100	600–800	(2080)	20.5	12–15	1176 (100)	3
19	Kern River, CA	1964	Kern River sand	155	700	35	—	80	—	(1900)	20	12.6	1540 (80)	9

Abbreviations: PV, pore volume; OIP, oil in place.

[a] From Ref. 1. © 1974 *J. Can. Pet. Tech.*, CIM.

[b] Oil content in bbls/acre-ft shown in parentheses. $S_o = (\text{bbl/ac-ft})/0.7758\phi$, where S_o and ϕ are in %.

Table 5.3 Steam Stimulation Field Experience: Operating Parameters[a]

Project Number	Location	Total Steam Injected per Well (STB water) per cycle	Injection Pressure (psi)	Injection Time (days)	Soak Period (days)	Oil Production Rate (before) (bbl/d)	Oil Production Rate (after) (bbl/d)	Production Period	Increm. Oil per Well (STB)	Number of Cycles	Bbl Oil[b] per bbl Steam	Notes
1	Guadalupe field, CA	—	—	—	—	45–50	250	11–12 months	—	4	—	c
2	Midway-Sunset field, CA	3194, 3588, 3995	140–160	8, 8, 7	5, 3, 10	5	59, 60, 57	5.5, 7, 5.5 months	10,974, 12,710, 9407	3	(8.8, 9.07, 6)	d
3	Midway-Sunset field, CA	3240, 5361, 6501	270–300	10, 14, 10	3, 2, 5	15	93, 51, 61	2, 7, 4 months	5763, 10,762, 8187	3	(4.2, 4, 2.7)	d
4	Midway-Sunset field, CA	3900, 4769	450, 300	6, 9	8, 7	5	15, 10	9, 6 months	4330, 1756	2	(2.02, 0.48)	e
5	Wilmington field, CA	8100–32,000	650–1400	6–17	5–8	30	240	~9 months	—	1	—	f
6	Wilmington field, CA	8100–32,000	650–1400	6–17	5–8	25	95	~9 months	—	1	—	g
7	Cat Canyon field, CA	2731–19,685	1250–2500	5–16	1–6.5	—	37–241	5–6 months	0–15,200	—	0–2.24	h
8	Midway-Sunset Section 5A, CA	12,549, 14,475, 9971	—	—	9, 29, 24	24	125–134	4–5 months	—	3	0.37	i
9	Midway-Sunset Section 23A, CA	12,843, 8504, 3795	—	—	89, 7, 3	—	—	4–5 months	—	3	0.10	j
10	West Newport, CA	—	800–1000	—	—	—	—	—	—	1–5	—	k
11	Huntington Beach, CA	—	—	—	1–10	850[a]	3250[a]	—	—	1–4	0.50	l
12	Santa Barbara, Venezuela	~11,000	—	—	2–9	1075[a]	3500[a]	—	—	2–4	3.40	m
13	Emlichheim, West Germany	3430	617	—	1–7	12.6	25–88	—	—	1	1.3	n
14	Tia Juana, Venezuela	40,000	400–800	—	—	25	250	353, 416, 190 days	—	3	3.0	o
15	Oxnard field, CA	4124, 1848, 4895, 2514	800–1150	12, 5, 14, 7	3, 4, 3, 0.5	—	80	1, 0, 8, 5 days	150, 0, 919, 18	4	0.04, 0, 0.2, 0.01	p
16	Oxnard field, CA	22,839, 14,756, 7500 10,671, 9315	800–1600	18, 11, 6, 8, 7	3, 3, 3, 3, 4	—	—	45, 101, 112, 176, 28 days	5153, 13,192, 11,497, 13,125 911	5	0.23, 0.89, 1.53, 1.23	q
17	Quiriquire field, Venezuela	—	1000	44, 49	—	143	213, 174	443, 305 days	—	2	1.3–1.6	r
18	Huntington Beach, CA	9850, 8400, 9750, 11,900	—	—	10	2–30	9–175	14, 17, 16, 14 days	30,000, 30,000, 27,500, 27,500	1–5	3, 3.7, 2.5, 2.4	s
19	Kern River, CA	10,854, 14,072	406, 392	18, 25	7, 9	3.9	59	151 days	8704, 480	2	0.8	t

[a] From Ref. 1. © 1974 *J. Can. Pet. Tech.*, CIM.

Notes for field experience table: a, Total rate for project; otherwise, rate is average per well. b, Figures in parentheses are bbl oil per million Btu's injected. c, Total 48 wells; average first-cycle recovery, 30,000–33,000 bbl. Economic success. d, Economic success. e, Failure. f, Total 67 wells; 6 on second cycle. Quality 50–85%. g, Total 5 wells. h, Several failures; hot-water soak most effective; double tubing. i, 73 wells: one cycle; 24: two cycles; 14: third cycle; 2: fourth cycle; 1: fifth cycle. j, Poor response. k, 166 treatments to 100 wells; a few in A & C zones. l, 259 steam stimulations on 125 tar sand wells; 96: two cycles; 34: three cycles; 3: four cycles. m, Economic success. Naphthenic-base oil; two zones 16° & 19° API. n, Total 17 wells. o, 73 wells, 700-ft spacing; 90% quality. p, Partial success; peak rate 690 bbl/d. q, Flowed in four cycles; pumping in fifth; peak rate 389 bbl/d. r, Best well in project. s, 30 wells: one cycle; 22: two cycles; 17: three; 9: four; 3: five. t, 7 wells' average; 80% quality.

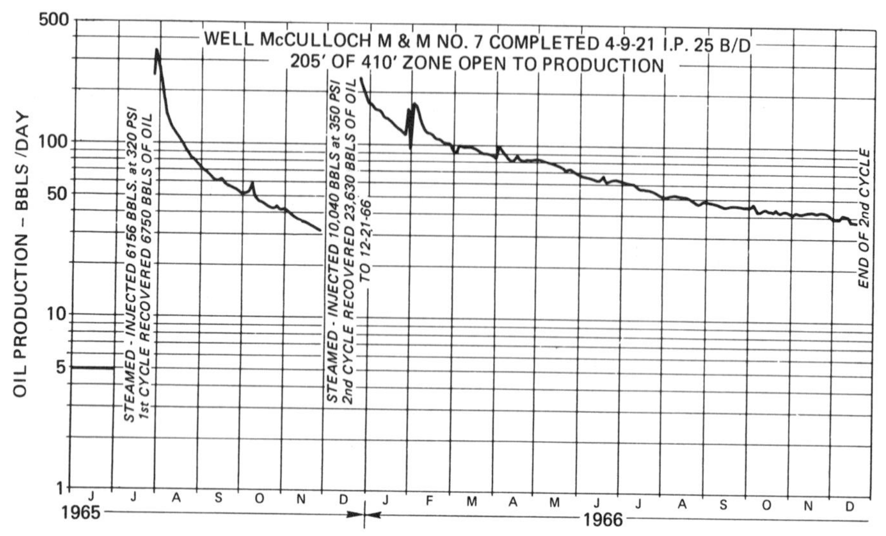

Figure 5.1 Typical response to cyclic steam injection, Midway–Sunset field.[10] © 1969 SPE-AIME.

drawn from these projects are reasonably general. Performance of a typical California well in the Midway–Sunset field, where gravity drainage is the principal producing mechanism, is shown in Figure 5.1.

STEAM STIMULATION IN M-6 PROJECT, TIA JUANA FIELD, VENEZUELA

The Tia Juana field operated by Maraven is one of the large heavy oil fields in the Bolivar coastal region of Venezuela.[3,4] It is located along the eastern shore of Lake Maracaibo and was discovered in 1936. Primary development in the M-6 area began in 1948. By early 1969, when steam stimulation operations began in this area, 67.8 Mbbl, or 11.7% of the original oil in place had been produced. Extrapolation of the cold primary trend indicated an ultimate primary recovery of 105.2 Mbbl, or 18.3%, of the stock tank original oil in place (STOOIP). Ultimate recovery estimated for steam stimulation and preceding cold primary is 136.9 Mbbl, or 23.8%, of the STOOIP. Additional recovery is expected from steam drive operations.

The structure of the field is a gentle homocline. The dip averages 4° and trends northeast to southwest. The four producing sands are in the lower Lagunillas formation, which is Miocene in age. The sands are unconsolidated and have excellent porosity and permeability. The crude-oil gravity is

12°API and varies in viscosity between 1000 and 9000 cp at 100°F. Other reservoir characteristics and project performance to the beginning of 1977 are displayed in Table 5.4. Performance of the first three cycles of a typical well is shown in Figure 5.2. The predominant entry of steam into the upper part of the open formation is illustrated in Figure 5.3.

The project area is 1831 acres. There are 145 wells in the project drilled on 231 m spacing. The installed capacity of the steam plant is 3480 tonne/d (320,000 lb/hr). As of February 1977, 143 of the 145 wells had been steam

Table 5.4 Reservoir and Project Performance Data—M-6 Project, Tia Juana Field[a]

Reservoir Data	
Project area	1831 acres
Average depth top sand	1600 ft sub-sea
Average net oil sand	125 ft
Porosity	38.1%
Oil saturation	0.85
Water saturation	0.15
Oil formation volume factor	1.04 RB/STB
Temperature	113°F
Oil gravity	12°API
Dead oil viscosity at 100°F	1000–9000 cp
Original reservoir pressure at 1600 ft	780 psig
Stock tank oil initially in place (STOOIP)	575×10^6 bbl/d
Project Performance, January 1, 1977	
Oil production rate	8714 bbl/d
Cumulative oil production	115.7×10^6 bbl
Cumulative oil as percentage of STOOIP	20.1%
Gas–oil ratio (instantaneous)	422 ft^3/bbl
Cumulative gas production	44.9×10^9 ft^3
Producing water cut	30.1%
Cumulative water production	9.9×10^6 bbl
Cumulative steam injected	11.9×10^6 bbl
Project Reserve Estimates	
Primary recovery	105.2×10^6 bbl
Percentage of STOOIP	18.3%
Steam soak recovery	31.7×10^6 bbl
Percentage of STOOIP	5.5%
Steam drive recovery	120.1×10^6 bbl
Percentage of STOOIP	20.9%
Total recovery	257.0×10^4 bbl
Percentage of STOOIP	44.7%

[a] From Ref. 4. © 1977 CIM.

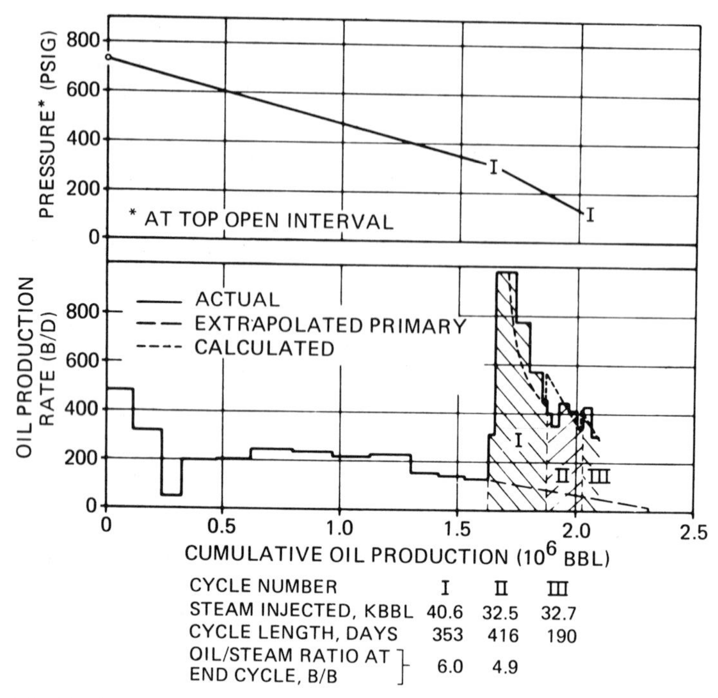

Figure 5.2 Typical performance of soaked well (LSJ-622).[3] © 1969 SPE-AIME.

stimulated, of which 7 were in the first cycle, 82 in the second, 53 in the third, and 1 in the fourth cycle. From the initiation of steam stimulation in March 1969 to the end of December 1976, 48.1 Mbbl of oil had been produced, of which 30.7 Mbbl is attributable to steam stimulation. The incremental oil–steam ratio is 2.56 bbl/bbl. In early 1970, the production by steam stimulation was about five times that under cold conditions extrapolated to the same cumulative recovery.

Maraven has found that steam preferentially enters the upper zones. This makes uniform depletion of the zones difficult. To alleviate this difficulty, selective injection into lower zones is often used. Wells have routinely been completed with gravel-packed liners. To accomplish selective injection, a plugging fluid[5] is squeezed into the gravel pack, and the liner is equipped with a special collar. The lower zones may then be effectively steamed.

An illustration of the effectiveness of this method is provided in Figure 5.4. This figure shows wellhead injection pressure plotted against steam injection rate for two cycles of a well in East Tia Juana. For both cycles, the wellhead pressure is extrapolated to zero injection rate. This pressure gives an indication of the pressure in the zone taking steam. For the first cycle, steam was injected conventionally, and the extrapolated pressure was

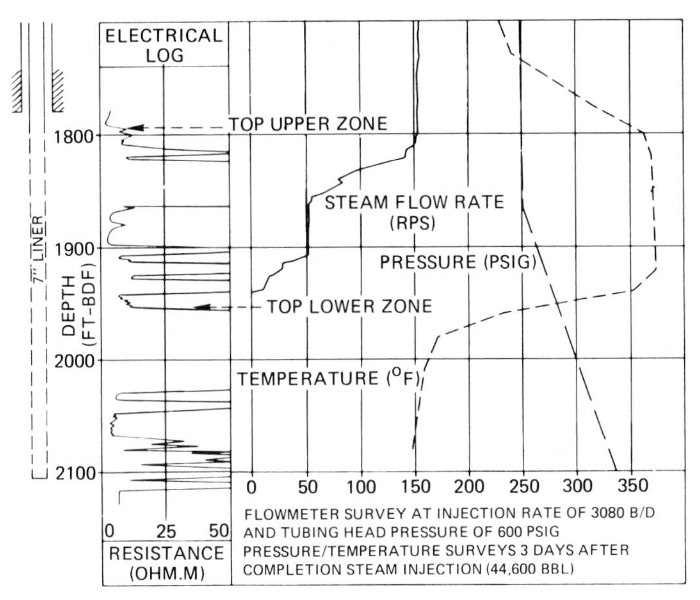

Figure 5.3 Typical steam injection, pressure and temperature profiles (well LSJ-2793).[3] © 1969 SPE-AIME.

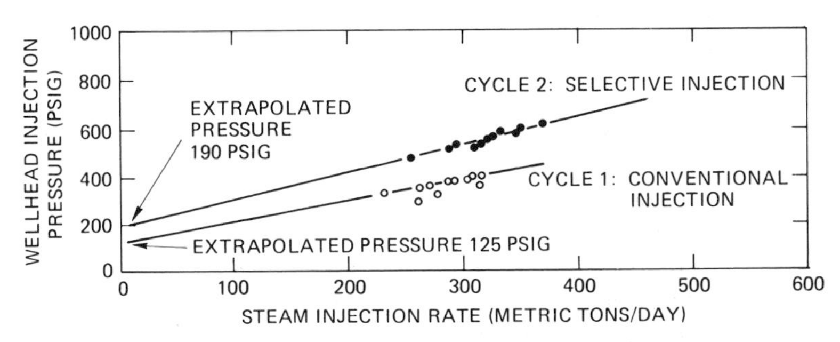

Figure 5.4 Steam injectivity, well LSE-1789, East Tia Juana field.[4] © 1977 CIM.

125 psig. Because the uppermost zone took most of the first cycle steam, the second cycle used selective injection into a lower zone. The extrapolated pressure for the second cycle is 190 psig, indicating that steam was successfully directed to the deeper high pressure zone.

The main producing mechanism at Tia Juana is sand compaction. As reservoir pressure decreases, the formation compacts, and its pore volume becomes smaller. This compaction is exhibited at the surface as subsidence. In addition to the compaction mechanism, a limited solution gas drive is present.

This is illustrated clearly in Figure 5.5, which plots cumulative subsidence

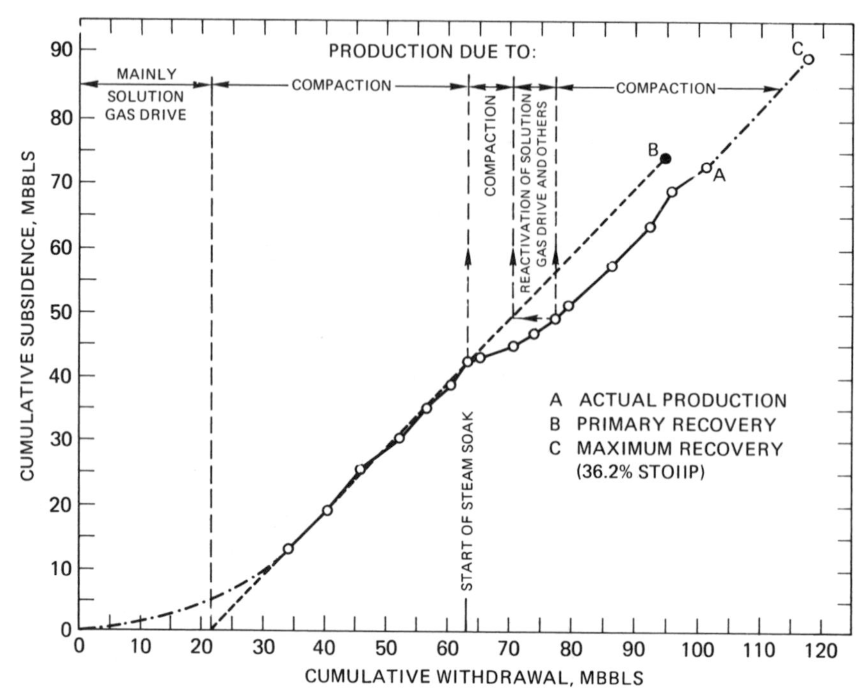

Figure 5.5 Influence of different reservoir drive mechanisms, D2/E2 project, Tia Juana field.[21] © 1977 CIM.

against fluid withdrawals for the D2/E2 project (an earlier Maraven project). After an initial period dominated by solution gas drive, the slope of the line approaches 45°, and the dominant mechanism becomes compaction. When steam stimulation is initiated, the solution gas drive mechanism is reactivated briefly; later the dominant compaction drive mechanism resumes. It is anticipated that steam stimulation will allow more complete use of the compaction mechanism (by obtaining a lower abandonment pressure) and hence provide higher ultimate recovery.

Maraven expects to convert the entire area to steam drive in the future and expects that the ultimate recovery from this area will be about 45%.

STEAM STIMULATION AT ESSO'S COLD LAKE PROJECT

The Cold Lake oil sands[7,8] are located 300 km northeast of Edmonton, Alberta, and cover an area of over 6500 km^2. They contain approximately 33×10^9 m^3 OOIP (208×10^9 STB). Esso Resources Canada Limited has conducted pilot operations in the sands since 1964. The Clearwater formation contains sands having a thickness of about 164 ft. It is the richest reservoir, and it has been the target of most pilot operations at Cold Lake to date. It occurs at depths of 1400–1480 ft. A summary of reservoir and oil characteristics is presented in Table 5.5. In the Clearwater formation, the average porosity and permeability are 35% and 1.5 darcies, respectively. At a depth of 1500 ft, reservoir temperature is 55°F, and the pressure is 450 psia. Cold Lake bitumen has a gravity of 10.2°API and a viscosity of 100,000 cp at the reservoir temperature. Oil mobility is extremely low at initial reservoir conditions despite the high formation permeability, and primary production is nil.

The immobility of the cold bitumen requires initial steam injection above fracturing pressures of about 1300 psi to achieve practical injection rates. Depending on the in situ stress state, fractures may be either horizontal or vertical. Fracture orientation depends on whether it is easier to lift the formation resulting in a horizontal fracture or to separate the formation along a vertical plane. Vertical fractures appear to trend southwest to northeast.

The Ethel pilots were initiated at Cold Lake in 1964 and operated until 1970. They showed that many cycles of steam stimulation would be required before steamflooding would be practical on commerical well spacings.

Based on the experience gained, a new project, the May pilot, was initiated in 1972 about 1 km to the northwest. It initially consisted of 23 wells in three enclosed 5-spot patterns with a well spacing of 2 ha. In this pilot, a number of steam stimulation tests were conducted, including the use of gas and solvent with steam to improve oil recovery. Early data indicated that the technique was beneficial in accelerating oil production, but the economic benefits were uncertain. Figure 5.6 shows the well locations in the Ethel and May pilots.

The reservoir quality improves to the northwest of the May pilot, with higher bitumen saturations and the absence of bottom water in most areas.

Table 5.5 Cold Lake Reservoir and Fluid Properties

Reservoir depth	460 m (1509 ft)
Reservoir thickness	50 m (164 ft)
Initial reservoir pressure	3100 kPa (450 psi)
Initial reservoir temperature	13°C (55°F)
Average reservoir permeability	1.5 darcies
Oil gravity	10.2°API
Viscosity at reservoir temperature	100,000 cp
Solution gas–oil ratio	9.85 sm^3/m^3 (55 scf/STB)
Average oil saturation	70%

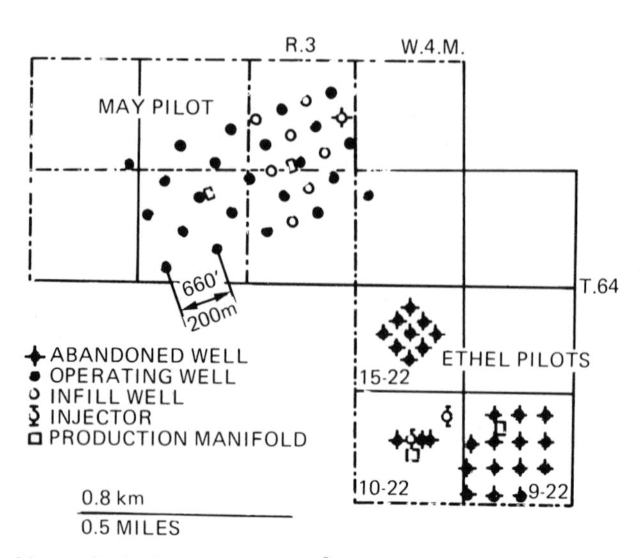

Figure 5.6 May–Ethel pilot configuration.[8] © 1984 The World Petroleum Congress, published by John Wiley & Sons Ltd.

A third pilot, Leming, was started in 1975 in this area about 4.5 km northwest of the May pilot, with 56 wells in eight clusters (Fig. 5.7). Leming wells were drilled on 2.9 ha well spacing. Later well configurations were designed to delay well-to-well communication in the on-fracture-trend direction for steam stimulation. All the wells were directionally drilled from central pads in order to reduce steam distribution line lengths and to minimize environmental impact. A variety of steam stimulation experiments were conducted, and by 1977, the results were sufficiently encourging to lead to the start-up of a large commercial project in 1985.

A major expansion of the Leming pilot was undertaken in 1980 with the addition of seven new pads containing 112 wells. This expansion also included installation of commercial prototype steam generators, pipeline systems, and a water reuse system. The Leming pilot reached a production output of 2600 m^3/d (16,350 STB/d) during 1984. The calendar day oil rate (CDOR) is the cumulative bitumen production divided by the total days for the entire cycle, including steam injection, production, and shut-in days. Figure 5.8 shows the May and Leming CDOR performances for average wells.

The Leming performance has been appreciably better than at May because of the higher quality of the reservoir and the design improvements incorporated in the later Leming project. Figure 5.9 shows the average oil–steam ratio (OSR) behavior of the May and Leming wells for each

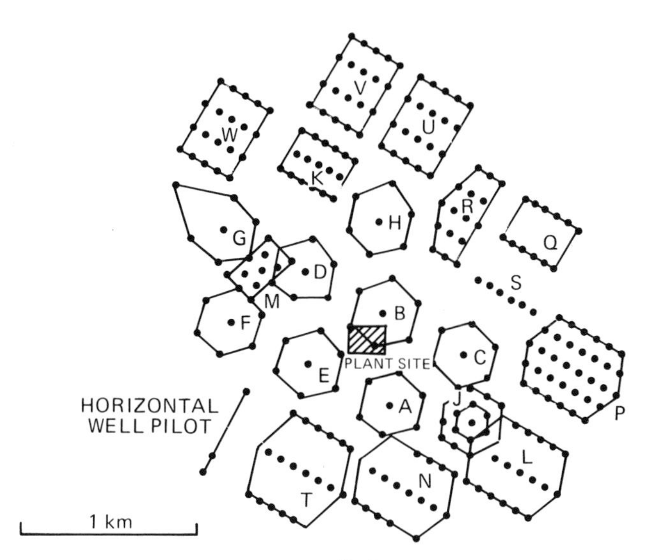

Figure 5.7 Leming pilot well location map.[8] © 1984 The World Petroleum Congress, published by John Wiley & Sons Ltd.

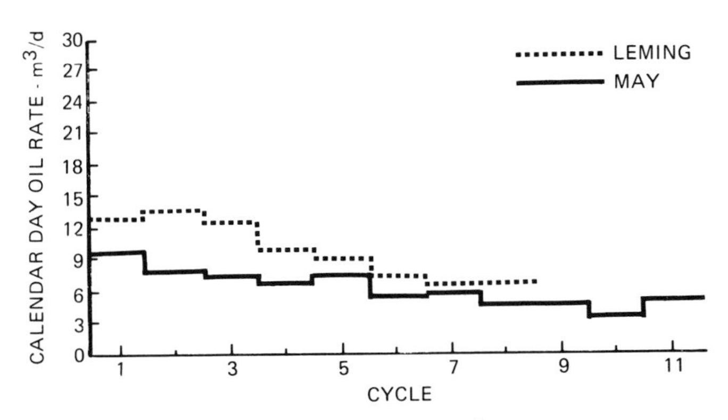

Figure 5.8 Cold Lake pilot calender day oil rates.[8] © 1984 The World Petroleum Congress, published by John Wiley & Sons Ltd.

stimulation cycle. The average OSR performance has varied from 0.30–0.45 at Leming compaed to 0.20–0.30 at May. The first-cycle OSR is relatively low in both pilots, with marked improvement for the next two cycles, and then exhibits a gradual decrease. The initial OSR can also be strongly affected by the volume of steam injected during a cycle. Large steam treatments result in the OSR being depressed at first. However, the

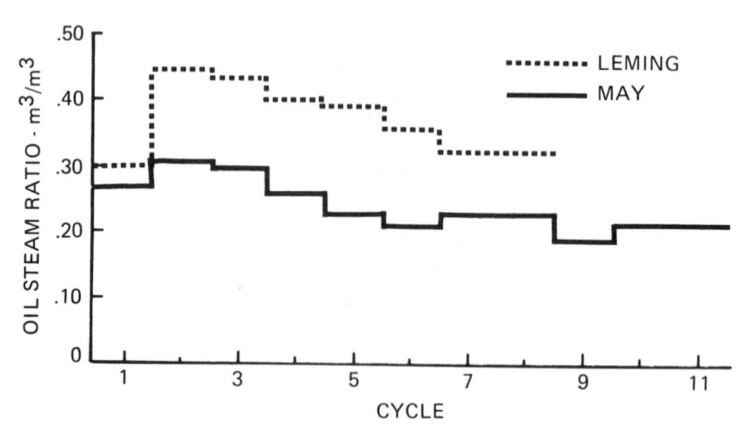

Figure 5.9 Cold Lake pilot oil–steam ratios.[8] © 1984 The World Petroleum Congress, published by John Wiley & Sons Ltd.

long-term OSR's approach a common value. Table 5.6 outlines the 1985 design basis for the commerical project.

In the commerical operation, wells are operated for about eight cycles of steam stimulation over about an 8-year period. After this, either cyclic steam stimulation will be continued or an alternative recovery method, probably steamflooding, will ensue.

The wells are completed over an interval of about 10 m at the bottom of the good oil sand and above any shale or tight streaks. A well spacing of 1.6 ha (4 acres), with an aspect ratio of 1.7, is used. The aspect ratio is the ratio of the distance between wells in the on-fracture trend direction to that in the off-fracture trend direction. This appears to be a good combination for Cold Lake to achieve good recovery and reduce premature well-to-well communication. A systematic technique of row steaming is employed where simultaneous injection into a row of wells aligned in the off-trend direction is used to help avoid premature breakthrough in that direction. Pilot results have shown that recovery of better than 20% OOIP, or 36.4×10^3 m^3/well, for wells on a 4-acre spacing can be achieved by steam stimulation. For design purposes in the Leming area, the average values of the CDOR at 8.4 m^3/d (53 bbl/d), OSR at 0.38, and WOR at 2.6, assume that new wells will be added each year to maintain a steady production rate.

Since desalination of produced water is uneconomical, the project utilizes once-through steam generators capable of making high-pressure wet steam from softened brine. The Leming project expansion of 1980–1981 incorporated testing prototypes of large steam generators of 1600-m^3/d (10-kbbl/d) capacity. These generators, fueled by gas, have three parallel, once-through tube systems and are operated to deliver steam at 80% quality and 14 MPa (2030 psia).

Table 5.6 Commercial Design Parameters for Recovery of Cold Lake Crude[a]

Recovery method	Cyclic steam stimulation
Number of cycles	8
Well life	8 years
Well completion	Bottom of good sand
Well spacing	1.6 ha (4 acres)
Aspect ratio	1.7
Steam volume[b]	11,000 m^3, first cycle (7000, 8000, 8500, 9000, etc. m^3)[c]
Average steam rate	230 m^3/d (1447 bbl/d)
Average CDOR	8.5 m^3/d (53 bbl/d)
Steady-state OSR	0.35
Steady-state WOR	2.6
Recovery	20%
Produced water recycle	100%

[a] From Refs. 8 and 9.
[b] 1 m^3 = 6.29 bbl.
[c] For second and subsequent cycles.

The Leming project also includes an experimental water reuse pilot plant. While recycling of produced water has been practiced for many years in California heavy-oil operations, the operating conditions at Cold Lake are more difficult. The high specific gravity of the bitumen increases the difficulty of total oil removal, and the higher operating temperatures and pressures of the steam generators call for better quality feedwater. The basic process employed consists of oil skimming and induced gas flotation for oil removal, hot lime softening to remove most of the calcium and magnesium and reduce silica, followed by ion exchange to complete the hardness removal.

SUMMARY OF OBSERVATIONS OF INDIVIDUAL PROJECT DATA

1. The first or second cycle is usually the best (see Figs. 5.1 and 5.2).
2. Thick gravity-drainage reservoirs with low decline rates can support numerous cycles. Solution gas drive reservoirs can support usually three to four cycles depending on initial conditions and crude value.[3,10]
3. Steam tends to enter the upper part of thick sands (see Fig. 5.3).
4. Treatment sizes range from 4400 to 70,000 bbl steam. (30–430 bbl/ft net sand; see Tables 5.3 and 5.6).

REASONS FOR STEAM STIMULATION PROJECT FAILURES (Tables 5.2–5.3)

1. Excessive pressure depletion (Coalinga, Tulare).[10]
2. Low oil saturation per acre foot of oil sand (Quiriquire,[11] Tulare,[10] Potter B[10]).
3. Wells too high on structure; in gas cap (Coalinga, White Wolf).[10]
4. High shale content; too many thin sands (Coalinga,[10] Tulare,[10], Quiriquire[11]).
5. Presence of thief water sands (Poso Creek).[10]
6. High clay content, low permeability (Tulare, White Wolf).[10]
7. Inadequate well completions in reservoirs requiring high-pressure steam injection.[10]

INCREMENTAL COSTS FOR STEAM STIMULATION

Costs for steam are a major item in a steam stimulation project and vary widely as expense for fuel and water treating vary.[10] Steam is commonly generated using once-through units with water pumped through a series of fired tubes and flashed to not greater than 80% quality steam. This method requires that only scale-forming ions (calcium and magnesium) be removed by ion exchange and that oxygen be removed before the water enters the steam generator. The water that passes through the steam generator still contains soluble salts, and these conventionally pass with the steam into the well. In projects where the produced water is reutilized, expense to deoil the water will be incurred.

A barrel (350 lb) of 80% quality steam will carry with it approximately 350,000 Btu.† Assuming 80% thermal efficiency of the steam generator, the fuel cost per barrel of steam generated will be:

1. Using gas:

$$\text{Cost of fuel} = \frac{350{,}000\ \text{Btu/bbl}}{0.8 \times 10^{6}\ \text{Btu/kcf}} C_{g}$$
$$= 0.438 C_{g} \quad (\text{cents/bbl})$$

where C_g is the cost of gas in cents per thousand cubic feet.

† This will vary with injection pressure and quality. For conditions remote from 800 psi and 0.80 quality, the following formula is used to estimate the Btu content of a barrel of steam:

$$\text{Btu/bbl steam} = 350(XH_{wv} + H_{ws} - H_{wr})$$

2. Using lease crude:

$$\text{Cost of fuel} = \frac{350{,}000 \text{ Btu/bbl steam}}{0.8(340)(18{,}000) \text{ Btu/bbl fuel}} C_o \times 100$$
$$= 7.1 C_o, \text{ cents/bbl §}$$

where C_o is the crude value in dollars per barrel.

Reference to the above numbers indicates that a break-even point (on a Btu basis) is reached at an incremental oil–steam ratio of

$$\frac{350{,}000}{0.8(340)(18{,}000)} = 0.071 \text{ bbl oil/bbl steam}$$

for the case of burning lease crude and generating 80% quality steam at 800 psig injection pressure. The reciprocal of this number is also useful to remember—about 14 bbl of steam are generated for each barrel of lease crude burned.

Other increased costs will be incurred in any steam operation:

1. Additional well costs:
   (a) Workovers for improved sand control plus packers and tubing insulation in high-temperature applications.
   (b) New well drilling; stronger casing and slotted liners may be required (N-80 and P-110 grades must frequently be used where elevated temperatures are encountered).

2. Water treating costs; can become significant if fresh water is scarce.

3. Produced oil dewatering; emulsion and sand problems can increase costs of steaming operations.

4. Steam distribution line costs; this can become a major item if large stationary steam generators are used.

5. Produced water disposal.

Cost reduction is being obtained by[12]

1. using larger generators and

2. reuse of produced water in areas where fresh water is scarce.

REFERENCES

1. Farouq Ali, S. M. "Current Status of Steam Injection as a Heavy Oil Recovery Method," *J. Can. Pet. Tech.* (January–March 1974), p. 115.

§ Assumes crude gravity 14°API (340 lb/bbl) with a heating value of 18,000 Btu/lb.

2. Leonard, J. "Increased Rate of EOR Brightens Outlook," *Oil Gas J.* (April 14, 1986), p. 71.
3. de Haan, J. H. "Early Results of the First Large-Scale Steam Soak Project in the Tia Juana Field, Western Venezuela," *J. Pet. Tech.* (January 1969), p. 101.
4. Herrera, A. J. "The M-6 Steam Drive Project Design and Implementation," *The Oil Sands of Canada–Venezuela, 1977*, CIM Special Vol. 17, p. 551.
5. Burkill, G. C. C. "How Steam is Selectively Injected in Open Hole Gravel Packs," *World Oil* (January 1982), pp. 127–136.
6. Bowman, C. H., and Gilbert, S. "A Successful Cyclic Steam Injection Project in the Santa Barbara Field, Eastern Venezuela," *J. Pet. Tech.* (December 1969), p. 1531.
7. Buckles, R. S. "Steam Stimulation Heavy Oil Recovery at Cold Lake, Alberta," SPE 7994, Presented at the California Regional Meeting of SPE, Ventura, CA, April 1979.
8. Mainland, G. G., and Lo, H. Y. "Technological Basis for Commercial In Situ Recovery of Cold Lake Bitumen," Eleventh World Petroleum Congress, London, Wiley, Chichester, July 1984, pp. 235–242.
9. Kemp, E. M. Presentation at the Saskatchewan Heavy Oil Conference, November 28–30, 1984, Saskatoon, Saskatchewan.
10. Burns, J. "A Review of Steam Soak Operations in California," *J. Pet. Tech.* (January 1969), p. 25.
11. Payne, R. W., and Zambrano, G. "Cyclic Steam Injection in the Quiriquire Field of Venezuela," SPE 1157, Presented at Society of Petroleum Engineers Meeting, Caracas, 1965.
12. Bursell, C. G. "Steam Displacement—Kern River Field," *J. Pet. Tech.* (October 1970), p. 1225.
13. Adams, R. H., and Khan, A. M. "Cyclic Steam Injection Project Performance Analysis and Some Results of Continuous Steam Displacement Pilot," *J. Pet. Tech.* (1969), pp. 95.
14. Boberg, T. C., and Lantz, R. H. "Calculations of the Production Rate of a Thermally Stimulated Well," *J. Pet. Tech.* (1966), p. 1613.
15. Owens, W. D., and Suter, V. E. "Steam Stimulation—Newest Form of Secondary Petroleum Recovery," *Oil Gas J.* (April 26, 1965), p. 82.
16. Teberg, J. A. "Wilmington Field Steam Operations," SPE 1494, Dallas Fall Meeting of SPE-AIME, October 1966.
17. Long, R. J. "Case History of Steam Soaking in the Kern River Field, California," *J. Pet. Tech.* (September 1965), p. 989.
18. Bott, R. C. "Cyclic Steam Project in a Virgin Tar Reservoir," *J. Pet. Tech.* (May 1967), p. 585.
19. de Haan, H. J., and Schenk, L. "Performance Analysis of a Major Steam Drive Project—The Tia Juana Field, Western Venezuela," *J. Pet. Tech.* (January 1961), p. 111.
20. Farouq Ali, S. M., and Meldau, R. F. "Current Steam Flood Field Experience," *J. Pet. Tech.* (October 1979), p. 1332.

21. Borregales, C. J. "Steam Soak on the Bolivar Coast," *The Oil Sands of Canada—Venezuela, 1977*, CIM Special Vol. 17, p. 561.
22. Rivero, R. T., and Hentz, R. C. "Resteaming Time Determination—Case History of a Steam-Soak Well in Midway–Sunset," *J. Pet. Tech.* (June 1975), p. 665.

Chapter 6

CALCULATION OF STEAM STIMULATION PERFORMANCE

ABSTRACT

A simplified, analytical model for calculating multicycle steam stimulation behavior, the Boberg–Lantz method, is presented in this chapter. In this method, the Marx–Langenheim equation is used to calculate the initial radius heated to steam temperature during the injection phase of the process. Then analytical solution of the equations for temperature decline as conduction occurs both vertically and radially is employed. An approximate stepwise approach is used to account for further temperature reduction owing to withdrawal of energy with the produced fluids. Since the method permits a calculation of the average temperature of the originally heated region, it lends itself well to the application of steady-state or pseudo-steady-state flow equations as presented in Chapter 4 for the estimation of the stimulated oil rate. A more sophisticated approach would be required for the case of a reservoir having very low or nonexistent oil mobility prior to steaming. The use of steady-state equations for flow provides an overly pessimistic prediction of stimulated rates for the immobile oil situation. An example hand calculation demonstrating the Boberg–Lantz method is provided. Because of the laborious nature of the calculations, they are usually programmed for a computer.

INTRODUCTION

Several techniques have been developed to calculate steam stimulation performance.[1–4] The method presented here is similar to that of Martin,[2] but it is easier to use and accounts for vaporization of produced water. Extensions of the model have been made to the gravity drainage case,[5] and other treatments of gravity drainage have appeared.[6] In addition, others have substituted a transient fluid flow analysis for the steady-state formulation presented here.[7] These extensions will not be covered in this book.

For very thick sands, numerical two-dimensional approaches may be required to obtain accurate results. Closmann et al.[4] discuss a partial two-dimensional numerical treatment of the stimulation problem. A more sophisticated three-phase numerical treatment, which accounts for heat transfer two dimensionally but assumes no vertical communication between layers, was presented by Weinstein.[9] Shutler's[10] method, which is a full two-dimensional, three-phase numerical treatment, can also be used for steam stimulation calculations but is more expensive computationally. These methods are covered in Chapter 11, where numerical methods for steamflooding are discussed. Numerical methods are usually required when computing cyclic steam injection recovery behavior in viscous oil sands where a large fraction of the ultimate oil recovery comes from the region heated around the wellbore.

BOBERG–LANTZ METHOD

Although the Boberg–Lantz[1] method describes the stimulation process in a simplified fashion, it has been used successfully to match the production history of a number of steam-stimulated wells. It is more applicable to a thin sand or a multiplicity of thin sands (Fig. 6.1) separated by shale stringers. It

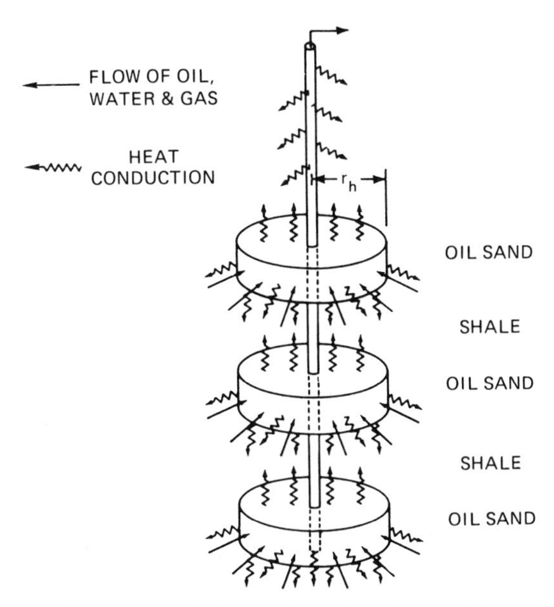

Figure 6.1 Schematic representation of heat transfer and fluid flow calculated by mathematical model.[1] © 1966 SPE-AIME.

is not applicable to thick sands, where gravity segregation of steam and oil will result in highly nonuniform heating of the oil around the wellbore. Nor is it applicable to cases where a large portion of the oil recovery comes from the region heated, as it neglects oil recovery from this region. It is best suited for widely spaced wells producing moderately viscous oils. It is most accurate for the first cycle of stimulation and becomes progressively less accurate for succeeding cycles because of the simplifying assumptions made.

The Boberg–Lantz method assumes that the reservoir is uniformly heated to a calculated distance from the wellbore. The size of the heated region is determined by accounting for heat losses to over and underburden while one-dimensional radial fluid flow is assumed to calculate fluid production history.

The mathematical model is developed about a coupled energy balance, fluid flow (oil, gas, and water), and time relationship. The first step in the calculation is to establish the temperature and size of the heated region; next, the well is allowed to soak; and finally, it is placed on production. A continuous energy balance on the heated region is conducted throughout the calculation.

Size of Heated Region

During steam injection, the oil sand near the wellbore is at condensing steam temperature T_s, the temperature of saturated steam at the sandface injection pressure. Pressure fall-off away from the well during injection is neglected in this analysis. Here T_s is assumed to exist out to a distance r_h where the temperature falls sharply to T_r, the original reservoir temperature. In reality, the temperature falls more gradually to reservoir temperature because of the presence of the hot-water bank ahead of the steam, but this is neglected to simplify the calculation. The heated zone radius is calculated by the Marx–Langenheim equation (see Chapter 3), assuming that each sand in a multisand reservoir is invaded equally and uniformly:

$$r_h^2 = \frac{\hat{h} M_s (\bar{X}_i H_{wv} + H_{ws} - H_{wr}) \xi_s}{4 K_h \pi (T_s - T_r) t_i N_s} \tag{6.1}$$

The function ξ_s is plotted as a function of dimensionless time $[\tau = 4K_h t_i / \hat{h}^2 (\rho c)_{R+F}]$ in Fig. 6.2. The ratio θ, equal to $(\rho c)_{ob}/(\rho c)_{R+F}$, is taken as unity ($\theta = 1$). The use of this equation for multisand reservoirs assumes that injection times are sufficiently short and that interbedded shale is sufficiently thick that no heating occurs at the midplane of the shale during the injection period.

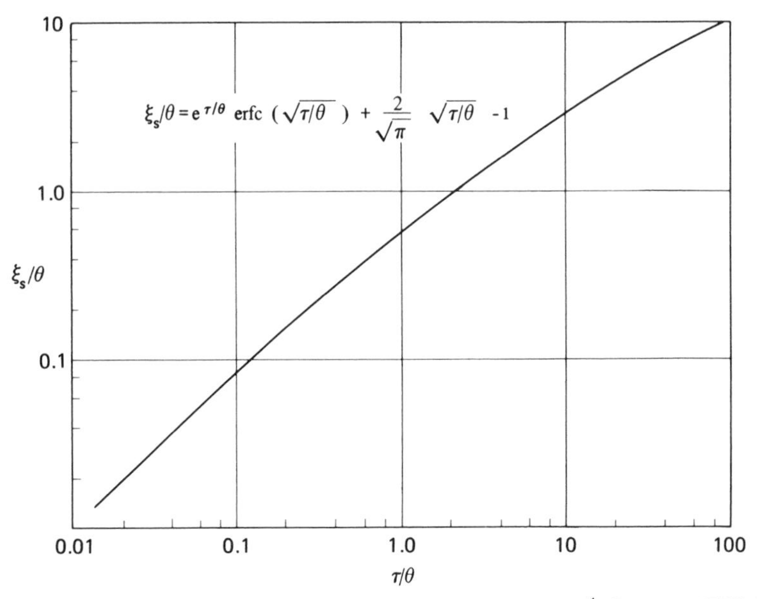

Figure 6.2 Dimensionless position of "steamed-out" region.[1] © 1966 SPE-AIME.

Temperature History of the Heated region

The average temperature T_{avg} of the heated region or regions is calculated by the following equation at any time after termination of steam injection; the equation is based on an approximate energy balance around the heated region $r_w < r < r_h$:

$$T_{avg} = T_r + (T_s - T_r)[\bar{v}_r \bar{v}_z (1-\delta) - \delta] \tag{6.2}$$

where $\bar{v}_r$ and $\bar{v}_z$ are dimensionless quantities that account for conduction of heat from the heated oil sand into unproductive rock, and δ is a correction term that accounts for the energy removed from the oil sand by the produced oil, gas, and water.

Radial Conduction Losses. The conduction correction for energy losses radially away from the heated region[1] is given by

$$\bar{v}_r = 2\int_0^\infty \frac{e^{-\bar{\theta}_r y^2} J_1^2(y)\,dy}{y} \tag{6.3}$$

where $\bar{\theta}_r = \alpha(t - t_i)/r_h^2$. A plot of $\bar{v}_r$ versus $\bar{\theta}_r$ is given in Figure 6.3; $\bar{v}_r$ may also be evaluated by

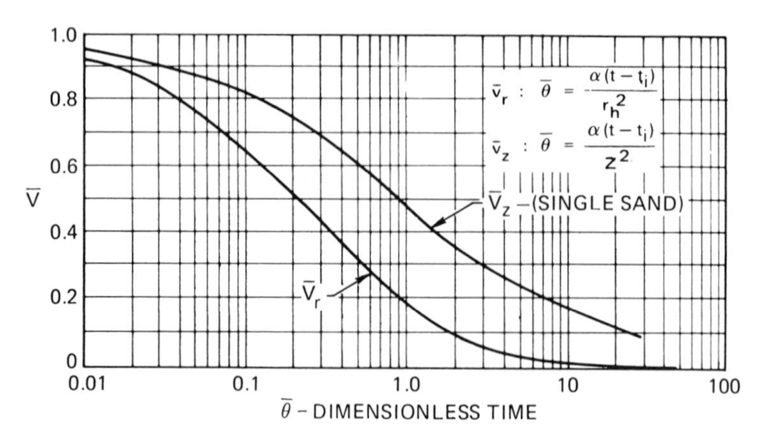

Figure 6.3 Solution for $\bar{v}_r$ and $\bar{v}_z$ (single sand).[1] © 1966 SPE-AIME.

$$\bar{v}_r = \frac{1}{\bar{\theta}_r} \sum_{k=0}^{\infty} s_k \tag{6.4}$$

where $s_o = \frac{1}{4}$ and

$$s_{k+1} = \left[\left(-\frac{1}{\bar{\theta}_r} \right) \frac{(k+1.5)}{(2+k)(3+k)} \right] s_k .$$

Vertical Conduction Losses. The heat transfer model for conduction in the vertical direction consists of approximating the sand–shale sequence by a stack of variable-thickness, equal-radii sand cylinders. These cylinders, initially at constant temperature, conduct heat as a function of time to variable-thickness interbedded shales and to semi-infinite overburden and underburden. Figure 6.1 shows a geometric sequence of an alternating sand–shale sequence depicting schematic fluid flow and heat loss. The general expression for $\bar{v}_z$ with this heat transfer model can be written[1]:

$$\begin{aligned} \bar{v}_z = \left(2 \sum_{m=1}^{N_s} \bar{h}_m \right)^{-1} \sum_{m=1}^{N_s} \sum_{n=1}^{N_s} & \times \left\{ W_1 \operatorname{erf} \frac{W_1}{\sqrt{w}} + W_2 \operatorname{erf} \frac{W_2}{\sqrt{w}} - W_3 \operatorname{erf} \frac{W_3}{\sqrt{w}} - W_4 \operatorname{erf} \frac{W_4}{\sqrt{w}} + \sqrt{\frac{w}{\pi}} \right. \\ & \left. \times \left[\exp\left(-\frac{W_1^2}{w} \right) + \exp\left(-\frac{W_2^2}{w} \right) - \exp\left(-\frac{W_3^2}{w} \right) - \exp\left(-\frac{W_4^2}{w} \right) \right] \right\} \end{aligned} \tag{6.5}$$

where $W_1 = B_m + \bar{h}_m - B_n$
$W_2 = B_n + \bar{h}_n - B_m$
$W_3 = B_m - B_n$

$$W_4 = B_n + \bar{h}_n - B_m - \bar{h}_m$$
$$w = 4\alpha(t - t_\mathrm{i})$$
$$B_j = B_{j-1} + \bar{h}_{j-1} + \bar{l}_{j-1}$$
$$B_1 = 0$$
$$\bar{h}_j = h_j + z$$
$$\bar{l}_j = l_j - z$$
h_j = thickness of sand j

l_j = thickness of shale j

$$z = \left[M_\mathrm{s} \frac{\bar{X}_\mathrm{i} H_\mathrm{wv} + H_\mathrm{ws} - H_\mathrm{wr}}{[\pi r_\mathrm{h}^2 (\rho c)_{\mathrm{R+F}} (T_\mathrm{s} - T_\mathrm{r}) N_\mathrm{s}]} \right] - \hat{h}$$

Here z is a hypothetical thickness that, when added to the individual sand thickness, accounts for all the energy injected, including that lost to the shale during the injection phase.

For a thick formation (>300 ft), it will be adequate at short times (<2 years) to approximate the sand–shale sequence as an infinite stack of sands and shales of uniform, but not necessarily equal, thicknesses. For this case,

$$\bar{v}_\mathrm{z} = \frac{Z}{L} + \frac{2L}{\pi^2 Z} \sum_{k=1}^{\infty} \left[\frac{1}{k^2} \sin^2 \frac{k\pi Z}{L} \exp\left(\frac{-k^2 \pi^2 \alpha (t - t_i)}{L^2} \right) \right] \tag{6.5a}$$

where $\hat{h} = \Sigma h_j / N_\mathrm{s}$

$l = \Sigma l_\mathrm{j} / (N_\mathrm{s} - 1)$

$L = (\hat{h} + l)/2$

$Z = (\hat{h} + z)/2$

For this case, $\bar{v}_\mathrm{z}$ can be estimated using Figure 6.4. For a single-sand reservoir, Eq. (6.5) becomes

$$\bar{v}_\mathrm{z} = \mathrm{erf} \frac{\bar{h}_1}{\sqrt{4\alpha(t - t_\mathrm{i})}} - \sqrt{\frac{4\alpha(t - t_\mathrm{i})}{\pi \bar{h}_1^2}} \left[1 - \exp\left(- \frac{\bar{h}_1^2}{4\alpha(t - t_\mathrm{i})} \right) \right] \tag{6.6}$$

This result is graphically presented in Figure 6.3.

Energy Removed with the Produced Fluids. The dimensionless quantity δ in Eq. (6.2) accounts for the energy removed from the formation with the produced fluids and is defined by

$$\delta = \frac{1}{2} \int_{t_\mathrm{i}}^{t} \frac{H_\mathrm{f}^* \, dx}{\pi r_\mathrm{h}^2 (\rho c)_{\mathrm{R+F}} (T_\mathrm{s} - T_\mathrm{r}) \sum_{i=1}^{N_\mathrm{s}} \bar{h}_\mathrm{i}} \tag{6.7}$$

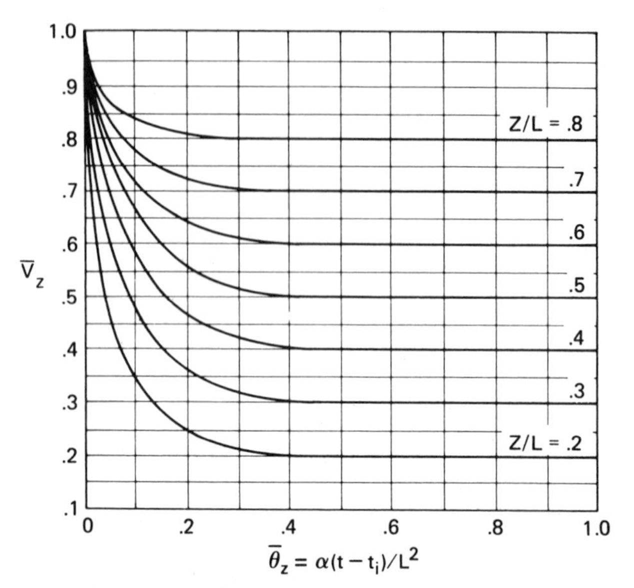

Figure 6.4 Solution for $\bar{v}_z$ (multiple sands).

Equation (6.2) must be solved in a stepwise manner since H_f^*, the energy removal rate, is a function of T_{avg}. Because approximations are used in deriving Eq. (6.2), as δ approaches unity, T_{avg} calculated by Eq. (6.2) can become less than T_r. This is physically impossible. When it happens, T_{avg} should be taken as equal to T_r.

For cases where negligible gas and water vapor are produced,

$$H_f^* = 5.61 q_{oh}[(\rho c)_o + R_w(\rho c)_w](T_{avg} - T_r) \quad \text{(Btu/d)} \tag{6.8}$$

and for the case where significant gas and water vapor are produced,

$$H_f^* = q_{oh}(H_{og} + H_w) \quad \text{(Btu/d)} \tag{6.9}$$

where

$$H_{og} = [5.61(\rho c)_o + R_g c_g](T_{avg} - T_r) \quad \text{(Btu/STB oil)} \tag{6.10}$$

$$H_w = 5.61\rho_w[R_w(H_{wt} - H_{wr}) + R_{wv}H_{wv}] \quad \text{(Btu/STB oil)} \tag{6.11}$$

Because of the elevated temperatures, a fraction of the produced water will be produced as water vapor. (This will always occur if gas is produced.)

The water–oil ratio for the produced water vapor is given by

$$R_{wv} = (0.0001356)\left(\frac{P_{wv}}{P_w - P_{wv}}\right)R_g \quad \text{(bbl liquid water at 60°F/STB)}$$

when $P_w > P_{wv}$ and $R_{wv} < R_w$.

The R_{wv} calculated by the above formula should never be greater than R_w. If $P_{wv} \geq P_w$, then $R_{wv} = R_w$. Here, P_{wv} is the saturated water vapor pressure at T_{avg} in psia and P_w is the producing bottom-hole pressure in psia.

In order to evaluate H_f^* accurately, one must know the values of R_w and R_g, the water–oil and gas–oil ratios following stimulation. In most cases, the water–oil ratios are higher immediately following injection than before injection. Later, R_w will stabilize close to or below the prestimulation values. Gas–oil ratios are frequently lower immediately following injection but soon return to the prestimulation level. A more sophisticated numerical calculation would be needed to accurately predict the behavior of R_w or R_g immediately following injection. For this simplified method, it is reasonable to neglect the change in R_g. For an optimistic evaluation, R_w can be assumed equal to or slightly below the normal water–oil ratio.

For a pessimistic evaluation, it could be assumed that only water would be produced ($R_w \rightarrow \infty$) until all of the water injected as steam is produced. An intermediate behavior (i.e., an average observed in one field case) is given in Figure 6.5. This figure plots the change in the water–oil ratio, ΔR_w, as a function of the ratio of the cumulative water produced to the cumulative water (as condensate) injected.

Calculation of Oil Production Rate. For most conventional heavy-oil reservoirs, the use of steady-state radial flow equations appears to be adequate for predicting the oil production rate response to steam injection. The reservoirs must have sufficient reservoir energy and sufficiently low oil viscosity to produce oil under cold conditions and a decline curve of cold production history must be available (or predictable).

For tar sands and pressure-depleted reservoirs, the procedure as outlined here will not be applicable, as it will predict an unrealistically low rate response. In these cases, the bulk of the stimulated oil production comes from oil sand that is actually heated (rather than from the portion of the reservoir that is still cold).

This method does not account for oil saturation depletion of the heated region. It assumes that cold oil moving in from outside the heated region replenishes the oil in the heated zone.

Steady-State Approximation for Productivity Index. For a given stage of reservoir depletion, the ratio of the oil productivity index in the stimulated

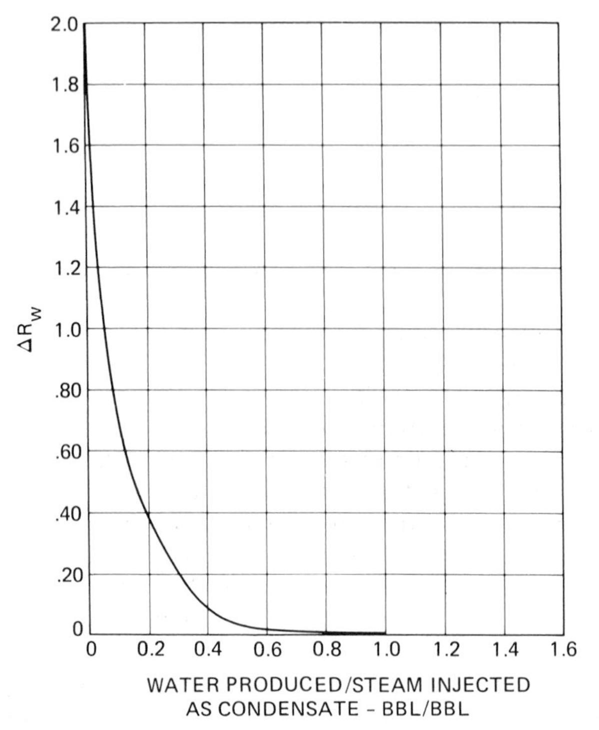

Figure 6.5 Increase in water–oil ratio following steam injection; average for Quiriquire wells Q-594 and Q-599.[1] © 1966 SPE-AIME.

condition ($J_h = q_{oh}/\Delta P$) to that in the unstimulated condition, J_c, can be estimated by an equation of the form

$$\bar{J} = \frac{J_h}{J_c} = \frac{1}{(\mu_{oh}/\mu_{oc})c_1 + c_2} \quad \text{(dimensionless)} \tag{6.12}$$

where μ_{oh} = kinematic oil viscosity at temperature T_{avg} (cs)
μ_{oc} = kinematic oil viscosity at temperature T_r (cs)

Here c_1 and c_2 are geometric factors that include the pattern geometry and the skin factor of the well. For preliminary engineering calculations, suitable values can be calculated from the formulas given in Table 6.1.

In Table 6.1, r_w is the effective well radius in feet. The effect of perforated casing, slotted liner, plugging by drilling mud, paraffins, and so on, is like that of creating a skin at the wellbore. To relate r_w to $\bar{r}_w$, the actual well radius, we must know the well skin factor s. The defining

Table 6.1 Expressions for c_1 and c_2 Used in Eq. (6.12)

System	c_1	c_2
Radial, P_e constant	$\dfrac{\ln(r_h/r_w)}{\ln(r_e/r_w)}$	$\dfrac{\ln(r_e/r_h)}{\ln(r_e/r_w)}$
Radial, P_e declining	$\dfrac{\ln(r_h/r_w) - r_h^2/2r_e^2}{\ln(r_e/r_w) - 1/2}$	$\dfrac{\ln(r_e/r_h) - 1/2 + r_h^2/2r_e^2}{\ln(r_e/r_w) - 1/2}$

relationship between r_w, $\bar{r}_w$, and s is given as

$$r_w = \bar{r}_w e^{-s} \quad \text{(ft)} \tag{6.13}$$

Also in Table 6.1, r_e is the drainage radius of the well in feet. This quantity is never accurately known but is usually taken as one-half the well spacing in a pattern development. As long as $r_e >> r_h$, fairly large variations in r_e have little effect on $\bar{J}$.

Implicit in the assumptions used to derive Eq. (6.12) is the assumption that heating and fluid injection have a negligible effect on the permeability to oil. When swelling clays are present, steam might reduce the permeability, although cases of plugging attributable to steam injection in permeable unconsolidated sands are rare.

Influence and Significance of Skin Factor. The skin factor s has a large effect on the stimulation response. The greater the skin factor, the greater c_1 will be relative to c_2. For this reason, s has a strong inverse relationship with J_c; but for the case of a stimulated well where μ_{oh}/μ_{oc} is small, s has only a weak inverse relationship with J_h.

Alteration of Skin Factor by Heating. The prior discussion assumes that the skin factor is not altered by heating. In some wells, asphaltene precipitation may be a cause of high skin factors. This and similar situations can be alleviated by heating, and suitable modification of c_1 and c_2 can be made to take this into account.

For the case where the skin factor s_c is reduced to s_h following stimulation, Eq. (6.12) for constant P_e becomes

$$\bar{J} = \frac{s_c + \ln(r_e/\bar{r}_w)}{(\mu_{oh}/\mu_{oc})[s_h + \ln(r_h/\bar{r}_w)] + \ln(r_e/r_h)} \tag{6.14}$$

A similar form can be written for the case where P_e is declining.

To obtain the oil rate as a function of time, it is necessary to know the unstimulated productivity index J_c and the static reservoir pressure P_e as a function of the cumulative fluid withdrawals. Then the stimulated oil rate, q_{oh}, can be calculated:

$$q_{oh} = \bar{J} J_c \, \Delta P \quad (\text{STB/d}) \tag{6.15}$$

where $\bar{J}$ is determined from Eq. (6.12) or (6.14), J_c is obtained from an extrapolated plot of the productivity index history of the well prior to stimulation, and ΔP, the pressure drawdown, is $P_e - P_w$ during the period of stimulated production. The static pressure P_e can be estimated from an extrapolation of a plot of P_e versus cumulative oil produced prior to stimulation for those cases where the pressure maintenance benefits of the injected fluids can be neglected.

Equations (6.2), (6.12), and (6.15) must be solved jointly in a stepwise manner.

Estimation of Heat Losses for Succeeding Cycles. In order to calculate the performance of cycles following the first, some account should be made for the residual heat left in the reservoir during preceding cycles. The energy remaining in the oil sand after preceding cycles can be approximated by

$$\text{heat remaining} = \pi r_h^2 (\rho c)_{R+F} \hat{h} N_s (T_{avg} - T_r) \tag{6.16}$$

This energy can then be added to that injected during the succeeding cycle. Since the maximum formation temperature is limited by the steam temperature, this additional energy will result in a larger heated radius for the new cycle. Consequently, for two cycles having identical injection conditions, this procedure will result in an improved productivity increase for the second cycle relative to the first.

The major assumption involved in this method is that interbedded shale and overburden and underburden are at initial reservoir temperature at the beginning of each cycle. This is a conservative assumption since calculated heat losses for cycles after the first will be higher than those actually observed.

Limitations of the Boberg–Lantz Method. The Boberg–Lantz method was designed for moderately viscous oil reservoirs that produce from multiple, relatively thin sands and for which extrapolatable cold production decline curves and water cut curves are available. It is not suitable for highly viscous oil reservoirs that have essentially zero cold production or for cases where well spacing is so close that depletion of oil from the heated region is a large fraction of the total oil recovered.

In these cases, effects of saturation changes, formation and fluid compressibilities, and thermal expansion become important mechanisms affecting stimulation response. Gros et al.[8] have extended the Boberg–Lantz method and incorporated a material balance that accounts for compressibility and thermal expansion effects. Approximate agreement of their model with observed stimulation response for a highly viscous oil was reported.

APPENDIX: EXAMPLE HAND CALCULATION, QUIRIQUIRE WELL Q-594

The calculation of the stimulation production history of Quiriquire well Q-594 is illustrated in the following example. Data for this well are given in Table 6A.1. Additional basic data are found in Figures 6A.1 and 6A.2, which give the estimated unstimulated variation of water–oil ratio, static pressure, and cold productivity index with cumulative oil production. On these plots, zero cumulative oil production corresponds to the start of stimulation. Oil viscosity as a function of temperature is given in Fig. 6A.3.

Table 6A.1 Stimulation Test Data for Quiriquire Well Q-594

Depth, ft	4050
Section thickness, ft	470
Net sand thickness, ft	183
Original reservoir temperature, °F	119
Original reservoir oil viscosity, cp	133
Oil gravity, °API	14.5
Oil specific heat, Btu/lb-°F	0.469
Formation thermal diffusivity, ft^2/d	0.631
Formation thermal conductivity, Btu/d-ft-°F	24.0
Sand-shale ratio	0.64
Average individual sand thickness, ft	11.43
Formation depth–section thickness ratio	8.5
Well radius, ft	0.292
Skin factor (before and after heating)	5.1
Effective drainage radius, ft	570
Normal producing bottom-hole pressure, psia	100
Static formation pressure, psia	490
Producing gas–oil ratio, scf/bbl	980
Prestimulation	
Oil rate, bbl/d	135
WOR, bbl/bbl	0.83
GOR, scf/bbl	985

Table 6A.1 (contd.)

First Stimulation Cycle	
Injected steam, $\times 10^6$ lb	18.13
Wellhead injection pressure, psig	770
Injection time,[a] days	46
Shut-in time following injection, days	2
Ratio of maximum pumping capacity to original lifting requirement	3
Water–oil ratio behavior following injection:	Fig. 6A.1 used in all three cases
Duration of cycle,[b] days	487
Stimulated producing days	378
Actual oil production, bbl	80,803
Calculated oil production bbl	84,000
Theoretical cold production, bbl	50,841

[a] Includes shut-in time following injection.
[b] Total calendar days including injection time.

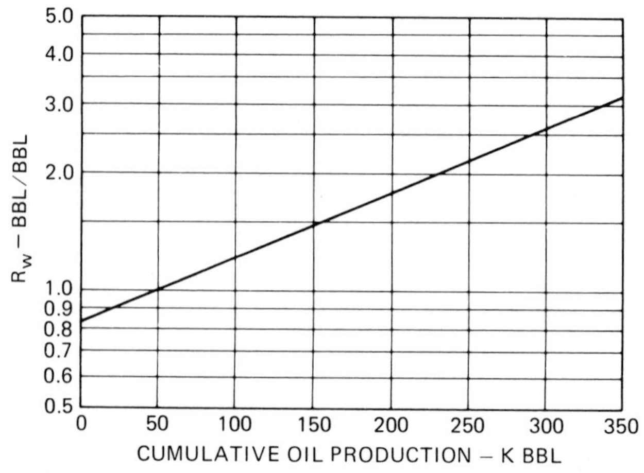

Figure 6A.1 Normal produced water–oil ratio, Quiriquire well Q-594.

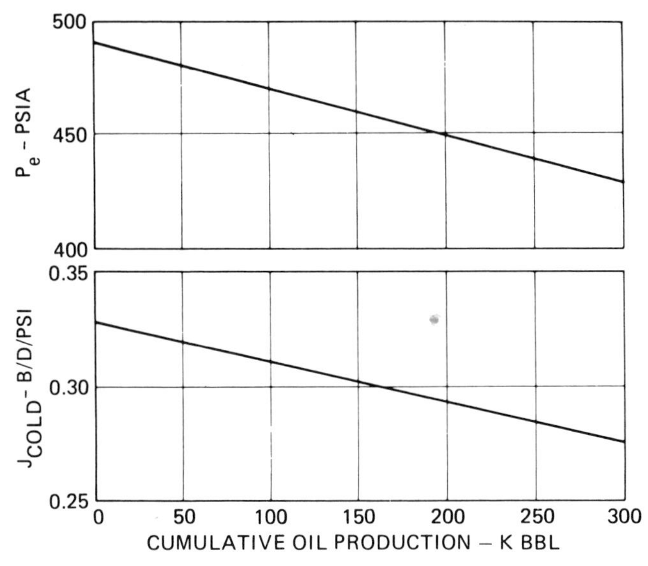

Figure 6A.2 Static pressure and unstimulated productivity index, Q-594.

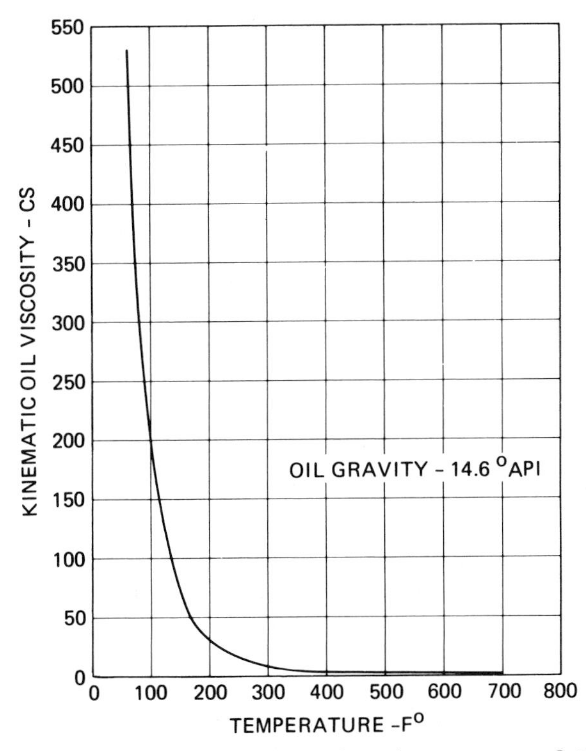

Figure 6A.3 Oil viscosity as function of temperature, Q-594.

Wellbore Heat Loss during Steam Injection

The average wellbore heat loss rate can be determined from Figure 2.9. Injection time is 46 days; hence, bottom-hole quality will be read for an average time equal to 23 days. Injection pressure is 770 psig, and injection is down a $3\frac{1}{2}$-in. tubing and a 7-in. casing. Using the injection through the tubing line,

$$\text{heat loss rate/ft} = 824 \text{ Btu/hr-ft}$$

The downhole quality (average) is calculated as

$$\bar{X}_i = 0.95 - \frac{(824)(24)(46)(4050)}{(18.13 \times 10^6)(689)}$$
$$= 0.655$$

Size of Heated Region

The radius of the region around the wellbore that is heated to steam temperature is calculated using Eq. (6.1). From the definitions for τ and ξ_s,

$$\tau = \frac{4(0.631)(44)}{(11.43)^2} = 0.850$$

$$\xi_s = \exp(0.850) \operatorname{erfc}(0.922) + \frac{2}{\sqrt{3.14}}(0.922) - 1.0$$
$$= 0.490 \quad \text{(see Fig. 6.2)}$$

Well Q-594 produces from 16 individual sands. Hence, $N_s = 16$, and Eq. (6.1) yields

$$r_h^2 = \frac{(0.490)(11.43)(18.13 \times 10^6)(0.655(689) + 509.5 - 86.9)}{4(24)(3.14)(518 - 119)(44)(16)}$$
$$r_h = 32.4 \text{ ft}$$

Next we calculate the hypothetical half-thickness Z, which accounts for vertical heat losses occurring by conduction to the shale during steam injection. Here, Z corresponds to the average half thickness of the individual sands, $\frac{1}{2}\hat{h}$. Using the formula for Z [see Eqs. (6.5) and (6.5a)],

$$Z = \frac{18.13 \times 10^6(0.655(689) + 509.5 - 86.9)}{2(3.14)(32.4)^2(38.0)(518-119)(16.0)} = 9.90$$

Temperature History of the Heated Region

To solve Eq. (6.2) for the average temperature T_{avg} of the heated region, a plot of $\bar{v}_r \bar{v}_z$ versus $t - t_i$ is constructed. Values of $\bar{v}_r$ are read from Figure 6.3 for values of $\bar{\theta}_r$, where

$$\bar{\theta}_r = \frac{0.631(t - t_i)}{(32.4)^2} = 0.0006011(t - t_i)$$

Values of $\bar{v}_z$ are read from Figure 6.4 for values of $\bar{\theta}_z$ and Z/L, where

$$\bar{\theta}_z = \frac{0.631(t - t_i)}{(15.28)^2} = 0.002702(t - t_i)$$

[*Note*: $L = (470 - 183)/2(15) + 183/2(16) = 15.28$ ft.] Then

$$Z/L = 9.9/15.28 = 0.65$$

At early time (up to $t - t_i = 30$ days), $\bar{v}_z$ is more accurately read from Figure 6.3 for values of $\bar{\theta}_z$:

$$\bar{\theta}_z = \frac{0.631(t - t_i)}{(9.9)^2} = 0.0064(t - t_i)$$

The product $\bar{v}_r \bar{v}_z$ is plotted versus $t - t_i$ in Figure 6A.4.

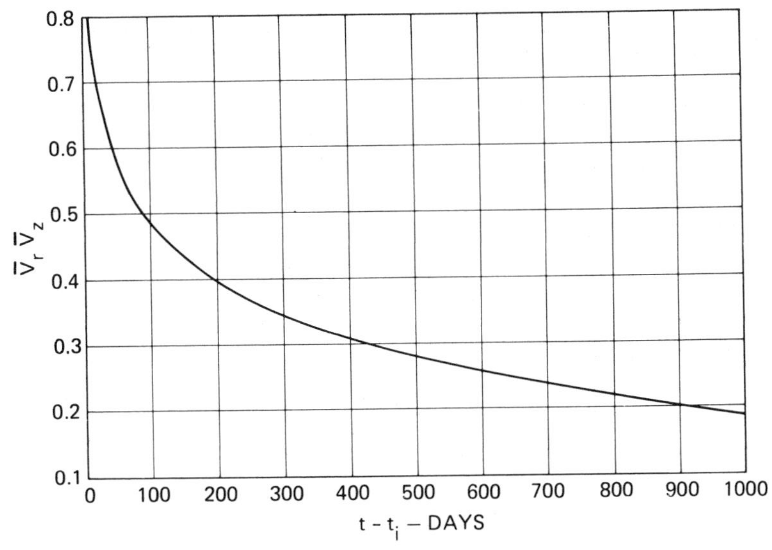

Figure 6A.4 $\bar{v}_r \bar{v}_z$ product for well Q-594.

Following steam injection, well Q-594 was shut in for 2 days. At the start of oil production, $t - t_i = 2$ days, and $\bar{v}_r\bar{v}_z = 0.922$. Since no fluids have been produced during this period, $\delta = 0$, and T_{avg} for the first time step of production is given by Eq. (6.2):

$$T_{avg} = 119 + (518 - 119)(0.922) = 487°F.$$

For future time steps, δ can be evaluated approximately as

$$\delta = \frac{1}{4} \frac{\sum_{i=1}^{N} H_{f_i}^* \Delta t_{p_i}}{Z\pi r_h^2 (\rho c)_{R+F}(T_s - T_r)N_s}$$

$$= \frac{\sum_{i=1}^{N} H_{f_i}^* \Delta t_{p_i}}{4(9.9)(3.14)(32.4)^2(38.0)(399)(16)}$$

$$= \frac{\sum_{i=1}^{N} H_{f_i}^* \Delta t_{p_i}}{3.166 \times 10^{10}}$$

where $H_{f_i}^*$ is the energy removal rate at the beginning of the time step, Δt_{p_i} is the number of producing days in the time step, and N is the number of time steps. Relatively short time steps must be used for this approximation to be valid.

Determination of the Oil Production Rate

The oil production rate is calculated using Eq. (6.15), which requires a knowledge of the pressure drawdown, the cold productivity index, and the ratio of the stimulated to unstimulated productivity index. Table 6A.2 gives the schedule of producing days and bottom-hole pressure during field testing of well Q-594 for the first 260 days following steam injection.

Figure 6A.2 gives the estimated static pressure P_e and cold productivity index J as a function of cumulative oil production following steam injection. The ratio of hot to cold productivity index is calculated by Eq. (6.12).

Table 6A.1 is used to evaluate the constants c_1 and c_2. To do this, r_w must be evaluated. Well Q-594 has an estimated skin factor $s_c = 5.1$. It is expected that s_h, the skin remaining after stimulation, will also have this value.

The effective well radius r_w is calculated as

$$r_w = \bar{r}_w e^{-s_c}$$

$$= 0.292e^{-5.1} = 0.00176 \text{ ft}$$

Table 6A.2 Estimated Bottom-Hole Pressures and Producing Days for Well Q-594 Following Steam Injection

$t - t_i$ (days)	Calendar days (Δt)	Producing days (Δt_p)	P_w (psia)
2			
	8	8	403
10			
	10	10	346
20			
	10	10	294
30			
	10	10	258
40			
	20	20	222
60			
	20	20	168
80			
	20	20	179
100			
	20	14	203
120			
	20	13	109
140			
	20	16	93
160			
	20	17	93
180			
	20	17	97
200			
	20	16	89
220			
	20	20	143
240			
	20	5	118
260			
	20	20	107
280			

Based on maps of the area, Q-594 is estimated to have a drainage radius r_e of about 570 ft. Hence, c_1 and c_2 are evaluated as

$$c_1 = \left[\ln\left(\frac{32.4}{0.00176}\right) - \frac{1}{2}\left(\frac{32.4}{570}\right)^2\right] \Big/ \left[\ln\left(\frac{570}{0.00176}\right) - 0.5\right] = 0.8056$$

and

$$c_2 = \left[\ln\left(\frac{570}{32.4}\right) - 0.5 + \frac{1}{2}\left(\frac{32.4}{570}\right)^2\right] \Big/ \left[\ln\left(\frac{570}{0.00176}\right) - 0.5\right] = 0.1944$$

For the first time step, $T_{avg} = 487°F$; hence, $\mu_{oh} = 1.96$ cp. Since original reservoir temperature is 119°F, $\mu_{oc} = 134$ cp. Hence, $\bar{J}$ is evaluated using Eq. (6.12):

$$\bar{J} = \left(\frac{1.96}{134}(0.8056) + 0.1944\right)^{-1} = 4.85$$

Since Q_o, the cumulative oil production since steam injection, is zero, $J_c = 0.329$ STB/d-psi, and $P_e = 491$ psia, both read from Figure 6A.2. By Eq. (6.15), the heated oil rate is

$$q_{oh} = 4.85(0.329)(491 - 403) = 141 \text{ STB/d}$$

Energy Removed With Produced Fluids

To calculate the value of T_{avg} for the second time step, it is necessary to evaluate $H_f^* \Delta t_p$ for the first time step. To evaluate H_f^*, the energy removal rate, by Eqs. (6.10) and (6.11), H_{og} and H_w must be calculated. We assume the produced gas has a specific heat of 0.02 Btu/°F-scf. Then H_{og}, in Btu/STB oil produced with the oil and gas, is

$$H_{og} = [5.61(0.469)(62.4)(0.971) + (985)(0.02)](487 - 119)$$
$$= 0.659 \times 10^5 \text{ Btu/STB}$$

To calculate H_w, the energy per stock tank barrel oil contained in the produced water (both vapor and liquid), we must know R_w, the produced water–oil ratio. Since Q_o is zero, the normal water–oil ratio $R_{wo} = 0.83$ from Figure 6A.1. Since Q_w/Q_{wi}, the ratio of barrels of water produced following steam injection to barrels of water injected (steam expressed as condensate), is zero, ΔR_w read from Figure 6.5 is 2.0. Hence,

$$R_w = 0.83 + 2.0 = 2.83 \text{ bbl/bbl}$$

Next, R_{wv}, the ratio of stock tank barrels of condensed water produced in the vapor state to stock tank barrels of oil, is calculated. For the first time step, $T_{avg} = 487°\text{F}$. Hence, the saturated water vapor pressure, $P_{wv} =$ 601 psia, which is greater than $P_w = 388$ psia. Thus, all the produced water for the first time step is assumed to be entirely vaporized at downhole conditions, and $R_{wv} = 2.83$. Had P_w been greater than P_{wv}, R_{wv} would have been calculated by the relation

$$R_{wv} = (0.0001356)\left(\frac{P_{wv}}{P_w - P_{wv}}\right)R_g \quad \text{(bbl/bbl)}$$

provided the value of R_{wv} so calculated is less than R_w. Now H_w can be calculated:

$$H_w = 5.61(62.4)[2.83(472 - 87) + 2.83(731)]$$
$$= 11.056 \times 10^5 \text{ Btu/STB}$$

Then H_f^* is calculated by Eq. (5.9):

$$H_f^* = 141(0.659 \times 10^5 + 11.056 \times 10^5)$$
$$= 1.652 \times 10^8 \text{ Btu/d}$$

and the value of the integral is

$$\int_{t_i}^{t} H_f^* \, dt = (1.652 \times 10^8)(8.0) = 1.322 \times 10^9 \text{ Btu}$$

Then δ for the second time step can be calculated:

$$\delta = \frac{1.322 \times 10^9}{3.166 \times 10^{10}} = 0.0418$$

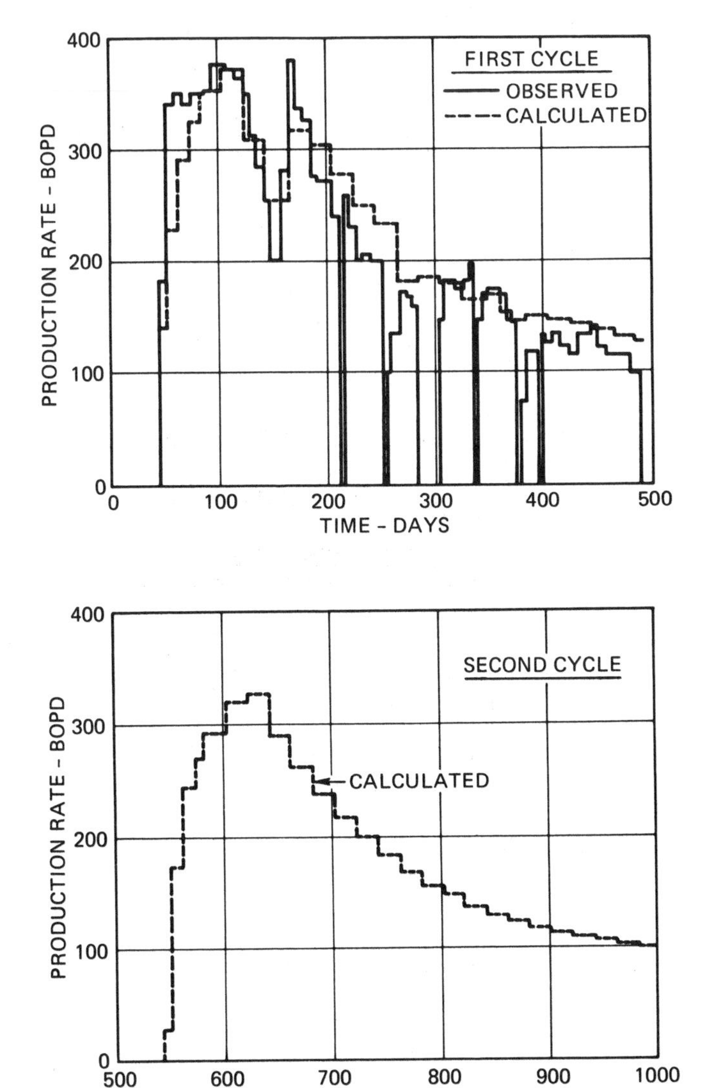

Figure 6A.5 Steam stimulation; comparison of calculated and observed oil production rate history (Quiriquire well Q-594).

Table 6A.3 Intermediate Steps in Calculating Steam Stimulation History of Q-594

$t - t_i$ (days)	$\bar{v}_r \bar{v}_z$	$\Sigma H_f^* \Delta t_p$ (10^{-10} Btu)	δ	T_{avg} (°F)	P_{wv} (psia)	μ_{oh} (cp)	$\bar{J}$	Q_o (10^{-3} STB)	J_{cold} (STB/d-psi)	ΔP (psi)	q_{oh} (STB/d)	Q_w (10^{-3} STB)	R_w (bbl/bbl)	R_{wv} (bbl/bbl)	H_{og} (10^{-6} Btu/STB)	H_w (10^{-6} Btu/STB)	H_f^* (10^{-10} Btu/d)
2	0.920	0.0	0.0	487	601	1.96	4.85	0.0	0.329	88	141	0.0	2.83	2.83	0.0675	1.1056	0.0165
10	0.800	0.132	0.0418	408	270	2.46	4.78	1.13	0.329	144	227	2.0	1.77	0.474	0.052	0.327	0.0086
20	0.714	0.218	0.0688	357	148	3.90	4.59	3.39	0.328	194	292	5.19	1.41	0.134	0.0426	0.159	0.0059
30	0.655	0.277	0.0875	322	92	5.90	4.35	6.32	0.328	228	325	8.76	1.22	0.073	0.0364	0.115	0.0049
40	0.612	0.326	0.103	297	64	8.03	4.12	9.57	0.327	260	351	12.27	1.08	0.053	0.0319	0.085	0.0041
60	0.553	0.408	0.129	260	35	12.98	3.67	16.59	0.326	310	371	18.94	0.95	0.034	0.0252	0.058	0.0031
80	0.512	0.470	0.148	234	22	18.84	3.25	24.01	0.325	293	309	25.84	0.93	0.019	0.0206	0.044	0.0020
100	0.487	0.510	0.161	218	16.5	23.48	2.98	30.18	0.324	263	254	31.70	0.95	0.012	0.0177	0.037	0.0014
120	0.464	0.529	0.167	207	13.3	27.72	2.77	33.74	0.323	354	317	35.12	0.96	0.019	0.0158	0.038	0.0017
140	0.449	0.551	0.174	197	10.8	32.38	2.57	37.87	0.322	367	304	39.04	0.97	0.017	0.0140	0.032	0.0014
160	0.428	0.574	0.181	187	8.8	37.55	2.38	42.73	0.321	363	277	43.85	0.99	0.013	0.0122	0.028	0.0011
180	0.414	0.593	0.187	179	7.3	43.27	2.20	47.44	0.320	355	250	48.56	1.00	0.011	0.0107	0.024	0.00088
200	0.398	0.608	0.192	171	6.1	49.60	2.03	51.70	0.320	360	234	52.91	1.02	0.010	0.0093	0.023	0.00075
220	0.387	0.620	0.196	165	5.3	56.14	1.88	55.44	0.319	303	182	56.76	1.03	0.005	0.0082	0.016	0.00045
240	0.373	0.629	0.199	159	4.6	61.64	1.77	59.08	0.318	324	182	60.58	1.05	0.005	0.0072	0.015	0.00040
260	0.364	0.631	0.199	156	4.3	66.67	1.68	59.99	0.318	335	180	61.54	1.05	0.006			

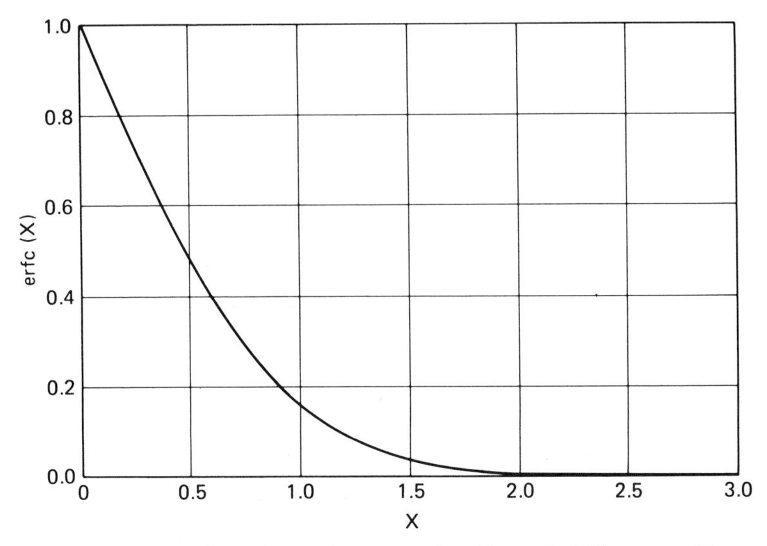

Figure 6A.6 Complementary error function erfc (X) versus X.

Next, $\bar{v}_r\bar{v}_z = 0.800$ for the second time step is read from Figure 6A.4 for $t - t_i = 10$ days. Then T_{avg} for the second time step is calculated using Eq. (6.2):

$$T_{avg} = 119 + (399)[0.800(1 - 0.0418) - 0.0418]$$
$$= 408°F$$

In this manner, the step-by-step solution is carried out through 260 days of oil production following steam injection. Table 6A.3 gives the intermediate steps in this computation.

Figure 6A.5 gives the calculated rate versus time for this case. The calculation result is compared to the actual field history for this well.

Figure 6A.6 is a plot of the complementary error function erfc (X) as a function of X for reference. The function erfc (X) is used in the evaluation of terms in Eqs. (6.5), (6.5a), and (6.6). [*Note*: erf $(X) = 1 - \text{erfc}(X)$.]

Because of the laborious nature of the calculations, they have been programmed for both mainframe and personal computers.

NOMENCLATURE

B_j Constant appearing in Eq. (6.5), ft

c_1, c_2 Constants appearing in Eq. (6.12), dimensionless; for definitions, see Table 6.1

- c_g Average specific heat of gas, over temperature interval T_r to T_{avg}, Btu/scf-°F
- c_o, c_w Average specific heats of oil and water over temperature interval T_r to T_{avg}, Btu/lb-°F
- D Depth of producing formation, ft
- $\hat{h}$ Average thickness of individual sand members, ft
- H_{wt} Enthalpy of liquid water at T_{avg} above 32°F, Btu/lb
- H_f^* Rate of energy removal with produced fluids (above T_r), Btu/d
- H_{wr} Specific enthalpy of liquid water at T_r, Btu/lb
- H_{ws} Specific enthalpy of liquid water at T_s, Btu/lb
- h_t Gross formation thickness, ft
- h_j Individual sand thicknesses, ft
- $\bar{h}_j$ Artificially increased sand thickness used in Eq. (6.5), ft
- H_{og} $(5.61(\rho c)_o + R_g c_g)\ (T_{avg} - T_r)$, Btu/STB oil
- H_w $5.61\rho_w[R_w(H_{wt} - H_{wr}) + R_{wv}H_{wv}]$, Btu/STB oil
- H_{wv} Specific enthalpy of vaporization of water at T_{avg}, Btu/lb
- $J_1(y)$ First order Bessel function, dimensionless
- $\bar{J}$ Ratio of stimulated to unstimulated productivity indices, dimensionless
- J_h, J_c Stimulated (hot) and unstimulated (cold) productivity indices, respectively, STB/d-psi (additional subscripts d and ud refer to damaged and undamaged, respectively)
- J_{cs} Cold specific productivity index, STB/d-psi-ft
- k Formation permeability, darcy
- K_h Overburden thermal conductivity, Btu/ft-d-°F
- k_d Damaged permeability, darcy
- k_o Oil permeability, darcy
- l_j Thickness of individual interbedded shale, ft
- $\bar{l}_j$ Artificially reduced shale thickness used in Eq. (6.5), ft
- l Average interbedded shale thickness, ft
- L Half average individual sand plus shale thickness [see Eq. (6.5a)], ft
- M_s Total mass of steam plus condensate injected in current cycle, lb
- m_s Mass steam injection rate, lb/hr
- N_s Number of individual sand members
- P_e Static formation pressure existing at distance r_e from wellbore, psia
- P_w Producing bottom-hole pressure, psia
- P_{wv} Saturated vapor pressure of water at T_{avg}, psia

q_h Rate at which energy is removed from formation with produced fluids at time t, Btu/d

q_o Oil production rate, STB/d

q_{oh} Oil production rate during stimulated production, STB/d

Q_o Cumulative oil production, STB

r Radial distance from wellbore, ft

r_d Radius of region of damaged permeability k_d, ft

r_e Drainage radius of well, ft

R_g Total produced gas–oil ratio, scf/bbl at stock tank conditions

r_h Radius of region originally heated, ft

r_w Effective wellbore radius, ft

$\bar{r}_w$ Actual wellbore radius, ft

R_w Total produced water–oil ratio, bbl/bbl at stock tank conditions

ΔR_w $R_w - R_{wo}$, bbl/bbl at stock tank conditions

R_{wo} Normal (unstimulated) water–oil ratio, bbl/bbl at stock tank conditions

R_{wv} Water produced in vapor state per stock tank bbl oil produced, bbl water vapor (as condensed liquid at 60°F)/STB

s Skin factor of well, dimensionless (subscripts c, h refer to cold and heated conditions, respectively)

s_k kth term in series solution for v_r, dimensionless

t Time elapsed since start of injection for current cycle, days

t_i Time of injection (current cycle), days

Δt_{pi} Number of producing days in a time step, days

T_{avg} Average temperature of originally heated oil sand at any time t, °F

T_r Original reservoir temperature, °F

T_s Condensing steam temperature at sandface injection pressure, °F

$\bar{v}_r, \bar{v}_z$ Temperature correction for radial and vertical conduction, respectively, dimensionless

w Constant appearing in Eq. (6.5), $4\alpha(t - t_i)$, ft^2

W_j Constant appearing in Eq. (6.5), ft

$\bar{X}_i$ Average downhole steam quality during injection phase, lb vapor/lb liquid plus vapor

X_{surf} Wellhead steam quality, lb vapor/lb liquid plus vapor

z Hypothetical thickness used in Eq. (6.5), ft

Z Half thickness of average heated sand [defined as $\frac{1}{2}(\hat{h} + z)$] for Eq. (6.5a), ft

α Overburden thermal diffusivity, ft^2/d
δ Quantity defined in Eq. (6.7), dimensionless
$\bar{\theta}$ Dimensionless time
$\bar{\xi}_s$ $e^{\tau} \operatorname{erfc}(\sqrt{\tau}) + (2\sqrt{\tau}/\sqrt{\pi}) - 1$, dimensionless
$(\rho c)_{R+F}$ Volumetric heat capacity of reservoir rock including interstitial fluids, Btu/ft^3-°F
ρ_o, ρ_w Stock tank fluid density of oil and water, respectively, lb/ft^3
τ $4K_h t_i/\hat{h}^2(\rho c)_{R+F}$, dimensionless

REFERENCES

1. Boberg, T. C., and Lantz, R. B. "Calculation of the Production Rate of a Thermally Stimulated Well," *Trans. AIME*, **237**, I-1613 (1966).
2. Martin, J. C. "A Theoretical Analysis of Steam Stimulation," *J. Pet. Tech.* (March 1967), p. 411.
3. Davidson, L. B., Miller, F. G., and Muller, T. D. "A Mathematical Model of Reservoir Response During the Cyclic Injection of Steam," *Soc. Pet. Eng. J.* (June 1967), p. 174.
4. Closmann, P. J., Ratliff, N. W., and Truitt, N. E. "A Steam-Soak Model for Depletion-type Reservoir," *J. Pet. Tech.* (June 1970), p. 757.
5. Towson, D. E., and Boberg, T. C. "Gravity Drainage in Thermally Stimulated Wells," *J. Can. Pet. Tech.* (October–December 1967), pp. 1–6.
6. Seba, R. D., and Perry, G. E. "A Mathematical Model of Repeated Steam Soaks of Thick Gravity Drainage Reservoirs," *J. Pet. Tech.* (January 1969), p. 87.
7. Bentsen, R. G., and Donahue, D. A. T. "A Dynamic Programming Model of the Cyclic Steam Injection Process," *J. Pet. Tech.* (December 1969), p. 1582.
8. Gros, R. P., Pope, G. A., and Lake, L. W. "Steam Soak Predictive Model," SPE 14240, Presented at the Sixtieth Annual Technical Conference and Exhibition of the SPE, Las Vegas, Nevada, September 1985.
9. Weinstein, H.G., Wheeler, J. A., and Woods, E. G. "Numerical Model for Steam Stimulation," *Soc. Pet. Eng. J.* (February 1977), p. 65.
10. Shutler, N. D. "Numerical Three-Phase Model of the Two-Dimensional Steamflood Process," *Soc. Pet. Eng. J.* (December 1970), p. 405.

Chapter 7

SCREENING CALCULATIONS FOR SELECTING STEAM STIMULATION WELLS

ABSTRACT

Frequently it is not necessary to make a full prediction using the Boberg–Lantz or a more sophisticated method to pick stimulation candidates. Simplified calculation procedures are illustrated in this chapter to assist in making well selections for steam stimulation. Two approaches are discussed. One, applicable for low-rate but mobile oil wells, is a simplified calculation method. The second, a more generally useful tool—the Boberg–West correlation—is a correlation of field-observed oil–steam ratios as a function of the product of original-condition oil transmissibility and formation pressure.

LOW-RATE WELL CASE

In general, if a selection has to be made among a group of wells as to which should be the next to be steamed, a general rule might be to pick that well that will produce the most incremental oil for a given volume of steam injected. For a *low-rate* well where the length of the cycle is essentially independent of the rate of production, the incremental oil recovery will be directly proportional to the incremental improvement in oil production rate. The engineer should stimulate first the well that will give most improvement in oil rate. For low-rate wells:

1. First calculate r_h for two (or more) wells for a given M_s.
2. For some average temperature, say $T_{avg} = \frac{1}{2}(T_s + T_r)$, calculate J_h/J_c for the wells.
3. Calculate $\Delta q_o = q_{oh} - q_{oc}$ for the various wells, where

$$q_{oh} = (J_h/J_c)J_c(P_e - P_w)_h = \bar{J}J_c\,\Delta P_h$$

$$q_{oc} = J_c(P_e - P_w)_c = J_c\,\Delta P_c$$

As an illustration, the following example is given. A selection is to be made between the following two wells for stimulation:

	Well 1	Well 2
Average individual sand thickness ($\hat{h}$), ft	25	12.5
Number of sands (N_s)	4	4
Thermal diffusivity (α), ft^2/d	0.883	0.883
Thermal conductivity (K_h), Btu/hr-ft-°F	1.4	1.4
Steam injected (M_s), kbbl	10	10
Injection time (t_i), days	10	10
Average bottom-hole steam quality ($\bar{X}_i$)	0.6	0.6
Temperature of steam in formation (T_s), °F	500	500
Original reservoir temperature (T_r), °F	100	100
Oil viscosity (μ_o)		
At 100°F, cp	500	50
At 300°F, cp	10	5
Oil formation volume factor (B_o), RB/STB	1.0	1.0
Cold productivity index (J_c), STB/d-psi	0.2	0.5
Producing bottom-hole pressure (P_w), psia	50	50
Static pressure (P_e), psia	500	300
Drainage radius (r_e), ft	300	300
Actual well radius ($\bar{r}_w$), ft	0.3	0.3
Cold skin factor (s_c), dimensionless	4.6	2.3
Heated skin factor (s_h), dimensionless	4.6	0.0

1. Calculation of r_h, Radius of the Heated Zone. Dimensionless time τ is calculated for well 1:

$$\tau = 4\alpha t_i/\hat{h}^2 = 4(0.883)(10)/(25)^2 = 0.0565$$

The function ξ_s is obtained from Eq. (3.7) or the table in Appendix A, Chapter 3:

$$\xi_s = 0.045$$

From Eq. (3.6), it follows that

$$r_h^2 = \frac{\hat{h}M_s(\bar{X}_i H_{wv} + H_{ws} - H_{wr})\xi_s}{4K_h\pi(T_s - T_r)t_i N_s} = \frac{(350\times 10^4)[0.6(714) + 488 - 68]\hat{h}\xi_s}{4(1.4\times 24)(3.14)(500-100)(10)(4)}$$

$$= (0.04398)\hat{h}\xi_s \times 10^4 = 495 \text{ ft}^2$$

$$r_h = 22 \text{ ft}$$

Similarly, for well 2,

$$\tau = 4\alpha t_i/\hat{h}^2 = 4(0.883)(10)/(12.5)^2 = 0.226$$

$$\xi_s = 0.167$$

$$r_h^2 = (4.398 \times 10^2)\hat{h}\xi_s = 918 \text{ ft}^2$$

$$r_h = 30 \text{ ft}$$

2. Calculation of J_h/J_c. Using Eq. (4A.6),

$$\frac{J_h}{J_c} = \frac{s_c + \ln(r_e/\bar{r}_w)}{(\mu_{oh}/\mu_{oc})[s_h + \ln(r_h/\bar{r}_w)] + \ln(r_e/r_h)}$$

For well 1,

$$\bar{J} = \frac{4.6 + \ln(300/0.3)}{(10/500)[4.6 + \ln(22/0.3)] + \ln(300/22)} = 4.12$$

For well 2,

$$\bar{J} = \frac{2.3 + \ln(300/0.3)}{(5/50)[0.0 + \ln(30/0.3)] + \ln(300/30)} = 3.33$$

3. Calculation of Δq_o. In the absence of information to the contrary, we assume that $\Delta P_c = \Delta P_h$ for both wells:

$$\Delta q_o = (\bar{J} - 1)J_c\,\Delta P$$

For well 1,

$$\Delta q_o = (4.12 - 1)(0.2)(450) = 281 \text{ bbl/d}$$

For well 2,

$$\Delta q_o = (3.33 - 1)(0.5)(250) = 291 \text{ bbl/d}$$

Since Δq_o is greater for well 2, this well is the one that would be selected for stimulation by this approach, although for practical purposes the Δq_o's are the same.

HIGH-RATE WELL CASE

For high-rate wells, the rate of heat removal from the formation will be high, and this effect will significantly accelerate the decline of the stimulation effect. Hence, the above approach, which neglects this effect, will unduly favor higher rate wells for stimulation. For this reason, a calculation such as the Boberg–Lantz method, which accounts for this effect, probably should be made for wells producing over 300 bbl/d to make an accurate well selection.

BOBERG–WEST CORRELATION

An approximate correlation has been observed of steam stimulation field results with the product of oil transmissibility and static pressure. This correlation can be used as a rough screening tool for identifying potential candidates for steam stimulation.

Theoretical support for this correlation is obtained by observing that the incremental oil–steam ratio may be expressed as

$$\Delta Q_o/M_s = \frac{1}{M_s}\left[\int_{\theta_i}^{\theta_p} \bar{J}J_c\,\Delta P\,d\theta - \int_{\theta=0}^{\theta_p} J_c\,\Delta P\,d\theta\right] \tag{7.1}$$

where θ_i is the time at the end of the injection soak period, and θ_p is the time at the end of the production phase, when the stimulation effect is essentially nil. If constant average values for J_c and ΔP can be assumed,

$$\Delta Q_o/M_s = \frac{1}{M_s}\left[J_c\,\Delta P\int_{\theta_i}^{\theta_p} (\bar{J}-1)\,d\theta - J_c\,\Delta P\,\theta_i\right]$$

$$= \bar{J}_c\,\Delta\bar{P}\left[\int_{\theta_i}^{\theta_p} (\bar{J}-1)\,d\theta - \theta_i\right]\left(\frac{1}{M_s}\right) \tag{7.2}$$

Note that for $\theta_p >> \theta_i$ and $\bar{J} >> 1$, this simplifies to the approximate equation

$$\Delta Q_o/M_s \cong \bar{J}_c\,\Delta\bar{P}\,\frac{\bar{J}_{avg}}{M_s}\,\theta_p \tag{7.3}$$

From Chapter 6, it follows for the case of steady-state flow that

$$\frac{\bar{J}}{s_c + \ln(r_e/\bar{r}_w)} = \frac{1}{(\mu_{oh}/\mu_{oc})[s_h + \ln(r_h/\bar{r}_w)] + \ln(r_e/r_h)}$$

and

$$\bar{J}_c = \frac{2\pi(k_o h)/B_o\mu_o}{s_c + \ln(r_e/\bar{r}_w)}$$

Equation (7.3) becomes

$$\frac{\Delta Q_o}{M_s} \cong \left(\frac{k_o h}{\mu_o}\right)\Delta\bar{P}\left[\frac{2\pi\theta_p/B_o M_s}{(\mu_{oh}/\mu_{oc})_{avg}[s_h + \ln(r_h/\bar{r}_w)] + \ln(r_e/r_h)}\right] \tag{7.4}$$

The OSR will be a strong function of $(k_o h/\mu_o)\,\Delta\bar{P}$, where $\Delta\bar{P}$ is the average pressure drawdown over the producing cycle. For $P_e >> P_w$, the $\Delta\bar{P}$ can be approximated with P_e, and it can be seen that the incremental OSR will be a strong function of the product $(k_o h/\mu_o)P_e$. This would suggest a correlation between the incremental OSR and the product of transmissibility and static pressure. The relationship would not be linear, however. As $(k_o h/\mu_o)P_e$ increases, the prestimulation rate would be expected to increase. For a given treatment size, the stimulated rate would likewise increase, and consequently, the energy withdrawal rate from the formation would increase. Thus, θ_p, the stimulation producing time, will be inversely affected by an increased $(k_o h/\mu_o)P_e$. The effect will become more pronounced for highly productive wells.

Data on seven field projects were used to determine whether a correlation could be observed between the OSR and the transmissibility–pressure product. The average cumulative incremental OSR through approximately three cycles of steam stimulation was plotted. The pressure used in the $(k_o h/\mu_o)P_e$ term was that at project start. In the case of one project in which the drive mechanism was gravity drainage, the pressure used was that at the base of the perforated oil column.

The data for all but one of the seven projects plotted as an S-shaped curve on log–log coordinates are shown in Figure 7.1. The project that deviated was in a field where water saturation, shale content, and prestimulation water cut were substantially higher than for the other projects. These factors would reduce θ_p and thus the incremental OSR.

Hence, for wells that have had treatments of comparable size, comparable degrees of wellbore damage, and comparable WOR's and GOR's there should be a reasonable correlation between OSR and the prestimulation oil transmissibility–static pressure product.

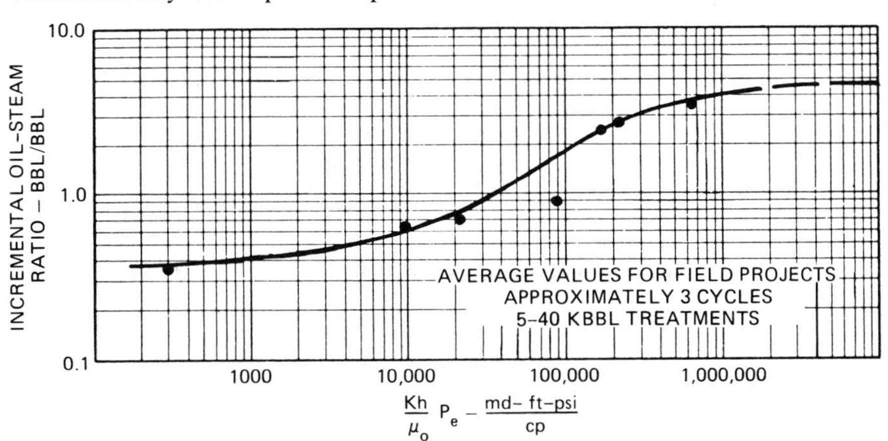

Figure 7.1 Boberg–West correlation of steam stimulation results.[1] © 1972 SPE-AIME.

Using Figure 7.1, we obtain the following for the same two wells discussed above:

For well 1,†

$$\frac{k_o h}{\mu_o} P_e = \frac{P_e J_c B_o[\ln(r_e/r_w) - 0.5]}{0.00708}$$

$$r_w = \bar{r}_w e^{-s} = 0.3e^{-4.6} = 0.003016$$

$$\frac{k_o h}{\mu_o} = \frac{0.2(1)[\ln(300/0.003015) - 0.5]}{0.00708} = 311\,.$$

$$\frac{k_o h P_e}{\mu_o} = (311.)(500) = 156{,}000$$

Incremental OSR = 2.3 bbl/bbl (from Fig. 7.1)

For well 2, $r_w = \bar{r}_w e^{-s} = 0.3e^{-2.3} = 0.030078$

$$\frac{k_o h P_e}{\mu_o} = (300)\frac{0.5(1)[\ln(300/0.030078) - 0.5]}{0.00708} = 184{,}000$$

Incremental OSR = 2.6 bbl/bbl (from Fig. 7.1)

Again, the two wells are quite close, with well 2 being marginally better.

NOMENCLATURE

- α Thermal diffusivity of overburden and underburden, ft²/d
- B_o Oil formation volume factor, RB/STB
- $\hat{h}$ Average individual sand thickness, ft
- h Total net sand thickness, ft
- H_{wv} Latent heat of vaporization of water at downhole injection temperature and pressure, Btu/lb
- H_{ws} Enthalpy of liquid water at downhole injection temperature and pressure, Btu/lb
- H_{wr} Enthalpy of liquid water at original reservoir conditions, Btu/lb
- $\bar{J}$ Ratio of stimulated to unstimulated productivity indices, dimensionless

†For a pressure-declining situation the cold oil P.I. is given by the pseudo-steady state equation:[2]

$$J_c = \frac{q_{oc}}{\Delta P} = \frac{2\pi k_o h}{B_o \mu_o(\ln(r_e/r_w) - 0.5)}$$

When millidarcy is used as the unit for permeability, the equation becomes

$$J_c = \frac{0.00708 k_o h}{B_o \mu_o(\ln(r_e/r_w) - 0.5)}$$

J_h, J_c Stimulated (hot) and unstimulated (cold) productivity indices, respectively, STB/d-psi (bar over J_h or J_c denotes average value)
K_h Thermal conductivity, Btu/hr-ft-°F
k_o Oil permeability, md
M_s Total steam plus condensate injected, bbl (as condensed liquid)
N_s Number of individual sands
P_e Static formation pressure existing at distance r_e from wellbore, psia
P_w Producing bottom-hole pressure, psia
ΔP $P_e - P_w$, psi
$\Delta \bar{P}$ $(P_e - P_w)_{avg}$ over time $\theta_p - \theta_i$, psi
ΔQ_o Cumulative incremental (over expected unstimulated) oil production, STB
q_{oh} Heated oil production rate, STB/d
q_{oc} Cold oil production rate, STB/d
r_e Drainage radius of well, ft
r_h Radius of region heated, ft
$\bar{r}_w$ Actual wellbore radius, ft
r_w Effective wellbore radius, ft
s Skin factor of well, dimensionless
θ_i Time of injection, days
θ_p Time to end of producing cycle, days
μ_o Oil viscosity, cp
τ Dimensionless time of steam injection, $=4\alpha t_i/h^2$
t_i Time of steam injection, days
T_r Original reservoir temperature, °F
T_s Downhole steam temperature, °F
$\bar{X}_i$ Average downhole steam quality during injection, lb/lb
ξ_s Dimensionless steam zone area, defined by Eq. (3.7)

Subscripts

c cold
h heated

REFERENCES

1. Boberg, T. C., and West, R. C. "Correlation of Steam Stimulation Performance," *J. Pet. Tech.* (November 1972), p. 1367.
2. Craft, B.C., and Hawkins, M.F. *Petroleum Reservoir Engineering*, Prentice-Hall, Inc., Englewood Cliffs, N.J. (1959) p. 286.

Chapter 8

INTRODUCTION TO STEAMFLOODING PRINCIPLES

ABSTRACT

Steamflooding is a continuous-injection process analogous to waterflooding. It can achieve higher oil recoveries than possible with cyclic steam stimulation but will generally have lower oil–steam ratios in the range of 0.1–0.4 bbl/bbl. This occurs because steamflooding involves eventually heating the entire reservoir, whereas steam stimulation concentrates heating locally at producing wells to achieve productivity increase. Steamflooding is capable of driving oil saturations down to a very low level in the parts of the reservoir that are swept by steam. The major problems with steamflooding are ones of sweep efficiency and heat losses to unproductive formations. This chapter serves to introduce the process and to provide preliminary guidance so that favorable steamflooding results can be achieved. The mechanisms of steam displacement and factors affecting displacement efficiency are discussed at the outset with guidance in estimating steam zone residual oil saturations. This is followed with a discussion of sweep efficiency and overall thermal efficiency. Screening criteria for selecting good steamflooding candidates are given, with brief descriptions of the effects of the following process variables: starting oil content, formation thickness, net–gross sand thickness ratio, depth, pressure level, absolute and relative permeability, steam injection rate, steam quality, well spacing, pattern type, and steam bank size. The status of current research on chemical additives for improving sweep is also discussed.

INTRODUCTION

The usual procedure for application of steam injection processes to a reservoir is to first steam stimulate and then begin flooding, after stimulation response has declined to the point that flooding is commercially more attractive. In reservoirs containing extremely viscous oils, an extended sequence of stimulation cycles is usually necessary before sufficient viscosity reduction is achieved to make flooding possible. Steam stimulation of production wells may be continued until the time of heat arrival from the production wells.

The economic success of a steamflood operation depends on a combination of oil recovery efficiency and the efficiency of steam utilization. Oil recovery is dependent on the displacement efficiency, or the degree to which oil is displaced from the contacted rock pores, and the volumetric sweep efficiency, or fraction of the total reservoir volume actually contacted.

DISPLACEMENT EFFICIENCY

Oil displacement during steamflooding is achieved through a number of different mechanisms. Figure 8.1 illustrates variations with distance of temperature and oil–water–steam concentrations for a one-dimensional steamflood. The oil is contacted in sequence first by cold water, then by hot water, and finally by steam. Oil is moved by cold water, just as in a conventional waterflood. For a moderately viscous oil, this is not a problem, but for a highly viscous oil, a severe problem of viscous blockage can develop as displaced oil cools ahead of the front after it has been displaced by steam. In such situations, the entire distance from injector to producer may have to be heated before the displaced oil can be produced.

Figure 8.2 is a series of photographs of a visual model of a steamflood of a highly viscous (40,000-cp) oil. An initial channel at the top of the sand pack has been provided so that the steam can enter the model. As the steam

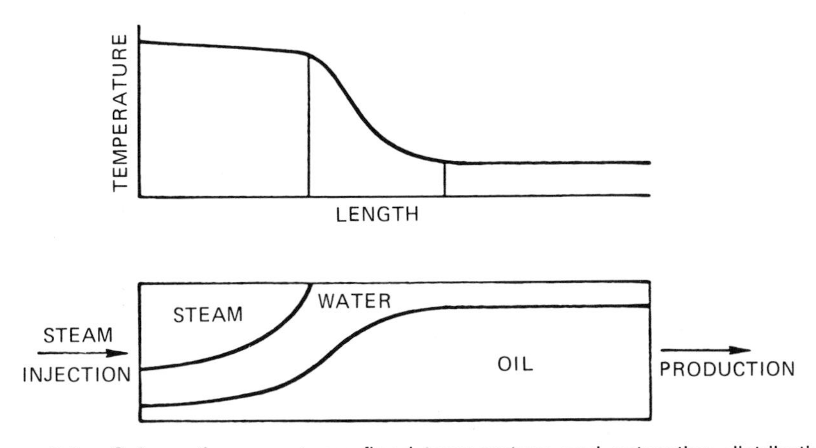

Figure 8.1. Schematic core steam flood temperature and saturation distribution.

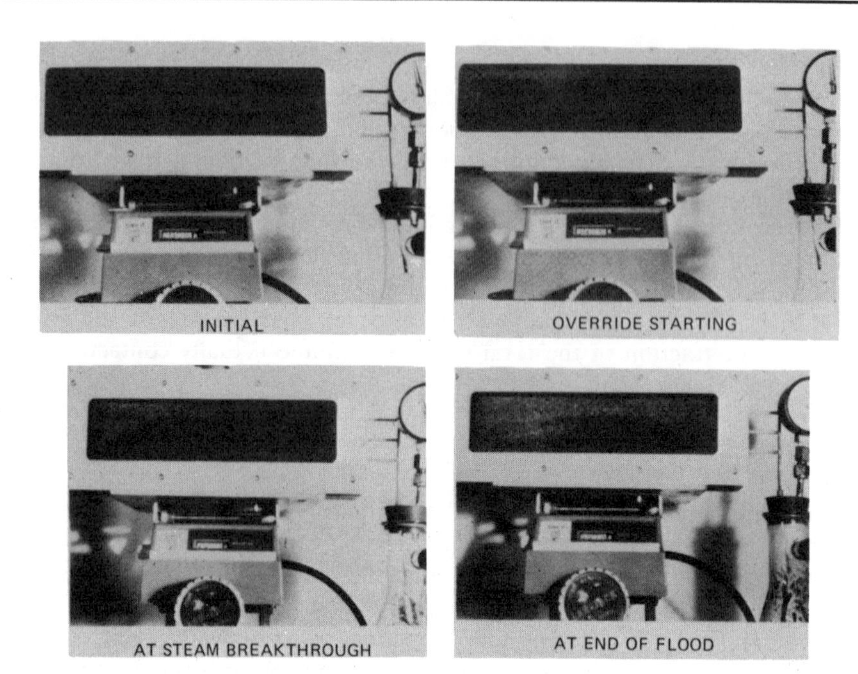

Figure 8.2. Visual laboratory model study of steamflood of highly viscous oil.

zone grows, a highly efficient displacement takes place as evidenced by the lightening of the sand grains despite the fact that the oil is over 3 million times more viscous than steam vapor at model conditions. What makes this efficient displacement possible is primarily a combination of mobility ratio improvement at the elevated temperature of the steam and a condensing gas drive displacement by the steam itself.

While steam distillation of the light ends of the crude and the solvent action of the recondensed light ends downstream of the heated zone have some influence, this effect must be minor in the case illustrated here for such a viscous oil. Steam vapor and condensed water moving through the steam zone are the major displacing agents. Why then does the oil saturation in the steam zone get so low that we can see the sand grains actually being cleaned? Application of Buckley–Leverett shock theory explains the effect; this is illustrated in Chapter 12, where the Shutler–Boberg analytical method is described.

Basically, oil displacement to a low S_{or} occurs because hundreds of pore volumes of steam vapor are traveling through the steam zone to the condensation front. An average steamflood will require 3–5 bbl of water flashed to 80% quality steam to move a barrel of oil. The volume of steam and water at 300 psia in an actual steamflood created by the flashing of these 3–5 bbl of water would be 232–386 bbl of wet steam. At the reduced pressure (essentially atmospheric) of the model in the illustration, those 3–5 bbl of 60°F water would be 4010–6683 bbl of wet steam. That is an enormous quantity of displacing agent to move just 1 bbl of oil.

Obviously, the gas drive effect of steam relative to other mechanisms becomes less effective as pressures increase. In order to maintain gas drive effectiveness, it may even be desirable to inject an inert gas with the steam at elevated pressures. At 2500 psia, the maximum practical pressure for conventional steam-generating equipment, our 3–5 bbl of water only occupies 21–34 bbl as wet steam. So there is much less effective displacing fluid at elevated pressures. Saturated steam temperature is however much hotter—668°F at 2500 psia versus 417°F at 300 psia and 212°F at 14.7 psia.

At temperatures over 500°F, cracking of the oil becomes a significant contributor to the displacement mechanism. In the cold- and hot-water zones ahead of the steam front, temperature effects on the relative permeability to oil and water become important factors. Mobility ratio improvement and thermal expansion of the oil contribute to improved oil recovery in these regions. At high pressures and for moderately viscous oils, these mechanisms become increasingly important relative to displacement within the steam zone itself.

The contributions of the various mechanisms to oil recovery vary with the reservoir system, particularly with the type of oil involved. Willman et al.[1] illustrated the importance of the distillation of the light ends in core floods of three refined oils containing, respectively, zero, 25, and 50% distillable materials (Table 8.1). For each oil, the core hot-waterflood recovery was as shown (54.8–58.0% of the oil in place (OIP)). The extra recovery attributed to the gas drive was 3%. Extra recovery caused by distillation was zero, 10.5, and 19.5%, respectively, for the nondistillable oil, 25% distillable oil, and 50% distillable oil. Additional recovery due to other mechanisms was estimated to range from 1.2 to 7.7%. It should be noted that in an

Table 8.1 Contributions to Steamflood Recovery[a]

	Recovery (% OIP)		
	Non-distillable Oil	25% Distillable Oil	50% Distillable Oil
Hot-waterflood recovery	54.8	54.8	58.0
Extra due to gas drive	3.0	3.0	3.0
Extra due to distillation	—	10.5	19.5
Other	1.2	7.7	3.4
Total steamflood recovery	59.0	76.0	83.9

[a]From Ref. 1.

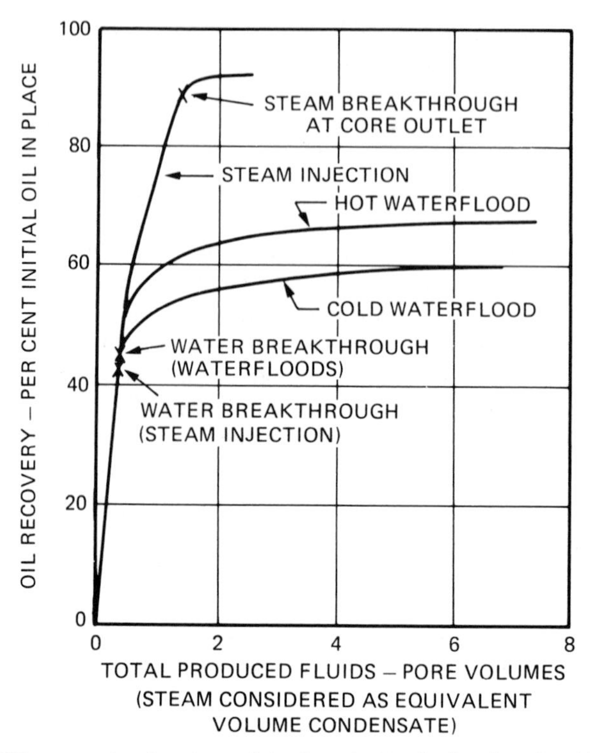

Figure 8.3. Oil recoveries by steam injection, hot waterflood and cold waterflood in core 9, initially containing 34.7° Homer crude and connate water.[1] © 1961 SPE-AIME.

an independent analysis of these same data, Farouq Ali[2] attributes 5% recovery to solvent extraction effects.

In another core flood, Willman et al.[1] compared the displacement efficiency of steam with that of hot and cold water (Fig. 8.3). For the crude under investigation, steamflood recovery was over 90%. Hot-waterflood recovery was less than 70%, and cold-waterflood recovery was about 60%.

Residual oil saturations in steamflooded regions can be quite low. For best values, experimental measurements should be made. For rough screening estimates, the National Petroleum Council[3,4] has used a value of 15 percent.

SWEEP EFFICIENCY

Channeling of injected steam and poor sweep are probably the most serious problems associated with steamflooding. High sweep efficiencies for steamfloods are prevented by the density and viscosity differences between the steam and the oil. There are strong tendencies for the steam to override the oil in the vertical plane (see Fig. 8.2) and to finger through it on an areal basis. Fingering and by-passing of oil are also strongly influenced by reservoir heterogeneities.

Despite the apparently adverse characteristics of a steamflood, volumetric sweep efficiencies are higher than might at first be expected. Correlation of volumetric sweep efficiency as a function of mobility ratio alone (Fig. 8.4) would suggest that steam would have extremely poor volumetric coverage. However, use of the mobility ratio as a correlating factor for reservoir sweep is an oversimplification of stability requirements for steamflooding. Stability requirements for a steamflood must include the effects of volume change due to steam condensation and the effects of heat transfer on the movement of the fluid front.

An analysis by Miller[5] points to factors tending to stabilize a steamflood. The first is the force of gravity, which tends to stabilize the steam front for down dip displacements. This force also tends to cause overriding for a horizontal displacement. The second is associated with fluid mechanical effects. The stability of the steam displacement front becomes greater as the ratio of the pressure gradient behind the front to that ahead of the front becomes greater. In other words, the steam front tends to be stable when values for the product of mobility ratio and ratio of liquid to steam velocities are less than 1. A small value for the velocity ratio tends to counteract the

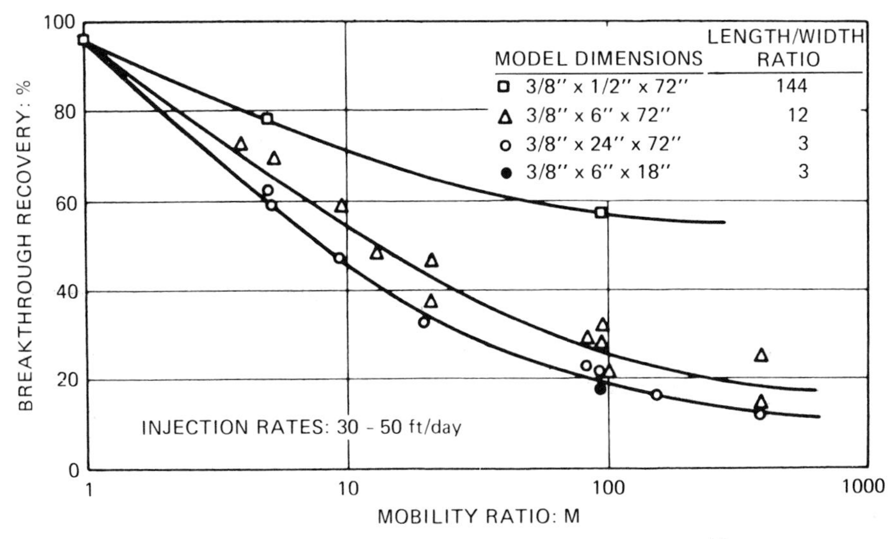

Figure 8.4. Effect of mobility ratio on recovery at breakthrough.[15] © 1959 SPE-AIME.

large value for the mobility ratio. The third factor is that lateral heat flow from fingers of steam tends to retard their growth and accelerate frontal advance between the fingers.

In the typical steamflood, with its overriding and channeling tendencies, oil recovery is obviously much more complicated than that represented by a one-dimensional model of the type discussed in Chapter 3. Condensation at the edges of the advancing steam zone provides a continuous and moving source of hot water. Parts of the reservoir are swept successively by cold water, hot water, and steam; parts are swept successively by cold and hot water; parts are swept only by cold water; and parts remain relatively untouched.

Sweep efficiencies for steamfloods vary over a wide range, and generalizations concerning expected values are obviously highly approximate. However, some guidance is given by values used in screening calculations by the National Petroleum Council (NPC)[3] (Table 8.2). With regard to vertical sweep, the NPC reported values for the maximum steam-swept zone

Table 8.2 Steamflood Recovery Projections (National Petroleum Council)[a]

Displacement Efficiency

Steam-swept regions	$S_{or} = 0.15$
Hot-water-swept regions	$S_{or} = 0.25$

Vertical Sweep

Steam Region		*Hot-Water Region*	
Well Spacing (acre)	Maximum Steam-Swept Zone (ft)	Well Spacing (acre)	Maximum Hot-Water-Swept Zone (ft)
2.5	20	2.5	40
5.0	15	5.0	35
10.0	10		

Areal Sweep

Steam Region		*Hot-Water Region*	
Well Spacing (acre)	Areal Sweep Efficiency (%)	Well Spacing (acre)	Areal Sweep Efficiency (%)
2.5	70	2.5	90
5.0	65	5.0	85
10.0	60	10.0	80

[a]From Ref. 3.

thickness ranging from 20 ft for a 2.5-acre spacing to 10 ft for a 10-acre spacing. The region swept by hot water ranged from 40 ft on a 2.5-acre spacing to 35 ft on a 5.0-acre spacing. The areal sweep efficiency for the steam zone was estimated to range from 70% on a 2.5-acre spacing to 60% on a 10-acre spacing. For the hot-water zone, areal sweep ranged from 90% on a 2.5-acre spacing to 80% on a 10-acre spacing.

THERMAL EFFICIENCY

Thermal efficiency is important in steamflooding. Heat losses from the surface distribution system, from the wellbore, and from the formation occur as the flood continues. Heat losses to overburden and underburden become increasingly important as the flood becomes advanced and more area is exposed for heat flow. Well spacing, sand thickness, and flooding rate are important factors that affect this type of heat loss.

One approach to conserving the heat that would otherwise be left behind the steam front is the use of thermal banks followed by water. Overall recovery utilizing a thermal bank will probably be less than that from continuous steam injection, but the oil–steam ratio will be higher. The decision on how large a steam bank to use is an economic one as well as a technical one.

SCREENING CRITERIA

Some idea of practical limits for certain key reservoir variables can be gained from the screening criteria utilized in recent studies of applications of enhanced recovery methods to domestic reservoirs. Table 8.3 lists such criteria for steam injection prospects as utilized by the NPC.[4]

Table 8.3 NPC Steam Drive Screening Criteria[a]

Oil gravity, °API	10–34
Oil viscosity, cp	≤ 1500
Depth, ft	≤ 3000
Zone thickness, ft	>20
Porosity, fraction	≥ 0.20
Transmissibility, md-ft/cp	≥ 5
Permeability, md	>250
Reservoir pressure, psia	≤ 1500
Minimum oil content at process start (ϕS_o), fraction	≥ 0.10
Rock type	Sandstone or carbonate

[a]From Ref. 4.

It should be noted that current steamfloods are usually being conducted at depths of less than 3000 ft. Still, it is not inconceivable that by using wellbore insulating techniques, reservoirs deeper than 5000 ft can be flooded provided that low pressure can be maintained. A minimum gross sand thickness of 20 ft is commonly accepted. The NPC study set a value of 5 md-ft/cp as the minimum value for transmissibility. Note that steam stimulation or hydraulic fracturing may be necessary to achieve sufficient transmissibility for very heavy oil reservoirs. The NPC gives a ϕS_o value of 0.1 or 776 bbl/acre-ft for the minimum acceptable starting oil concentration. A somewhat higher minimum value, on the order of 1000 bbl/acre-ft, may be required for an economic operation.

EFFECT OF RESERVOIR PARAMETERS

Oil Saturation and Porosity

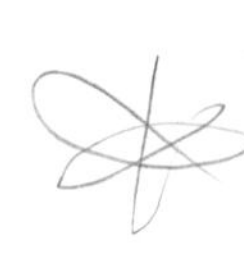

Oil saturation and porosity are key variables for the steam drive process. There is some minimum product of oil saturation and porosity for each reservoir below which an economically attractive process is highly unlikely. The latest NPC study[4] discussed previously suggests that the oil content expressed as ϕS_o be 0.1 or higher. Oil–steam ratios increase with increasing oil saturations and porosities.

Formation Thickness

As the bed thickness increases, the relative importance of the heat lost to overburden and underburden decreases, yielding an improved oil–steam ratio. For thin reservoirs, the fractional heat losses can become so great that the process is no longer economic. Excessive thickness, however, compared to that for a thinner formation, can lead to a poor vertical sweep and thus to a reduced recovery level.

Net–Gross Sand Thickness Ratio

Low values of net–gross sand thickness ratio reflect the presence of layers of barren rock that require heating but produce no oil. Consequently, a higher net–gross ratio will raise the oil–steam ratio.

Depth

Increasing depth adversely affects steamflooding because of increased heat losses in the injection wells and because deep wells are more expensive than shallow ones. Considerable infill drilling to arrive at acceptable (5–10-acre) spacings for steamflooding will be economically prohibitive in deep formations.

Pressure

There is lowered effectiveness of a given mass of steam at higher pressures. We already discussed the fact that steam vapor occupies more volume at a lower pressure, but there is also a very important thermal effect. With increasing pressure, the temperature of the injected steam is greater, requiring increased heating of the reservoir.

Since the heat of vaporization of water decreases with increasing temperature, the quantity of steam required for reservoir heating becomes even greater. The ratio $H_{wv}/(T_s - T_r)$ will be proportional to the steam zone size and hence proportional to the quantity of oil displaced by the steam zone. This ratio will be about four times greater at 300 psia than at 2500 psia for a reservoir originally at 100°F, indicative of the much greater steam displacement potential at low pressures.

Rock Properties

High values of rock transmissibility (md-ft/cp) enable flooding at high rates, thus reducing the relative effects of heat loss. Changes in rock absolute and relative permeabilities are known to occur with changes in temperature. The changes are usually in a favorable direction, but the effects on steamflood behavior usually are small.

EFFECT OF OPERATING PARAMETERS

Steam Injection Rate

A high steam injection rate is important in minimizing heat losses. However, securing higher injection rates through higher injection pressures creates greater heat losses in the injection wells as well as greater stresses in the casing. In addition, as discussed above, a high pressure in the steam zone has a negative effect. Channeling can also be more severe at high rates so the effect of steam injection rate is complex.

Steam Quality

The higher the steam quality, the higher the heat content per unit mass of steam. In general, oil recovery increases with the quantity of heat injected.

It is also easier to meter and control the flow of high-quality steam in surface installations.

Well Spacing

Close well spacing tends to reduce the time to steam breakthrough for a particular injection rate, thus improving the thermal efficiency of a steamflood by reducing the time for heat losses. Optimum spacing, however, involves many other factors, in particular, the steamflood sweep efficiency.

Well Pattern

The two most commonly used patterns for steamflooding are five-spots and inverted seven-spots. Differences in areal sweep efficiency are minor. Seven-spots have two producers per injector and thus provide greater productive capacity, an important consideration in viscous oil situations requiring high rates of injection and production for an effective flood.

Steam Bank Size

As has been indicated earlier, the use of steam banks increases the oil–steam ratio but usually leads to a lower overall recovery. The careful design of bank sizes has probably been the exception rather than the rule in field tests to date. Such designs require a careful and sophisticated engineering effort.

Chemical Flood Aids

Various methods are being investigated to improve the performance of steamfloods. In particular, methods have been designed for improving injection profiles and securing better vertical sweep. Shell[6] has developed an acrylic/epoxy emulsion gel system for plugging thief zones. Chevron[7] has utilized a one-step furfuryl alcohol process for formation plugging. They also report the successful use in Midway-Sunset field of an anionic sulfonate (SD-1000) at concentrations as low as 0.1% to generate foams in situ.[8] They recommend that a noncondensible gas such as nitrogen be injected to maintain the foam. There was a long sustained period of incremental oil following sulfonate injection. The use of steam foam was concluded to have favorable economics with surfactant usage between 0.25 and 1.0 lb per incremental barrel of oil.[8]

Fitch and Minter[9] have described field tests of Union Oil Company's foaming agent. Workers at Stanford University[10–12] are testing the effectiveness of foaming agents over extended periods of time at elevated temperatures. A successful test of a commercial surfactant for in situ foam generation (Suntech IV) injected with nitrogen was reported in the Kern River field.[12]

Reportedly successful tests have been made of a Phillips surfactant for reducing channeling in steam stimulation. Work has been performed at the University of Southern California[13] on the utilization of high-temperature surfactants to improve displacement efficiency during steamfloods. Getty[14] has recently described the results of a steam foam diverter to improve sweep efficiency in steamfloods in the Kern River field.

Simplified screening analysis methods are described in Chapter 9. Chapter 10 presents some actual steamflood case histories. Chapter 11 covers two- and three-dimensional numerical models of steamflooding and results from these models. Chapter 12 describes the Shutler–Boberg analytical model to give further insight into steam displacement mechanics.

REFERENCES

1. Willman, B. T., Valleroy, V. V., Runberg, G. W., Cornelius, A. J. and Powers, L. W. "Laboratory Studies of Oil Recovery by Steam Injection," *J. Pet. Tech.* (July 1961), pp. 681–690.
2. Farouq Ali, S. M., "Steam Injection," *Secondary and Tertiary Oil Recovery Processes*, Interstate Oil Compact Commission, Oklahoma City, 1974, Chapter VI, pp. 127–182.
3. National Petroleum Council. *Enhanced Oil Recovery: An Analysis of the Potential for Enhanced Oil Recovery from Known Fields in the United States—1976 to 2000,* National Petroleum Council, Washington, DC, 1976, pp. 169–202.
4. National Petroleum Council. *Enhanced Oil Recovery*, National Petroleum Council, Washington, DC, June 1984.
5. Miller, C. A. "Stability of Moving Surfaces in Fluid Systems With Heat and Mass Transport, III. Stability of Displacement Fronts in Porous Media," *AIChE J.* (May 1975), pp. 474–479.
6. Knapp, R. H., and Welbourn, M. E. "An Acrylic/Epoxy Gel System for Formation Plugging: Laboratory Development and Field Testing for Thief Zone Plugging," SPE 7083, Presented at the Fifth Symposium on Improved Methods for Oil Recovery of the Society of Petroleum Engineers of AIME, Tulsa, Oklahoma, April 16–19, 1978.
7. Hess, P. H. "One-Step Furfuryl Alcohol Process for Formation Plugging," *J. Pet. Tech.* (October 1980), pp. 1834–1842.
8. Ploeg, J. F., and Duerksen, J. H. "Two Successful Steam/Foam Field Tests, Sections 15A and 26C Midway-Sunset Field," SPE 13609, Presented at the 1985 California Regional Meeting, Bakersfield, California, March 27–29, 1985.
9. Fitch, J. P., and Minter, R. B. "Chemical Diversion of Heat Will Improve Thermal Oil Recovery," SPE 6172, Presented at the Fifty-First Annual Fall

Meeting of the Society of Petroleum Engineers of AIME, New Orleans, October 3–6, 1976.

10. Progress Review. Quarter Ending December 31, 1978. Contracts for Field Projects and Supporting Research on Enhanced Oil and Gas Recovery and Improved Drilling Methods, United States Department of Energy, No. 17, p. 93.
11. Chiang, J. C., Sanyal, S. K., Castnaier, L. M., and Brigham, W. E. "Foam as a Mobility Control Agent in Steam Injection Processes," SPE 8912, Presented at the Fiftieth Annual California Regional Meeting of the Society of Petroleum Engineers of AIME, Los Angeles, California, April 9–11, 1980.
12. Brigham, W. E., Marcou, J. A., Sanyal, S. K., Malito, O. P., and Castanier, L. M. "A Field Experiment of Improved Steam Drive With In Situ Foaming," SPE 12784, Presented at the 1984 California Regional Meeting, Long Beach, California, April 11–13, 1984.
13. Elson, T. D., and Marsden, S. S., Jr. "The Effectiveness of Foaming Agents at Elevated Temperatures Over Extended Periods of Time," SPE 7116, Presented at the 1978 Regional Meeting of the Society of Petroleum Engineers of AIME, San Francisco, California, April 12–14, 1978.
14. Greaser, G. R., and Shore, R. A. "Steamflood Performance in the Kern River Field," SPE 8834, Presented at the First Joint SPE/DOE Symposium on Enhanced Oil Recovery, Tulsa, Oklahoma, April 20–23, 1980.
15. Blackwell, R. J., Rayne, J. R., and Terry, W. M. "Factors Influencing the Efficiency of Miscible Displacement," *Trans. AIME* **217,** 5 (1959).

Chapter 9

SIMPLIFIED SCREENING CALCULATIONS OF STEAM DRIVE OIL RECOVERY

ABSTRACT

Screening and preliminary calculations to predict steamflooding behavior can be made with a number of analytical models that greatly simplify the heat and fluid mechanics of steamflooding. Their successful use depends largely on the user's ability to understand the underlying assumptions and to use the most realistic model for a given situation. A failure to do so will often lead to overly optimistic or pessimistic estimates of oil recovery and steam requirements. All of the models discuss only continuous steam injection and are essentially one dimensional with respect to heat flow. The methods covered are summarized below.

The *Marx–Langenheim method* assumes pistonlike advance of the steam front and uses the Marx–Langenheim heat transfer equation to calculate the steam zone volume. This method assumes all oil in the steam-invaded region is displaced to an assumed steam zone residual oil saturation.

The *Myhill–Stegemeier method* is similar to the Marx–Langenheim method but uses the more advanced Mandl–Volek equations to calculate the steam zone volume. It defines the size of the steam-invaded region somewhat more optimistically than the Marx–Langenheim method described here.

The *van Lookeren model* can be used to obtain an approximate estmate of the vertical sweep efficiency.

The *Vogel* and *Miller models* assume complete steam override of a thick oil sand. The oil column is heated by downward conduction from the overlying steam zone.

The *Butler* and *Edmunds models* describe the rate at which highly viscous oil or bitumen can be heated by steam and mobilized for two specific cases. Butler's equation is specific for calculating the rate of production of bitumen from a steam zone, where mobilized oil is produced by gravity drainage. Edmund's equation considers the case of stripping mobilized bitumen from the walls of a steam-heated fracture.

INTRODUCTION

Numerous analytical steam drive models have appeared in the literature.[1-23] The models discussed in this chapter are quite simplified and, for the most part, do not calculate fluid flow aspects rigorously. A more rigorous analytical fluid flow treatment is provided by the Shutler–Boberg model discussed in Chapter 12.

The Shutler–Boberg model, as well as the first two models discussed in this chapter, the simplified Marx–Langenheim and the Myhill–Stegemeier, are similar. All three neglect gravity effects and assume 100% vertical sweep. They are basically best suited for thin sands where the well spacing–thickness ratio is at least 7 or greater. They are all based on a similar heat flow theory—that of Marx and Langenheim or the Mandl–Volek model already discussed in Chapter 3.

For thick or heterogeneous sands having highly viscous oils where gravity override can be expected and where thermal breakthrough occurs early in the life of the project, these models will be overly optimistic as they will not predict the bypassing of the oil column and channeling of steam to producing wells that occurs after breakthrough. The model of van Lookeren[11] provides a method for estimating the vertical sweep efficiency for steamflooding a homogeneous thick sand where gravity override is important.

The Vogel model[18] permits rough estimates of oil–steam ratios for the extreme situation where steam breakthrough at producing wells occurs very early in the life of the project. An extension of this model by Miller[19] permits rate-versus-time predictions. Finally, the Butler[20-22] and Edmunds[23] models are discussed as analytical tools for predicting oil–steam ratios when conducting a steam gravity drainage or steam-stripping type of operation in an extremely high viscosity oil sand where interwell communication would have to be achieved by means of a vertical or horizontal fracture.

SIMPLIFIED MARX–LANGENHEIM MODEL

Marx–Langenheim[1] assumptions include pistonlike displacement and 100% vertical sweep. Generally, these assumptions are not valid except for thin sands. Yet this model has good utility in many situations where these assumptions do not strictly apply, and it can be used for rough screening of reservoir candidates. In its simplest form, this model assumes that all additional oil recovery comes from the steam-invaded zone.

Oil Recovery Estimate at Steam Breakthrough

The oil recovery N_p at steam breakthrough is estimated as

$$N_p = \frac{A_s h \phi}{5.61 B_o}(S_{oi} - S_{ors}) \tag{9.1}$$

where A_s = area swept by steam (ft^2)
h = net sand thickness (ft^2)
ϕ = porosity (fraction)
S_{oi} = initial oil saturation (fraction)
S_{ors} = average steam zone oil saturation at breakthrough (fraction)

Oil–Steam Ratio at Steam Breakthrough

Defining

$$H_t = \bar{X}_i H_{wv} + H_{ws} - H_{wr} \tag{9.2}$$

Equation (3.6a) becomes

$$\frac{\xi_s}{\theta} = \frac{4K_{hob}A(T_s - T_r)(\rho c)_{ob}}{h_t H_t m_s (\rho c)_{R+F}}$$

where A differs from A_s in that A is the area swept by steam and hot water.

The oil–steam ratio is

$$R_{os} = \frac{N_p}{m_s t} = \frac{A_s h \phi (S_{oi} - S_{ors}) h_t H_t (\rho c)_{R+F} (\xi_s/\theta)(350)}{4K_{hob}(T_s - T_r)(\rho c)_{ob} t A (5.61) B_o} \tag{9.3}$$

and

$$\frac{\tau}{\theta} = \frac{4K_{hob}(\rho c)_{ob} t}{h_t^2 (\rho c)_{R+F}^2} \tag{9.4}$$

In Eq. (9.4), a distinction is made between the net sand thickness, h and h_t, the gross sand thickness (including shale stringers). It is assumed that the shale stringers are thin and are heated to steam temperature uniformly as the steam front passes.

Hence, substituting τ/θ in Eq. (9.3) yields

$$R_{os} = \frac{A_s h \phi (S_{oi} - S_{ors}) H_t (\xi_s/\theta)(62.4)}{A(T_s - T_r)(\rho c)_{R+F} h_t (\tau/\theta) B_o} \tag{9.5}$$

In the Appendix, it is shown that E_h, the thermal efficiency, may be calculated as

$$E_h = \frac{\xi_s/\theta}{\tau/\theta} \tag{9.6}$$

Combining Eqs. (9.5) and (9.6) gives

$$R_{os} = \frac{62.4\phi h E_h A_s [H_t(S_{oi} - S_{ors})]}{B_o h_t (T_s - T_r)(\rho c)_{R+F} A} \tag{9.7}$$

An empirical relation between A_s and A assumes that these areas are proportional to the following ratio:

$$\frac{A_s}{A} = \frac{\text{(total energy injected)} - \text{(energy injected as hot water)}}{\text{total energy injected}}$$

$$= \frac{H_t - (H_{ws} - H_{wr})}{H_t} = \frac{\bar{X}_i H_{wv}}{H_t} \tag{9.8}$$

Making this substitution,

$$R_{os} = \frac{62.4\phi h E_h [\bar{X}_i H_{wv}(S_{oi} - S_{ors})]}{B_o h_t (T_s - T_r)(\rho c)_{R+F}} \tag{9.9}$$

We shall later examine the utility of Eq. (9.9) in estimating the oil–steam ratio. The value of the residual oil saturation in the steam zone should be determined experimentally. It will generally be a low value even for very heavy oils ($S_{ors} \approx 0.15$).

Equation (9.9) will predict that the oil–steam ratio increases as ϕ, h/h_t, E_h, X_i, H_{wv}, and S_{oi} increase and as S_{ors}, $T_s - T_r$, and $(\rho c)_{R+F}$ decrease. Factors tending to increase the Marx–Langenheim thermal efficiency E_h are

1. short flood time resulting from a high injection rate or close well spacing to decrease t;
2. high steam-injection rate m_s; and
3. thick sand, having a large h_t

Estimating Steam Injection and Oil Production Rates

A common error in designing steam drive projects is to assume that oil displaced by the steam zone can be moved at a rate higher than is physically possible. In a pistonlike displacement, most of the oil moves ahead of the steam and hot-water zone through unheated oil sand. In a five-spot, the initial oil rate would be[24]

$$q_{oi} = \frac{0.003541 k_o h \, \Delta P}{B_o \mu_{oc}[\ln(d/r_w) - 0.619]} \tag{9.10}$$

where k_o = permeability to oil (md)
h = net sand thickness (ft)
ΔP = difference between bottom-hole pressure at injection and producing wells (psi)
μ_{oc} = cold oil viscosity (cp)
d = distance between injection and producing wells (ft)
r_w = effective wellbore radius (ft)

It would be pessimistic to assume that the oil rate calculated by Eq. (9.10) would apply for the entire flood. There is little pressure drop through the steam zone relative to that through the cold reservoir in the case of a high-viscosity oil. Neglecting this pressure drop, when the steam zone has reached the halfway point (midway between injector and producer), the oil rate is[24]

$$q_{o,1/2} = \frac{0.00708 k_o h \, \Delta P}{B_o \mu_{oc} \ln(d/2r_w)} \tag{9.11}$$

To obtain an improved estimate for k_o in Eq. (9.11), an estimate of the flowing water–oil ratio ahead of the steam zone should be made. Then k_w/k_o can be obtained[24]:

$$\frac{k_w}{k_o} = R_w \frac{\mu_w}{\mu_o} \frac{B_o}{B_w} \tag{9.12}$$

The producing water–oil ratio R_w is roughly equal to the reciprocal of R_{os}, the oil–steam ratio, as a first approximation. An initial guess of R_w can be made by estimating R_{os} in the range 0.2–0.4 bbl/bbl.

The steam injection rate m_s (in lb/d) is calculated as

$$m_s = 350 q_{o,1/2}/R_{os} \tag{9.13}$$

The calculation procedure is as follows:

1. Calculate m_s by Eqs. (9.11) and (9.13).
2. Calculate R_{os} by Eq. (9.9).
3. Calculate k_w/k_o by Eq. (9.12) and improve the estimate of k_o.
4. Calculate $q_{o,1/2}$ and m_s by Eqs. (9.11) and (9.13). Use this new estimate of m_s and repeat the calculation steps if necessary.

If the oil rate by Eq. (9.11) is below about 50 bbl/d, one should not estimate the oil–steam ratio using Eq. (9.9) for pistonlike displacement. The oil sand may have to be heated indirectly to lower the oil viscosity. For such a situation, steam requirements would be greater than would be predicted using Eq. (9.9). Oil sands that do not permit flooding under cold conditions to give reasonable rates are less desirable candidates than those that permit reasonable rates under conventional depletion.

An example of the Marx–Langenheim method is illustrated next to select the better of two sands as steamflooding candidates:

Sand 1	Sand 2
$k_o = 2$ darcies	$k_o = 1$ darcy
$h_t = h = 50$ ft	$h_t = h = 100$ ft
$\mu_o = 100$ cp	$\mu_o = 200$ cp
$P_s = 1000$ psi	$P_s = 500$ psi
$T_s = 545$°F	$T_s = 467$°F
$P_w = 100$ psi	$P_w = 100$ psi
Depth = 2000 ft	Depth = 1000 ft
$T_r = 200$°F	$T_r = 100$°F
$S_{oi} = 0.80$	$S_{oi} = 0.90$
$S_{ors} = 0.15$	$S_{ors} = 0.15$
$\phi = 0.30$	$\phi = 0.25$
$B_o = 1.0$	$B_o = 1.0$
$(\rho c)_{ob} = 38$ Btu/ft^3-°F	$(\rho c)_{ob} = 38$ Btu/ft^3-°F
$(\rho c)_{R+F} = 27$ Btu/ft^3-°F	$(\rho c)_{R+F} = 27$ Btu/ft^3-°F
$K_h = 1.4$ Btu/hr-ft-°F	$K_h = 1.4$ Btu/hr-ft-°F

Both sands are assumed to have 10-acre patterns. Assuming 70% areal sweep, $A = 0.70(10)(43{,}560) = 3.049 \times 10^5$ ft^2. Both candidates have injection wells equipped with a 7-in casing and a $2\frac{1}{2}$-in tubing with injection through the tubing. For a 10-acre five-spot, $d = 467$ ft. A well radius $r_w = 0.3$ ft is assumed. A first estimate of $R_{os} = 0.3$ is made for both cases.

1. First estimate of steam injection rate:

   (a) Sand 1:

$$q_{o,1/2} = \frac{0.00708(2000)(50)(1000-100)}{(1.0)(100)\ln[467/2(0.3)]} = 957 \text{ bbl/d}$$

$$m_s = 350(957)/(0.3) = 1.12 \times 10^6 \text{ lb/d}$$

(b) Sand 2:

$$q_{o,1/2} = \frac{0.00708(1000)(100)(500 - 100)}{(1.0)(200)\ln[467/2(0.3)]} = 213 \text{ bbl/d}$$

$$m_s = 350(213)/0.3 = 0.248 \times 10^6 \text{ lb/d}$$

2. First estimate of E_h, R_{os}:

(a) Sand 1:

$$\frac{\xi_s}{\theta} = \frac{4(1.4)(24)(3.049 \times 10^5)(545 - 100)(38)}{(50)(m_s)(\bar{X}_i H_{wv} + H_{ws} - H_{wr})(27)}$$

$$= \frac{5.132 \times 10^8}{m_s(\bar{X}_i H_{wv} + H_{ws} - H_{wr})}$$

For the injection conditions stated,

$$H_{wv} = 649 \text{ Btu/lb}$$

$$H_{ws} = 542 \text{ Btu/lb}$$

$$H_{wr} = 68 \text{ Btu/lb}$$

The average downhole steam quality is estimated assuming a 4-year pattern life. The wellbore heat loss rate is obtained at the midpoint injection time (2 years), from Figure 2.8, as 650 Btu/hr-ft. The surface steam quality is assumed to be 0.7 for both cases. The bottom-hole steam quality is calculated by Eq. (2.5):

$$\bar{X}_i = 0.7 - \frac{650(2000)(24)}{(1.12 \times 10^6)(649)} = 0.66$$

Hence,

$$\frac{\xi_s}{\theta} = \frac{5.132 \times 10^8}{(1.12 \times 10^6)[0.66(649) + 542 - 68]} = 0.508$$

Using the Appendix of Chapter 3, τ/θ is then estimated as

$$\tau/\theta = 0.884$$

The thermal efficiency for sand 1 is then calculated by Eq. (9.6):

$$E_h = 0.508/0.884 = 0.575$$

Hence,

$$R_{os} = \frac{62.4(0.3)(50)(0.575)}{50(545 - 100)(27)}[(0.66)(649)(0.8 - 0.15)] = 0.250 \text{ bbl/bbl}$$

A new estimate is made of m_s:

$$m_s = 350(957)/0.250 = 1.34 \times 10^6 \text{ lb/d}$$

This is within 20% of the previous estimate of m_s and is in acceptable agreement considering the approximate nature of the calculation.

(b) Sand 2;

$$\frac{\xi_s}{\theta} = \frac{4(1.4)(24)(3.049 \times 10^5)(467 - 100)(38)}{100(m_s)(\bar{X}_i H_{wv} + H_{ws} - H_{wr})(27)}$$

$$= \frac{2.117 \times 10^8}{m_s(\bar{X}_i H_{wv} + H_{ws} - H_{wr})}$$

For the conditions stated,

$$H_{wv} = 755 \text{ Btu/lb}$$

$$H_{ws} = 449 \text{ Btu/lb}$$

$$H_{wr} = 68 \text{ Btu/lb}$$

For sand 2, the downhole steam quality is estimated as

$$\bar{X}_i = 0.7 - \frac{650(1000)(24)}{(0.248 \times 10^6)(755)} = 0.617$$

Hence, the following are calculated as above:

$$\frac{\xi_s}{\theta} = \frac{2.117 \times 10^8}{(0.248 \times 10^6)[0.617(755) + 449 - 68]} = 1.008$$

Using this value of ξ_s/θ and either Figure 3.6 or the table in Appendix A of Chapter 3, τ/θ is estimated as

$$\tau/\theta = 2.22$$

Then,

$$E_h = 1.008/2.22 = 0.454$$

$$R_{os} = \frac{62.4(0.25)(100)(0.454)}{(100)(467 - 100)(27)}[0.617(755)(0.9 - 0.15)]$$

$$= 0.250 \text{ bbl/bbl}$$

A new estimate is made of m_s:

$$m_s = 350(213)/0.250 = 0.298 \times 10^6 \text{ lb/d}$$

Again, this is probably in close enough agreement with the previous estimate. Since both sands have nearly identical estimated oil–steam ratios, one would probably pick the sand that could be flooded at the higher rates (i.e., sand 1). However, if new well costs were a significant part of the overall cost of the operation, a more thorough economic study would be required to determine whether the shallower sand having the greater oil in place per unit of area (sand 2) would have the economic edge.

Comparisons of the simplified Marx–Langenheim estimates of oil–steam ratios to field-project-observed values are presented in the next chapter.

SIMPLIFIED MODEL OF MYHILL AND STEGEMEIER

The Myhill and Stegemeier[5] method is a more advanced version of the simplified Marx and Langenheim model.

Assumptions basic to this model are similar to that of Marx and Langenheim, but the more advanced model of Mandl and Volek[4] is used to compute the size of the steam zone.

Step 1. Estimate the Mandl and Volek critical time (t_{cD}) from the following equation†

$$e^{t_{cD}} \operatorname{erfc}\sqrt{t_{cD}} = 1 - f_{hD} \tag{9.14}$$

where erfc is the complementary error function[1] defined as

$$\operatorname{erfc} x = 1 - \frac{2}{\sqrt{\pi}} \int_0^x e^{-t^2}\, dt \tag{9.15}$$

and f_{hD} is the fraction of thermal energy injected in the latent form and is defined as

$$f_{hD} = \frac{\bar{X}_i H_{wv}}{c_w(T_s - T_r) + \bar{X}_i H_{wv}} \tag{9.16}$$

† As an approximation, Jones[6] suggested that the critical time can be estimated by

$$t_{cD} = 0.48\left(\frac{f_{hD}}{1 - f_{hD}}\right)^{1,71}$$

where H_{wv} = latent heat of steam (Btu/lb)
$\bar{X}_i$ = average quality of steam at bottom hole (lb/lb)
c_w = specific heat of water (Btu/lb-°F)
T_s = steam temperature (°F)
T_r = initial formation temperature (°F)

Step 2. Determine the dimensionless time t_D associated with a selected injection time t by

$$t_D = \frac{4K_h(\rho c)_{ob}}{h_t^2(\rho c)_{R+F}^2}\, t \tag{9.17}$$

where K_h = thermal conductivity of overburden and underburden (Btu/ft-hr-°F)
h_t = gross thickness of reservoir (ft)
$(\rho c)_{R+F}$ = average volumetric heat capacity of steam zone (Btu/ft^3-°F),
$= (1-\phi)\rho_r c_r + \phi[\rho_o c_o S_o + \rho_w c_w S_w + (S_g \rho_s/\Delta T)(H_{wv} + c_w\,\Delta T)]$
$(\rho c)_{ob}$ = average volumetric heat capacity of underburden and overburden (Btu/ft^3-°F)
t = time since start of steam injection (hr)

Step 3. Determine the thermal efficiency E_{hs}.

If the dimensionless time t_D defined in Eq. (9.17) is less than the critical time calculated from Eq. (9.14), heat transfer across the steam front is dominated by conduction, and thermal efficiency should be determined by

$$E_{hs} = \frac{1}{t_D}\left(e^{t_D}\,\mathrm{erfc}\sqrt{t_D} + 2\sqrt{t_D/\pi} - 1\right) \quad \text{if } t_D < t_{cD} \tag{9.18}$$

Values of the function $e^{t_D}\,\mathrm{erfc}\sqrt{t_D} + 2\sqrt{t_D/\pi} - 1$ are listed in Appendix A of Chapter 3.

On the other hand, if $t_D > t_{cD}$, heat transfer across the steam front is dominated by convection, and E_{hs} is given by[16]

$$E_{hs} = \frac{1}{t_D}\left\{G(t_D) + (1-f_{hD})\frac{U(t_D - t_{cD})}{\sqrt{\pi}} \times \left[2\sqrt{t_D} - 2(1-f_{hD})\sqrt{t_D - t_{cD}} - \int_0^{t_{cD}} \frac{e^x\,\mathrm{erfc}\sqrt{x}\,dx}{\sqrt{t_D - x}} - \sqrt{\pi}\,G(t_D)\right]\right\} \tag{9.19}$$

$$G(t_D) \equiv e^{t_D}\,\mathrm{erfc}\sqrt{t_D} + 2\sqrt{\frac{t_D}{\pi}} - 1 \tag{9.20}$$

and $U(t_D - t_{cD})$ is the unit function defined as

$$U(t_D - t_{cD}) = \begin{cases} 0 & \text{for } t_D - t_{cD} < 0 \\ 1 & \text{for } t_D - t_{cD} > 0 \end{cases} \tag{9.21}$$

The function E_{hs} is given in Figure 9.1 as a function of t_d and f_{hD}.

Step 4. Determine the volume of the steam zone V_S:

$$V_S = \frac{i_s t \rho_w [c_w(T_s - T_f) + \bar{X}_i H_{wv}] E_{hs}}{(\rho c)_{R+F}(T_s - T_f)} \tag{9.22}$$

where i_s is the steam injection rate at the cold-water equivalent (bbl/d).

Step 5. Determine the total amount of oil recovered Q_o and the oil–steam ratio R_{os}:

$$Q_o = V_S \phi (S_{oi} - S_{or})(h/h_t) \tag{9.23}$$

$$R_{os} = Q_o / i_s t \tag{9.24}$$

Step 6. Repeat steps 2–5 for another time step.

EXAMPLE. Given the following data for a steamflood project, estimate the oil recovery and the oil–steam ratio for years 1–5.

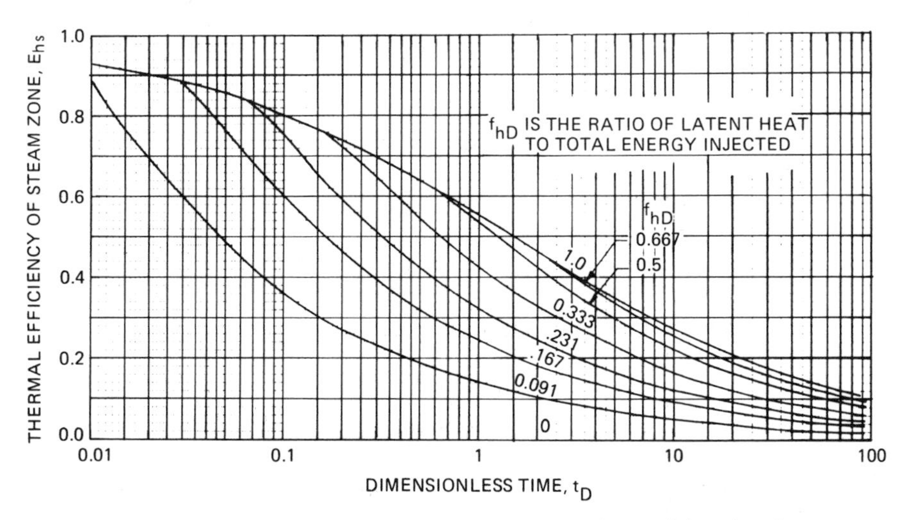

Figure 9.1. Fraction of heat injected in steamflood remaining in steam zone (f_{hD} = ratio of latent energy to total energy injected).[16] © 1982 SPE-AIME.

Area, acres	61.0
Porosity	0.35
Gross thickness, ft	100.0
Net thickness, ft	70.0
Oil saturation prior to steamflood	0.52
Oil saturation after steamflood	0.10
Reservoir temperature, °F	90.0
Thermal conductivity of overburden and underburden, Btu/ft-hr-°F	0.71
Volumetric heat capacity of heated reservoir, Btu/ft^3-°F	47.0
Volumetric heat capacity of underburden and overburden, Btu/ft^3-°F	47.0
Steam injection rate, bbl/d	6600.0
Steam injection pressure, psia	200.0
Steam quality at bottomhole	0.6

Step 1. Determine the critical time t_{cD}.
From steam tables, for $P = 200$ psia, we have

$$T_S = 381.86°\text{F} \qquad \text{and} \qquad H_{wv} = 843.7 \text{ Btu/lb}$$

Hence, the fraction of thermal energy injected in the latent form is

$$f_{hD} = \frac{\bar{X}_i H_{wv}}{c_w(T_s - T_r) + \bar{X}_i H_{wv}} = \frac{0.6 \times 843.7}{1.0(381.86 - 90) + 0.6 \times 843.7} = 0.635$$

From Eq. (9.14), we have

$$e^{t_{cD}} \operatorname{erfc}\sqrt{t_{cD}} = 1 - 0.635 = 0.365$$

From Appendix A, Chapter 3,

$$t_{cD} = (1.264)^2 = 1.597$$

Step 2. Calculate the dimensionless time t_D at 1 year from Eq. (9.17):

$$t_D = \frac{4 \times 0.71 \times 47 \times 24 \times 365 \times 1}{(100)^2 \times 47^2} = 0.053$$

Step 3. Calculate the thermal efficiency E_{hs}. Since t_D is less than t_{cD}, Eq. (9.18) is used. From the Appendix, Chapter 3,

$$e^{t_D} \operatorname{erfc} t_D + (2\sqrt{t_D/\pi}) - 1 = 0.045$$

Hence,

$$E_{hs} = 0.045/0.053 = 0.850$$

Alternatively, this value can be read from Figure 9.1.

Step 4. Determine the steam zone volume after 1 year of steam injection by Eq. (9.22):

$$V_S = \frac{6600 \times 365 \times 5.615 \times 62.4 \times [1.0 \times (381.68 - 90) + 0.6(843.7)] \times 0.850}{47(381.86 - 90)}$$

$$= 41.741 \times 10^6 \text{ ft}^3$$

Step 5. Determine the oil recovery and the oil–steam ratio at the end of 1 year:

$$Q_o = \phi V_S(S_{oi} - S_{or})(h/h_t) = 4.295 \times 10^6 \text{ ft}^3$$

$$= 0.765 \times 10^6 \text{ bbl}$$

$$R_{os} = Q_o/i_s t = 0.318$$

To find the percentage of oil in place recovered at the end of the first year, we need to know the oil in place at the start of the project (OOIP):

$$N = 61 \text{ acres} \times 43560 \text{ ft}^2/\text{acre} \times 70 \text{ ft} \times 0.35 \times 0.52/5.615 \text{ ft}^3/\text{bbl}$$

$$= 6.029 \times 10^6 \text{ bbl}$$

$$\text{Recovery} = \frac{0.765 \times 10^6}{6.029 \times 10^6} = 12.7\% \text{ OOIP}$$

Step 6. Repeat steps 2–5 for years 2–5. Results are listed in the following table.

Time (yr)	E_{hs}	Q_o (Mbbl)	Oil–Steam Ratio (bbl/bbl)	Recovery (% OOIP)
1	0.850	0.765	0.318	12.7
2	0.799	1.439	0.299	23.9
3	0.764	2.063	0.285	34.2
4	0.736	2.650	0.275	44.0
5	0.714	3.216	0.267	53.0

Jones[6] reports that the Myhill–Stegemeier method is usually somewhat optimistic and has modified the method to give results closer to those of 14 field projects by means of an empirical adjustment. Example results for the unmodified Myhill–Stegemeier model are compared to actual observed OSR's from a number of field projects in Chapter 10.

Other authors[25,26] consider the presence of hot water downstream of the vertical condensation front from the moment of injection. Although Willman et al.,[25] Farouq Ali,[26] and others consider the oil displacement by the hot-water condensate downstream of the condensation front, Myhill and Stegemeier[5] ignore the contribution as being small.

VAN LOOKEREN MODEL: ESTIMATING VERTICAL SWEEP

An approximate method developed by van Lookeren[11] and based on segregated flow principles is the only analytical method that considers the effects of dip, the ratio of gravity to viscous forces, and the liquid level in the injection well on the production response. The method also yields approximate shapes for the steam zone under the influence of buoyant forces.

Based on results for various special cases, van Lookeren suggests that the vertical sweep efficiency of steam flowing radially from an injection well in a homogeneous sand should be approximately equal to

$$E_{sv} = \bar{h}_{st}/h_t = 0.5A_R \tag{9.25}$$

where $\bar{h}_{st}$ = average steam zone thickness (m)
h_t = total sand thickness (m)

and

$$A_R = \left(\frac{\mu_{st} w_{st,i}}{\pi(\rho_o - \rho_{st}) g h_t^2 k_{st} \rho_{st}} \right)^{1/2}$$

where μ_{st} = steam viscosity at reservoir conditions [Pa-sec (kg/sec-m)]
$w_{st,i}$ = injected dry-steam mass flow rate at bottom hole (kg/sec)
ρ_o, ρ_{st} = density of oil, steam (kg/m^3)
g = acceleration by gravity (m/sec^2)
k_{st} = permeability to steam (m^2)

As an illustration, van Lookeren provides an example calculation for a case where the bottom-hole steam injection rate is 230 tonnes/d at 570 psia into a 98-ft (30-m) thick sand. The following additional data are given:

$\mu_{st} = 0.018$ cp (1.8×10^{-5} kg/sec-m)
$\rho_o = 840$ kg/m^3
$\rho_{st} = 20$ kg/m^3
$k_{st} = 250$ md (assumed to be one-quarter of the single-phase permeability), $= 0.25 \times 0.987 \times 10^{-12}$ m^2 $= 0.247 \times 10^{-12}$ m^2

Then

$$A_{R} = \left(\frac{1.8 \times 10^{-5} \times 230/(24 \times 3.6)}{3.14(840-20) \times 9.8 \times 30^{2} \times 0.247 \times 10^{-12} \times 20} \right)^{1/2} = 0.65$$

From Eq. (9.25), the vertical sweep efficiency would be

$$E_{sv} = h_{st}/h_{t} = 0.5 A_{R} = 0.33$$

Figure 9.2 gives a correlation of laboratory results giving vertical sweep efficiency of the steam zone, E_{sv}, as a function of the ratio of gravity to horizontal viscous forces for the case of steam injected into a water sand.

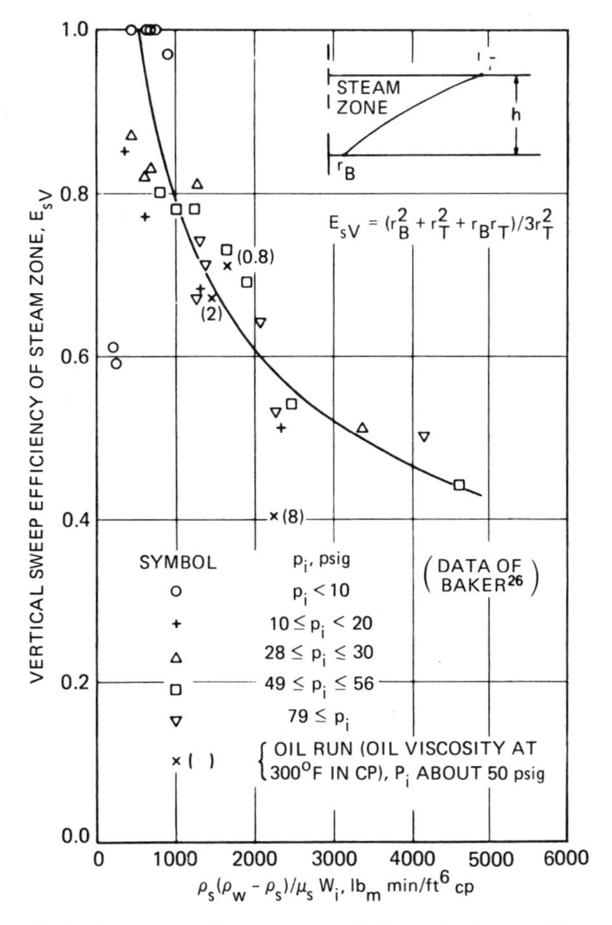

Figure 9.2. Vertical sweep efficiency correlation of steam drives (p_i = injection pressure).[16] © 1982 SPE-AIME.

For the case of more viscous oils, gravity override will be more severe as evidenced by the data for an 8-cp oil.

VOGEL MODEL: OVERRIDING STEAM ZONE

Vogel[18] assumes that the injected steam immediately overlays the formation and then conducts heat vertically upwards into the overburden and downward into the oil column. The cumulative heat stored in the formation, Q_T, then becomes

$$Q_T = Ah_s(\rho c)_s \, \Delta T + 2K_{h1} A \, \Delta T \sqrt{t/\pi\alpha_1} + 2K_{h2} A \, \Delta T \sqrt{t/\pi\alpha_2} \quad (9.26)$$

where
A = area overlain by steam zone (ft^2)
h_s = thickness of steam zone (ft)
$(\rho c)_s$ = volumetric heat capacity of steam zone (Btu/ft^3-°F)
ΔT = difference between steam and original reservoir temperature, $= T_s - T_r$ (°F)
K_{h1}, K_{h2} = thermal conductivity of overburden and oil sand, respectively (Btu/hr-ft-°F)
t = time of steam injection (hr)
α_1, α_2 = thermal diffusivity of overburden and oil sand, respectively (ft^2/hr)

The final steam zone thickness, h_s, could be estimated by the method of van Lookeren as a first approximation to make an overall estimate of the oil–steam ratio:

$$R_{os} = \frac{N_p}{m_s t} = \frac{Ah_s\phi(S_{oi} - S_{ors})}{m_s t} \quad (9.27)$$

If the injected heat equals that stored in the oil sand and overburden and heat removed by the produced fluids is negligible, the oil–steam ratio obtained by combining Eqs. (9.26) and (9.27) and including conversion factors is

$$R_{os} = \frac{62.4 h_s \phi \, \Delta S \, H_t}{\Delta T[h_s(\rho c)_s + 2(K_{h1}\sqrt{t/\pi\alpha_1} + K_{h2}\sqrt{t/\pi\alpha_2})]} \quad (9.28)$$

where

$$\Delta S = S_{oi} - S_{ors} \qquad H_t = \bar{X}_i H_{wv} + H_{ws} - H_{wr}$$

Using Eq. (9.26), the data for sand 2 discussed in connection with the

Marx–Langenheim method and further assuming:

$$h_s = 30 \text{ ft}$$

$$K_{h1} = K_{h2}$$

$$t = 4 \text{ yr} = 4(365)(24) = 35{,}040 \text{ hr}$$

$$\alpha_1 = \frac{1.4}{38} = 0.0368 \text{ ft}^2/\text{hr} \qquad \alpha_2 = \frac{1.4}{27} = 0.0518 \text{ ft}^2/\text{hr}$$

$$R_{os} = \frac{62.4(30)(0.25)(0.75)(847.0)}{367[30(27) + 2(1.4\sqrt{35040/3.14(0.0368)} + 1.4\sqrt{35040/3.14(0.0518)})]}$$

$$= 0.222 \text{ bbl/bbl}$$

This value for R_{os} compares to 0.25 bbl/bbl calculated by the Marx–Langenheim method for sand 2. Note that energy loss with the produced fluids would lower the oil–steam ratio still further.

MILLER MODEL

Miller et al.[19] have extended Vogel's model to permit a complete rate-versus-time prediction. This model is depicted in Figure 9.3. The Miller model assumes that the oil viscosity reduction is entirely due to conductive heating of the oil layer, which is overlain uniformly by the steam zone. The

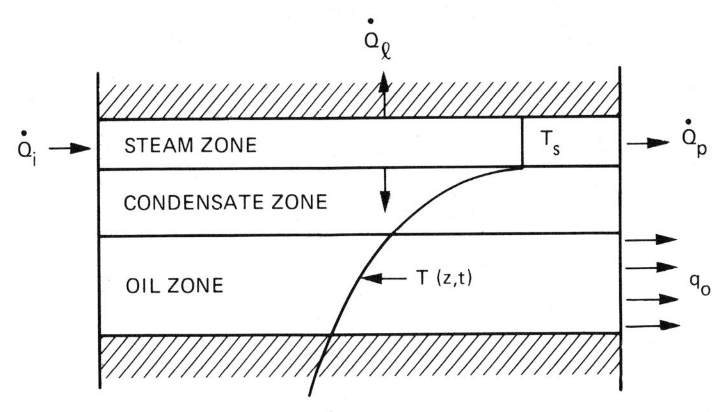

Figure 9.3. Conceptual schematic of steam override predictive model, Miller method.[19] © 1985 SPE-AIME.

following integration is carried out numerically:

$$\frac{q_o}{q_{oi}} = \frac{\mu_{oi}}{h} \int_{h_c}^{h_c+h_o} \frac{dz}{\mu_o} = \frac{\mu_{oi}}{h} \sum_{h_c}^{h_c+h_o} \frac{\Delta z}{\mu_o} \tag{9.29}$$

where q_{oi} = presteam oil production rate (STB/d)
q_o = stimulated oil production rate (STB/d)
μ_{oi} = initial oil viscosity (cp)
μ_o = heated oil viscosity at depth z (cp)
h = gross reservoir thickness (ft)
h_o = gross oil zone thickness (ft)
h_c = gross condensate zone thickness (ft)
z = depth below steam interface (ft)

The model does not account for pressurization of the reservoir or other effects considered to be unimportant. Vogel's assumption that heat is conducted from a stationary steam–oil zone interface at its initial position is considered to be adequate. The temperature can then be calculated by

$$T(z, t) = T_i + (T_s - T_i)\text{erfc}[z/\sqrt{4\alpha(t-\tau)}] \tag{9.30}$$

where $T(z, t)$ = temperature at depth z and time t (°F)
T_i = original reservoir temperature (°F)
T_s = steam zone temperature (°F)
α = thermal diffusivity (ft^2/d)
t = time since initiation of steam injection (days)
τ = lag time or time for steam zone to blanket oil zone (days)

The steps in the calculation are as follows:

1. The steam zone thickness h_s is set equal to zero or equal to the thickness of any thin gas cap that exists prior to steam injection.
2. After a time step Δt, the thickness of a condensate zone is calculated based on an estimate of f_{cp}, the fraction of the condensed steam that is produced from the reservoir:

$$h_c = \frac{(1 - f_{cp})2K_h(T_s - T_i)\sqrt{t-\tau}}{\rho_c H_{wv}(h_n/h)\phi(S_{oi} - S_{oc})\sqrt{\pi\alpha}} \tag{9.31}$$

where f_{cp} = fraction of condensed steam produced from reservoir
K_h = thermal conductivity (Btu/hr-ft-°F)
ρ_c = water density at T_s (lbm/ft^3)
H_{wv} = latent heat of vaporization (Btu/lb)

h_n = net oil sand thickness (ft)
ϕ = porosity (fraction)
S_{oi} = initial oil saturation (fraction)
S_{oc} = average oil saturation in condensate zone (fraction)

3. The oil zone thickness is calculated as

$$h_o = h - h_s - h_c \tag{9.32}$$

4. The oil rate is calculated by Eq. (9.29) in conjunction with Eq. (9.30) and a viscosity temperature equation or plot. The oil zone is divided into depth increments Δz of about 1 ft to perform the necessary integration.
5. Cumulative produced volumes of oil from the condensate zone, oil zone, and steam zone are calculated, and the new thickness h_s of the steam zone is calculated by

$$h_s = \frac{N_{ps}}{7758 A\phi(h_n/h)(S_{oi} - S_{os})} \tag{9.33}$$

where S_{os} = average oil saturation in steam zone
N_{ps} = cumulative oil production from steam zone (bbl)

Good agreement of this method with several field tests was reported. Predicted oil production rate and observed history are shown in Figures 9.4–9.7 for four California steamfloods. Agreement betwen observed and calculated history is good for the initial history, but the Miller model calculates conservatively for late times. An improvement on Miller's method would be to use an equation or numerical approach that accounts for the effect on the temperature distribution of the downward movement of the steam–oil zone interface with time. An equation derived by Carslaw and Jaeger[32] and presented by Closmann and Smith[33] permits calculation of the temperature distribution below a constant-temperature source (the steam zone) moving downward with a constant velocity v_s:

$$T(z, t) = T_i + \tfrac{1}{2}(T_s - T_i)\left[\operatorname{erfc}\frac{z}{2\sqrt{\alpha t}} + \left(\exp\frac{-v_s}{\alpha}(z - v_s t)\right)\operatorname{erfc}\frac{z - 2v_s t}{2\sqrt{\alpha t}}\right] \tag{9.34}$$

where $z - v_s t > 0$ and z is the distance below the initial position of the steam–oil interface. Replacement of Eq. (9.30) by Eq. (9.34) has been found to give improved calculation of late-time temperature distributions in the oil zone and should permit calculation of higher late-time oil production rates.

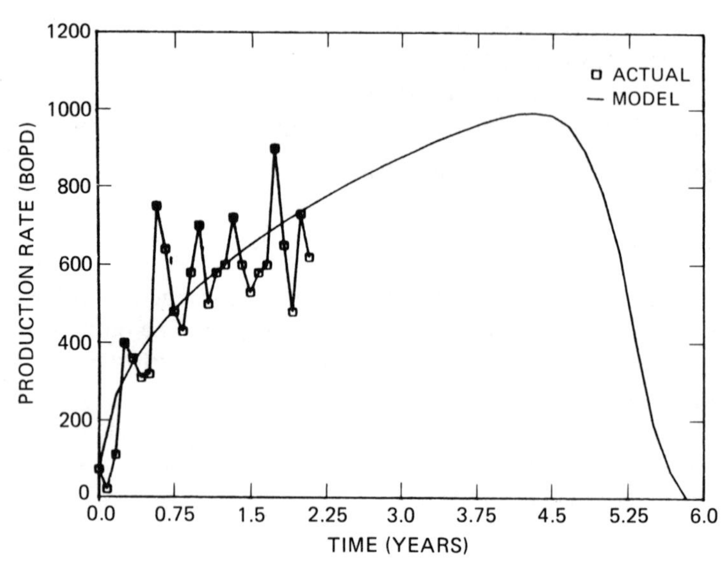

Figure 9.4. Match of Coalinga production data,[27] Miller model.[19] © 1985 SPE-AIME.

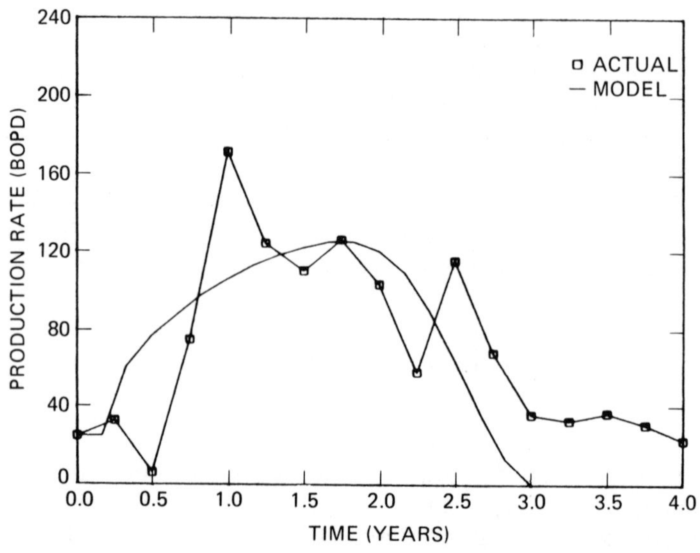

Figure 9.5. Match of Kern A production data,[28–30] Miller model.[19] © 1985 SPE-AIME.

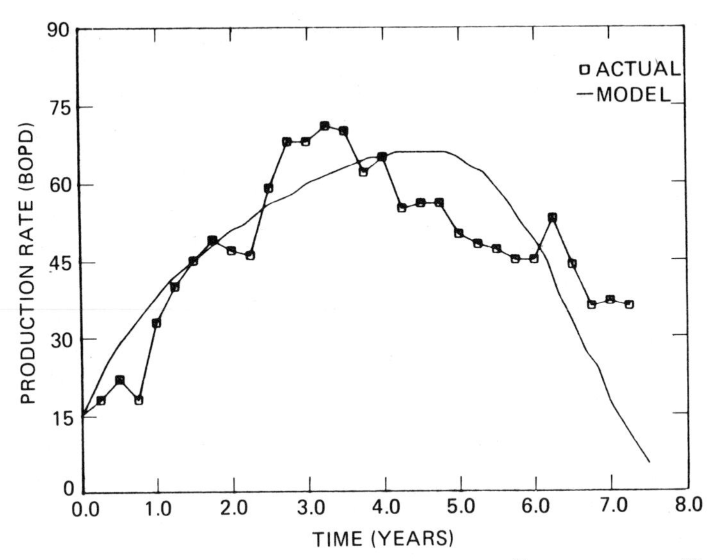

Figure 9.6. Match of Kern Canfield production data,[31] Miller model.[19] © 1985 SPE-AIME.

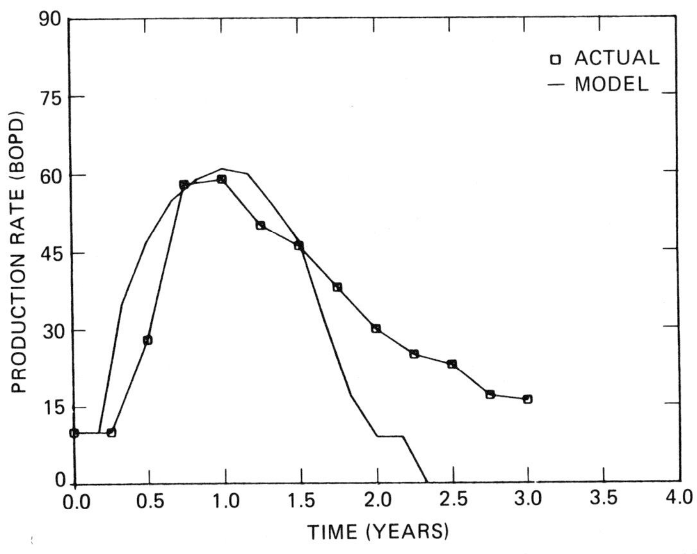

Figure 9.7. Match of Kern San Joaquin production data,[31] Miller model.[19] © 1985 SPE-AIME.

BUTLER MODEL: GRAVITY DRAINAGE TAR SAND CASE

Butler et al.[20] presented a simple model for a process of producing a highly viscous oil by gravity drainage from an expanding steam zone. In this tar sand process, the difficulty of communicating between vertical wells on commercial spacing in a tar sand is overcome by using a horizontal well located near the base of the sand as a production drain hole. Vertical wells that inject at several points 15–20 ft above the horizontal well are used to supply steam.

Initially, the steam may have to be injected above fracture pressure, but later, as heating in the vicinity of the horizontal well occurs, production of oil commences and injection pressures are lowered. As the steam rises, more oil sand is heated, and a conical-shaped steam zone forms (see Fig. 9.8). Eventually, a steady-state condition is reached where the flow rate of heated oil trickling down the walls of the steam zone can be approximated by a simple equation Butler has derived.

It is assumed that a steady-state condition is reached with the walls of the steam zone advancing at a constant velocity such that the temperature distribution by conduction of heat into the saturated tar sand is calculated by

$$\frac{T - T_r}{T_s - T_r} = \exp\left(-\frac{U\bar{x}}{\alpha}\right) \tag{9.35}$$

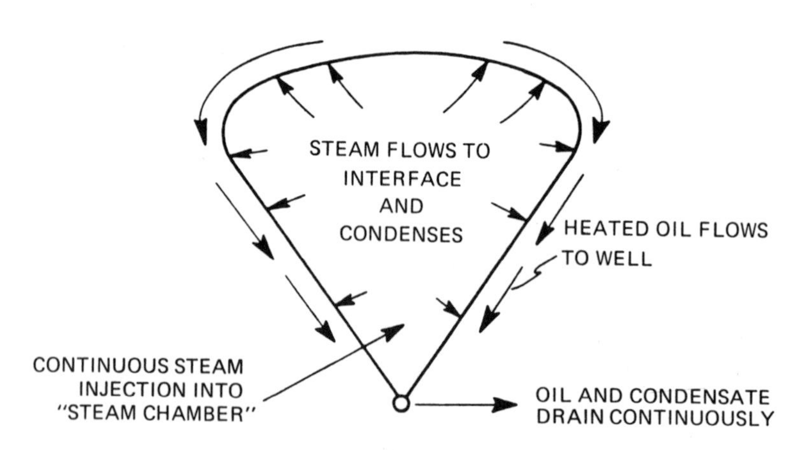

Figure 9.8. Butler model gravity drainage concept.[20] © 1981 *Canadian Journal of Chem. Engineering.*

where U = velocity of steam zone interface (m/d)
$\bar{x}$ = distance perpendicular to interface (m)
α = thermal diffusivity (m^2/d)
T_r = original reservoir temperature (°C)
T_s = steam zone temperature (°C)

An approximate equation relating the kinematic oil viscosity as a function of temperature is employed to obtain an analytical solution:

$$\frac{\upsilon}{\upsilon_s} = \left(\frac{T_s - T_r}{T - T_r}\right)^m \tag{9.36}$$

where υ = kinematic viscosity at temperature T (m^2/d)
υ_s = kinematic viscosity at temperature T_s (m^2/d)
m = viscosity exponent (usually in the range 3–4 for highly viscous oils) (dimensionless)

From Darcy's law, the oil drainage rate per unit length of steam zone interface is calculated as

$$\hat{q} = k_o g \sin\theta \int_0^\infty \frac{d\bar{x}}{\upsilon} \tag{9.37}$$

where $\hat{q}$ = oil drainage rate per unit horizontal length of interface (m^3/m-d)
g = gravitational acceleration (m/d^2)
θ = angle heated interface makes with horizontal (degrees)

Combination of Eqs. (9.35)–(9.37) and integrating results in

$$\hat{q} = \frac{k_o g \alpha \sin\theta}{m \upsilon_s U} \tag{9.38}$$

As the interface advances, oil over a differential height dy is mobilized to flow at the differential increases in rate $d\hat{q}$ in accordance with material balance:

$$d\hat{q} = U\phi(S_{oi} - S_{ors})\, dy/\sin\theta \tag{9.39}$$

where S_{oi} = initial oil saturation (fraction)
S_{ors} = residual oil saturation in steam zone (fraction)
ϕ = porosity (fraction)

Here U is related to the vertical distance of any point above the bottom of the formation, y, by

$$U = -\cos\theta\, dy/dt \tag{9.40}$$

Combination of Eqs. (9.38)–(9.40), rearranging, and integrating to obtain a total drainage flow rate $\hat{q}_{\mathrm{T}}$ over a total reservoir thickness H gives

$$\int_0^{\hat{q}_{\mathrm{T}}} \hat{q}\, d\hat{q} = \int_0^H \frac{\phi(S_{\mathrm{oi}} - S_{\mathrm{ors}}) k_{\mathrm{o}} g \alpha \, dy}{m v_{\mathrm{s}}}$$

which results in

$$\hat{q} = \sqrt{\frac{2\phi(S_{\mathrm{oi}} - S_{\mathrm{ors}}) k_{\mathrm{o}} g \alpha (H - y)}{m v_{\mathrm{s}}}} \tag{9.41}$$

where y is the vertical distance above the bottom of the formation in meters.

Eventually two interfaces that align with the well will develop. Therefore one should multiply the above rate by $2L$, where L is the length of the steam zone along the well to obtain the total horizontal well drainage rate.

Since heat losses to overburden and underburden are not considered, Eq. (9.41) is not valid as the top and bottom of the formation are reached, and it overpredicts rates of gravity drainage in these areas of the reservoir. The assumptions are most accurate at the midplane of the reservoir. Later work by Butler and associated workers[21,22] addresses these problems as well as the assumptions inherent in the steady-state temperature distribution to obtain a more exact solution for this complex fluid flow–heat transfer problem. Field test results for the horizontal well–gravity drainage approach have not yet been published.

EDMUNDS MODEL: TAR SAND CASE

In general, steamflooding is not possible in a highly saturated high-viscosity tar sand unless it has been preheated (with steam flowing through a vertical or horizontal fracture or through a thin-water or gas-saturated zone). Numerical methods and/or scaled physical models and field testing are usually required to estimate process performance in such a situation.

Once communication through an induced fracture or other flow channel has been achieved between injector and producer, bitumen recovery after breakthrough requires the steam to advance into the reservoir in a direction essentially normal to the flow. Edmunds[23] describes this as a "steam drag" process "where a thin layer of oil sand adjacent to the channel is heated, allowing the bitumen to be dragged away by the flowing steam." This author adopted from Butler et al.[20] a simple analytical model that neglects gravitational and convective effects but nevertheless gives a reasonable first estimate of rates of production from such a process.

The Edmunds model[23] predicts that the bitumen production rate should be proportional to the square root of the pressure gradient. The model oil rate is calculated as

$$q_o = h\sqrt{\frac{2\phi(S_{oi} - S_{or})k_a k_{ro}\,\Delta P\alpha}{\mu_{os} m}} \tag{9.42}$$

where h = height of steam front (vertical fracture case) (m)
k_a = absolute permeability (m^2)
k_{ro} = oil relative permeability (dimensionless)
m = bitumen viscosity exponent ($\mu_o = \mu_{os} T_d^m$)
$T_d = (T - T_r)/(T_s - T_r)$
μ_{os} = oil viscosity at steam temperature (Pa-sec)
ΔP = pressure drop (Pa)
α = thermal diffusivity of oil sand (m^2/sec)

The instantaneous oil–steam ratio is calculated by

$$\frac{q_o}{q_s} = \sqrt{\frac{2\phi(S_{oi} - S_{or})k_{ro} L\alpha h}{\mu_{os} m \lambda_s w q_s}} \tag{9.43}$$

where λ_s = effective two-phase steam mobility, $(\rho_g k_{rg})/(\bar{X}\rho_w \mu_g)$
$\bar{X}$ = average flowing steam quality
ρ_g, ρ_w = density of steam in gas and water phase, respectively
k_{rg}, μ_g = relative permeability (dimensionless) and viscosity (Pa-sec) respectively, of steam in gas phase
L = fracture length (m)
q_s = volume flow rate of steam (m^3/sec)
w = effective half-width of steam zone (m)

Equation (9.43) predicts that the oil–steam ratio will be inversely proportional to the square root of the oil viscosity at the steam temperature, which is in agreement with the physical model study of Doscher and Ghassemi.[34] Edmunds obtained reasonable agreement of the results calculated by Eqs. (9.42) and (9.43) with numerical model results for the steam drag process.† In general, oil–steam ratios for such a process are marginal, but optimization of steam rates can be undertaken to avoid excessive channeling of steam.

Possible steamflood process improvements cited by Edmunds would be to

1. employ heat recovery from the produced fluids so as to reduce net heat input;

† Inspection of Eq. (9.43) would apparently imply that as q_s approaches zero, the oil–steam ratio approaches infinity. This is obviously not correct. Low rates would invalidate the underlying assumptions of Edmund's model. Also, h would go to zero and the OSR would be indeterminate.

2. decrease steam mobility (increase ΔP with foams, etc.); and
3. increase bitumen mobility in the heated zone (with solvents, CO_2, etc., to lower μ_{os}).

GUIDELINES FOR APPLICABILITY OF STEAMFLOODING CALCULATIONS

One-Dimensional (Marx–Langenheim,[1] Myhill–Stegemeier,[5] or Shutler–Boberg[9]) Model

1. Use for thinner sands or wide spacings where injector–producer interwell distance–net sand thickness ratio is greater than 7 ($d/h > 7$). For cases with $d/h < 7$ or where the overriding steam zone is extreme, use the Vogel[18] or Miller[19] methods. The Shutler–Boberg[9] method should be used for cases where prior waterflooding or natural depletion has significantly lowered oil saturation prior to the start of the steamflood.
2. Do not use these methods for highly viscous oils requiring fracturing or other means of preheating prior to flooding. The Edmunds method[23] can be used for preliminary guidance, but numerical methods should be used for cases where complex heat and fluid flow distributions exist prior to or during steamflooding.

Two- or Three-Dimensional Models[36–39]

1. Generally required for thick, highly heterogeneous, or high-viscosity oil systems.
2. Effects of possible additional benefits such as oil distillation and thermal cracking should be accounted for.

The simplified methods can give reasonable results if caution is taken that they are not misapplied. For the cases where they do not apply, a more exact numerical method or a scaled physical model† should be employed. In general, however, owing to the effect of unknowns such as areal sweep and other critical reservoir parameters, it is always recommended that pilot testing in the field be conducted prior to initiating a commercial project. The calculations should serve to direct the engineer to the best candidates for piloting and guide him as to reasonable injection rates to employ.

† An excellent discussion of steamflooding and steam stimulation model scaling relationships is given by Niko.[35]

APPENDIX: DERIVATION OF EXPRESSION FOR MARX–LANGENHEIM THERMAL EFFICIENCY

The thermal efficiency is defined as

$$E_h = \frac{\text{energy remaining in oil sand}}{\text{total energy injected}} = \frac{h_t A(\rho c)_{R+F}(T_s - T_r)}{m_s t H_t} \tag{9A.1}$$

From Eq. (3.10),

$$A = \frac{(\xi_s/\theta) h_t H_t m_s (\rho c)_{R+F}}{4K_{hob}(T_s - T_r)(\rho c)_{ob}}$$

From Eq. (3A.4),

$$t = \frac{(\tau/\theta) h_t^2 (\rho c)_{R+F}^2}{4K_{hob}(\rho c)_{ob}}$$

Substituting these expressions for t and A in Eq. (9A.1) and canceling terms, we obtain

$$E_h = \frac{\xi_s/\theta}{\tau/\theta} = \frac{\xi_s}{\tau} \tag{9A.2}$$

REFERENCES

1. Marx, J. W., and Langenheim, R. N. "Reservoir Heating by Hot Fluid Injection," *Trans. AIME*, **216**, 312 (1959).
2. Ramey, H. J., Jr. "Discussion of Reservoir Heating by Hot Fluid Injection," *Trans. AIME*, **216**, 364–365 (1959).
3. Farouq Ali, S. M. "Effects of Differences in the Overburden and the Underburden on Steamflood Performance," *Producers Monthly* (December 1966), pp. 8–14.
4. Mandl, G., and Volek, C. W., "Heat and Mass Transport in Steam-Drive Processes," *J. Pet. Tech.* (March 1969), pp. 59–79.
5. Myhill, N. A., and Stegemeier, G. L., "Steam Drive Correlation and Prediction," *J. Pet. Tech.* (February 1978), pp. 173–182.
6. Jones, J., "Steam Drive Model for Hand-Held Programmable Calculators," *J. Pet. Tech.* (September 1981), pp. 1583–1598.
7. Yortsos, Y. C., and Gavalas, G. R., "Analytical Modeling of Oil Recovery by Steam Injection: Part 1. Upper Bounds," *Soc. Pet. Eng. J.* (April 1981), pp. 162–178.

8. Yortsos, Y. C., and Gavalas, G. R., "Analytical Modeling of Oil Recovery by Steam Injection: Part 2. Asymptotic and Approximate Solutions," *Soc. Pet. Eng. J.* (April 1981), pp. 179–190.
9. Shutler, N. D., and Boberg, T. C. "One-Dimensional, Analytical Technique for Predicting Oil Recovery by Hot Water or Steamflooding," *Soc. Pet. Eng. J.* (December 1972), pp. 489–498.
10. Neuman, C. H. "A Mathematical Model of Steam Drive Process—Applications," SPE 4757, presented at the Forty-Fifth Annual Regional Meeting of the Society of Petroleum Engineers, Ventura, California, April 2–4, 1975.
11. van Lookeren, J. "Calculation Methods for Linear and Radial Steam Flow in Oil Reservoirs," *Soc. Pet. Eng. J.* (June 1983), pp. 427–439.
12. Gomaa, E. E. "Correlations for Predicting Oil Recovery by Steamflood," *J. Pet. Tech.* (February 1980), pp. 325–332.
13. Aydelotte, S. R. and Pope, G. A. "A Simplified Predictive Model for Steam Drive Performance," *J. Pet. Tech.* (May 1983), pp. 991–1002.
14. Farouq Ali, S. M. "Steam Injection Theories—A Unified Approach," SPE 10746, presented at the Fifty-Second Annual Regional Meeting of the Society of Petroleum Engineers, San Francisco, California, March 24–26, 1982.
15. Effinger, A. W., and Wasson, J. A. "Applying Marx and Langenheim Calculations to the Prediction of Oil Recovery by Steamflooding in Venango Sands," USBM Information Circular 8432, U.S. Department of Interior, 1969.
16. Prats, M. *Thermal Recovery*, SPE Monograph Series, SPE of AIME, Dallas, TX, 1982.
17. Matthews, C. S. "Steamflooding," *J. Pet. Tech.* (March 1983), pp. 465–471.
18. Vogel, J. V. "Simplified Calculations for Steamfloods," *J. Pet. Tech.* (July 1984), pp. 1127–1136.
19. Miller, M. A., and Leung, W. K. "A Simple Gravity Override Model of Steamdrive," SPE 14241, Presented at the Sixtieth Annual Technical Conference and Exhibition of the Society of Petroleum Engineers of AIME, Las Vegas, Nevada, September 1985.
20. Butler, R. M., McNab, G. S., and Lo, H. Y. "Theoretical Studies on the Gravity Drainage of Heavy Oil During In Situ Steam Heating," *Can. J. Chem. Eng.* (August 1981), pp. 455–460.
21. Butler, R. M., and Stephens, D. J. "The Gravity Drainage of Steam-Heated Heavy Oil to Parallel Horizontal Wells," *J. Can. Pet. Tech.* (April–June 1981), pp. 90–96.
22. Butler, R. M. "A New Approach to the Modeling of Steam-Assisted Gravity Drainage, Paper No. 84-35-21, Presented at the Thirty-Fifth Annual Technical Meeting of the Petroleum Society of CIM, Calgary, June 10–13, 1984.
23. Edmunds, N. R. "An Analytical Model of the Steam Drag Effect in Oil Sands," Paper No. 83-34-48, presented at the Thirty-Fourth Annual Meeting of the Petroleum Society of CIM, Banff, May 1983.
24. Muskat, M. *Physical Principles of Oil Production*, McGraw-Hill, New York, 1949, pp. 236, 656.

25. Willman, B. T., Valleroy, V. V., Runberg, G. W., Cornelius, A. J., and Powers, L. W. "Laboratory Study of Oil Recovery by Steam Injection," *J. Pet. Tech.* (July 1961), pp. 681–690.
26. Farouq Ali, S. M. *Oil Recovery by Steam Injection*, Producer's Publishing, Bradford, PA, 1970.
27. Taschman, B. M. "Zone 1 Steam Project, Coalinga Field," *J. Pet. Tech.* (March 1982), pp. 511–516.
28. Bursell, C. G., and Pittman, G. M. "Performance of Steam Displacement in the Kern River Field," *J. Pet. Tech.* (August 1975), pp. 997–1004.
29. Chu, C., and Trimble, A. E. "Numerical Simulation of Steam Displacement—Field Performance Application," *J. Pet. Tech.* (June 1975) pp. 765–776.
30. Bursell, C. G. "Steam Displacement—Kern River Field," *J. Pet. Tech.* (October 1970), pp. 1225–1231.
31. Greaser, G. R., and Shore, R. A. "Steamflood Performance in the Kern River Field," SPE 8834, Presented at the First Joint SPE/DOE Symposium on Enhanced Oil Recovery, Tulsa, Oklahoma, April 1980.
32. Carslaw, H. S., and Jaeger, J. C. *Conduction of Heat in Solids*, 2nd ed., Oxford University Press, New York, 1959, p. 388.
33. Closmann, P. J., and Smith, R. A. "Temperature Observations and Steam-Zone Rise in the Vicinity of a Steam-Heated Fracture," *Soc. Pet. Eng. J.* (August 1983), pp. 575–586.
34. Doscher, T. M., and Ghassemi, F. "The Effect of Reservoir Thickness and Low Viscosity Fluids on the Steam Drag Process," SPE 9897, California Regional Meeting of the Society of Petroleum Engineers, Bakersfield, CA, 1981.
35. Niko, H., and Troost, P. J. P. M. "Experimental Investigation of the Steam-Soak Process in a Depletion-Type Reservoir," *J. Pet. Tech.* (August 1971), pp. 1006–1014.
36. Shutler, N. D. "Numerical Three-Phase Model of the Two-Dimensional Steamflood Process," *Soc. Pet. Eng. J.* (December 1970), pp. 405–417.
37. Shutler, N. D. "Numerical Three-Phase Model of the Linear Steamflood Process," *Soc. Pet. Eng. J.* (June 1969), pp. 232–246.
38. Coats, K. H. "Simulation of Steamflooding with Distillation and Solution Gas," *Soc. Pet. Eng. J.* (October 1976), pp. 235–247.
39. Coats, K. H., George, W. D., Chu, C., and Marcum. B. E. "Three-Dimensional Simulation of Steamflooding," *Soc. Pet. Eng. J.* (December 1974), pp. 573–592.

Chapter 10

STEAMFLOODING CASE HISTORIES

ABSTRACT

Results from a number of actual steamflood projects are presented in this chapter. Some general observations are taken from five early successful pilots at the outset. Some of these pilots were later expanded into successful commercial projects. The superior documentation of the five projects led to their selection for this review. The best performing pilots had high oil content, shallow depth, thick sands of low shale content, and good fluid transmissibility at original reservoir conditions. Operating conditions, pattern type, and size are also discussed. It is shown that oil–steam ratios calculated by the Marx–Langenheim method described in Chapter 9 generally agree well with those observed for these projects. A number of more recent projects are compared to similar results obtained using the Myhill–Stegemeier method, which usually predicts somewhat optimistically. Details are given for eight specific steamflood projects, most of which were still in operation as of 1986.

INTRODUCTION

Although a fairly large number of steam drive projects have been conducted, oil recovery to date from profitable commercial steamfloods is still probably less than the commercial oil recovery from steam stimulation projects. Undoubtedly, this will change with time. As of 1970, the U.S. production derived from steam drive projects was about 30,000 bbl/d, mostly in California. Only 5 years earlier, this production was probably less than 10,000 bbl/d, derived mostly from experimental pilots. In 1982, worldwide production from steam drive projects was estimated at about 270,000 bbl/d, mostly in the United States.[1] By 1986, steam processes (cyclic and drive) were estimated to be producing in excess of 650 kbbl/d worldwide.[2]

GENERAL OBSERVATIONS FROM EARLY PROJECTS

Reservoir Characteristics of Early Successful Projects

Table 10.1 lists the reservoir data for five early commercial steam drive projects, and Table 10.2 summarizes the producing characteristics of these projects.[3–7] Pattern type, size, and oil recovery information are given in Table 10.3. Four of the projects have reportedly been economic successes and have been expanded. The Tia Juana, Venezuela project, however, was not economic at the low crude price prevailing in Venezuela during the 1960s, despite the fact that it had higher per-well oil-producing rates and higher oil–steam ratios than the other projects.

Inspection of Table 10.1 shows that these reservoirs have the following favorable characteristics for a steam drive operation:

1. *High Oil Content.* All projects listed have a high oil content, exceeding 1400 bbl/acre-ft.
2. *Shallow Depth.* The formation depths range from 500 to 2800 ft. Injection pressures and temperatures are low, as are costs of new wells needed to complete flooding patterns.
3. *Thick Sands, Low Shale Content.* All of the projects have thick sands and, with the exception of the Winkleman Dome project, low shale contents. The sands range in thickness from 45 to 100 ft. The net sand figure for the Tia Juana project refers to two sands that were flooded simultaneously from dual-injection wells. The sands were separated by a 90-ft shale barrier that was not included in the gross sand interval number quoted in Table 10.1.
4. *Moderate Oil Viscosities under Cold Conditions.* The original oil viscosities for the projects range from 180 to 4000 cp. These viscosity levels are low enough that preheating of the oil sand is not required to obtain acceptable oil mobility as would be needed in the case of a tar sand.
5. *High Sand Permeabilities.* All the projects have a most important characteristic—high sand permeability. This is needed to permit reasonable ease of oil displacement of the high-viscosity crude. Steamfloods of heavy oils in tight sands have rarely been successful because of poor injectivity and productivity.

Another reservoir characteristic that has a beneficial influence is the presence of a thin bottom-water zone. When warming occurs from the bottom of the oil column, the sand heats quickly as steam rises through it under the influence of gravity.

In all the projects, steam tends to override and the steam zone acts as a

Table 10.1 Reservoir Characteristics of Early Successful Steam Drive Projects

Reference	Field	Oil Content (bbl/acre-ft)	Depth (ft)	Sand Thickness[a] (ft)		Oil Viscosity (cp)	Permeability (md)
				Gross	Net		
3	Kern River, CA[b]	1485	700	120	100	4000	4000
4	Schoonebeek, Holland	1920	2800	80	80	180	5500
5	Winkleman Dome, WY	1420	1220	185	85	900	1000
6	Tia Juana, Venezuela	1920	1450–1700	250	200	100–100,000	2000
7	Slocum, TX[c]	1470	500	60	45	2000	3500

[a] In flooded areas.
[b] Kern Project.
[c] Phase 1.

Table 10.2 Producing Characteristics of Early Successful Steam Drive Projects

Field	Number of Wells		Operating Life (yr)	Oil Production Rate (bbl/d)		Steam Injection Rate (bbl/d)	Oil–Steam Ratio (bbl/bbl)
	Injectors	Producers		Per Production Well	Project		
Kern River, CA	47	83	5.5[a]	53[b]	4306[b]	18,500[b]	0.23[b]
Schoonebeek, Holland	4	9	6.3[c]	192	1720	4550	0.38
Winkleman Dome, WY	6	12	3.0[d]	59	530	1380	0.21[e]
Tia Juana, Venezuela	7	24	5.3[f]	222	5320	8650	0.62
Slocum, TX	7	53	3.0[g]	28	1500	7200	0.21

[a] To January 1970.
[b] First half 1969 average.
[c] To January 1967.
[d] To January 1968.
[e] Estimated incremental over primary.
[f] To January 1967.
[g] To January 1970 (Phase 1 only).

Table 10.3 Pattern and Oil Recovery Data for Early Sucessful Steam Drive Projects[a]

Field	Type[b]	Area/Pattern Acres	Cumulative Oil Recovery			Date Reported
			kbbl/ Pattern[c]	Percentage of OOIP[d]	Total Percentage of OOIP[e]	
Kern River, CA	5-spot	2.5	81[d]	NA	NA	1/70
Schoonebeek, Holland	5-spot	16.4	1003	33	38	1/67
Winkleman Dome, WY	5-spot	10.0	65	7	7	1/68
Tia Juana, Venezuela	7-spot	19.6	1470	21	33	1/66
Slocum, TX	13-spot	5.65	171	35	36	1/70

[a]Abbreviations: OOIP, original oil in place; NA, not available.
[b]Inverted.
[c]Attributable to steam drive.
[d]Average of nine combined patterns.
[e]Oil recovery obtained by all methods.

hot, expanding gas cap. Hence, high permeability in the top of the sand is usually detrimental, as it may promote steam channeling. A high-permeability streak exists at the top at Schoonebeek, where steam channels to up-dip wells. Most of the oil production comes from the down-dip wells in that project.

Operating Characteristics of Early Projects

In the interest of controlling steam channeling through the numerous sands of the Kern River formation, the operator[3] has adopted the following procedures:

1. Set casing through the oil zone.
2. Selectively jet-perforate 50–60 ft of interval near the bottom of the zone. Inside casing slotted liners have been used to control sand where this has been a problem.

By using this technique, when the lower sands become depleted, the perforations can be squeezed and the upper sands selectively perforated later. The upper interval can then be efficiently flooded since it will have been warmed to some extent from the flood of the lower zone.

Table 10.2 gives general operating characteristics for the five projects in Table 10.1. With the exception of the Kern River project, the rates given are averages over the project life. Average rates for the last six months of 1969 are given for the Kern project.

Oil-producing rates per producing well for the various projects have ranged from 28 STB/d for the thinner Slocum sand to 222 STB/d for the massive dual-sand reservoir at Tia Juana. Owing to the higher sand permeability and lower oil viscosity than for the other projects, the Schoonebeek Valanginian sand yielded the highest oil production rates per foot of sand of any of these early steam drive projects.

Pattern Type and Size

Table 10.3 compares the pattern types and sizes for the five projects. The most popular pattern has been the 5-spot. However, the Tia Juana project employed an inverted 7-spot and the Slocum project an inverted 13-spot. The advantages of these latter patterns are that they have more producing wells per injection well, and pattern oil productivity is improved. Much less pressure drop occurs in the vicinity of the steam injection wells than near the producers. In addition to increasing the number of producing wells per injector, additional productivity improvement can be obtained by steam stimulating the producers.

The size of the patterns used in the commercial projects varies from 2.5 acres in the Kern project to 19.6 acres in the Tia Juana project, where the massive Lower Lagunillas sands are being flooded. The smaller patterns are generally used in the shallower, thinner sands to permit high flooding rates and to reduce heat losses.

COMPARISON OF FIELD RESULTS TO SIMPLIFIED SCREENING ESTIMATES

Marx–Langenheim Method

The oil–steam ratios from the early steam drive projects discussed in Tables 10.1–10.3 range from 0.21 to 0.62 bbl/bbl. Table 10.4 shows a comparison of the oil–steam ratios of these projects with values calculated by the Marx–Langenheim method presented in Chapter 9 [Eq. (9.9)]. Also shown are the data used for the calculations.

A reduced steam zone thickness occurring because of gravity override is not accounted for by this equation. Nevertheless, the agreement with field values is generally good for all of the cases except Winkleman Dome. Only a small fraction of the top sand had been flooded in this project, and it was still in an early stage at the time of these data, so the assumptions employed in Eq. (9.9) really do not apply in this case.

Myhill–Stegemeier Method

Table 10.5 gives more recent data for four of the above projects plus data for some other recent field projects. Oil–steam ratios calculated by the method of Myhill and Stegemeier presented in Chapter 9 [Eq. (9.24)] are compared to results from these projects in Table 10.6.

Except for Brea, Coalinga, and Yorba Linda, which are closely predicted by this method, the Myhill–Stegemeier method generally predicts optimistically. The Marx–Langenheim method predicts more conservatively because it calculates a smaller steam zone volume than does the Myhill–Stegemeier method.

SELECTED PROJECT DETAILS

Kern River, California

The largest steam drive operation in the United States is being conducted in the Kern River field near Bakersfield, California.[3,12,16–19] The oil production rate from these floods was reported at 107 kSTB/d in March of 1983.[20]

Table 10.4 Comparison of Marx–Langenheim Calculated and Actual Oil–Steam Ratios for Early Field Projects[a]

Field	ϕ	h_n (ft)	h_t (ft)	T_I (°F)	T_r (°F)	S_{o_i}	m_s ($\times 10^3$ lbs/hr)	D (ft)	t (yr)	R_{os}(Calculated) (bbl/bbl)	R_{os}(Actual) (bbl/bbl)
Kern River	0.384	100	120	450	90	0.50	5.7	700	4	0.27	0.23
Schoonebeek	0.3	80	80	486	100	0.85	15.1	2800	6	0.36	0.38
Winkleman Dome	0.248	85	185	551	85	0.75	3.5	1220	3.5	0.08	0.21
Tia Juana[b]	0.33	100	120	432	113	0.75	10.8	1410	5.3	0.56	0.62
Slocum	0.292	45	60	467	75	0.65	15.0	500	3	0.20	0.21

[a]For all cases;

$S_{ors} = 0.15$

$K_{h_{ob}} = 1.4$ Btu/hr-ft-°F (except Schoonebeek)

$K_{h_{ob}} = 1.27$ Btu/hr-ft-°F (for Schoonebeek)

$(\rho c)_{R+F} = 27$ Btu/ft^3-°F

$(\rho c)_{ob} = 31.4$ Btu/ft^3-°F

[b]Upper zone calculation only.

Table 10.5 More Recent Steam Drive Field Projects, Summary of Conditions Used to Apply Myhill–Stegemeier Method [a]

Field	Reference	Number of Injectors	Initial Temperature T_i (°F)	Petrophysical Properties[b] h_t (ft)	h/h_t	ϕ	ΔS_o[c]	Steam Parameters X_i	p_s (psig)	m_s (bbl/d)	A (acres/well)	t (yr)
Brea ("B" sand)	8	4	175	300	0.63	0.22	0.40	0.54	2000	500	10	8
Coalinga (Section 27, Zone 1)	9	40	96	35	1.0	0.31	0.37	0.55	400	500	9.2	4
El Dorado (northwest pattern)	10	4	70	20	0.85	0.26	0.20	0.45	500	200	1.6	1
Inglewood	11	1	100	43	1.0	0.37	0.40	0.7	400	1100	2.6	1
Kern River	12	85	90	55	1.0	0.32	0.40	0.5	100	360	2.5	5
Schoonebeek	4	4	100	83	1.0	0.30	0.70	0.7	600	1250	15	6
Slocum (Phase 1)	13	7	75	40	1.0	0.37	0.34	0.7	200	1000	5.65	2.5
Smackover	14	1	110	50	0.5	0.36	0.55	0.8	390	2500	10	1
Tatums (Hefner steam drive)	15	4	70	66	0.56	0.28	0.55	0.6	1300	685	10	5
Tia Juana	6	7	113	200	1.0	0.33	0.50	0.8	300	1400	12	5.3
Yorba Linda ("F" sand)	1	2	110	32	1.0	0.30	0.31	0.7	200	850	35	4.5

[a] From Ref. 1.
[b] $(\rho c)_{R+F} = 35$ Btu/ft^3-°F, $(\rho c)_{ob} = 42$ Btu/ft^3-°F, $K_h = 1.2$ Btu/ft-hr-°F.
[c] ΔS_o = oil saturation at start of steaming minus change in oil saturation from estimated primary during steam drive period minus 0.15. The term 0.15 represents an average value of the residual oil saturation after steam drive.

Table 10.6 Comparison of Field Results to Method of Myhill and Stegemeier[a,b]

Field	Dimensionless Quantity of Steam Injected W_{sD}	Dimensionless Steam Zone Size V_{sD}	Calculated Incremental Oil/Steam Ratio (bbl/bbl)	Field Incremental Oil/Steam Ratio (bbl/bbl)
Brea	0.5	0.15	0.13	0.14
Coalinga	0.94	0.45	0.16	0.18
El Dorado	1.6	0.315	0.05	0.02
Inglewood	1.26	1.256	0.41	0.28
Kern River	1.92	1.139	0.32	0.26
Schoonebeek	0.95	0.617	0.43	0.35
Slocum	1.41	1.202	0.29	0.18
Smackover	1.23	0.756	0.27	0.21
Tatums	1.54	0.397	0.13	0.10
Tia Juana	0.47	0.551	0.59	0.37
Yorba Linda "F"	0.54	0.280	0.16	0.17

[a]From Ref. 1.
[b]$W_{sD} = W_{s,eq}/(7758\ Ah_t)$; $V_{sD} = V_s/(7758\ Ah_t\phi)$. V_s is given in Eq. (9.22). $W_{s,eq}$ is the equivalent volume of injected steam based on a reference enthalpy of 10^3 Btu/lb, where

$$W_{s,eq} = (2.853 \times 10^{-6}\ \text{bbl/Btu})\,\frac{c_w(T_{sb} - T_A) + X_{sb}H_{wvsb}}{c_w(T_{idh} - T_i) + X_{sdh}H_{wvdh}}\,Q_i$$

and Q_i = injected heat (Btu)
c_w = water specific heat (Btu/lb-°F)
T = temperature (°F)
X = steam quality (lb/lb)
H_{wv} = steam enthalpy of vaporization (Btu/lb)
Subscripts: sb = at steam boiler; A = ambient; i = initial in formation; dh = injection downhole.

The Kern River field is a large, shallow California field that has seen intense thermal recovery activity during the last 25 years. A map and cross section of the early Kern River projects of Texaco (formerly Getty Oil Company) is shown in Figures 10.1 and 10.2. In general, the sands of all the projects are relatively clean and do not contain an abundance of swelling clays. The unconsolidated sands have permeabilities of 1–5 darcies and porosities of 28–30%. Oil viscosity averages 4000 cp at a reservoir temperature of 85°F.

This flood has been expanded several times since its inception. The steamflood in the original pilot was preceded by a period of hot-water injection during 1961–1964. The early production history of the Kern

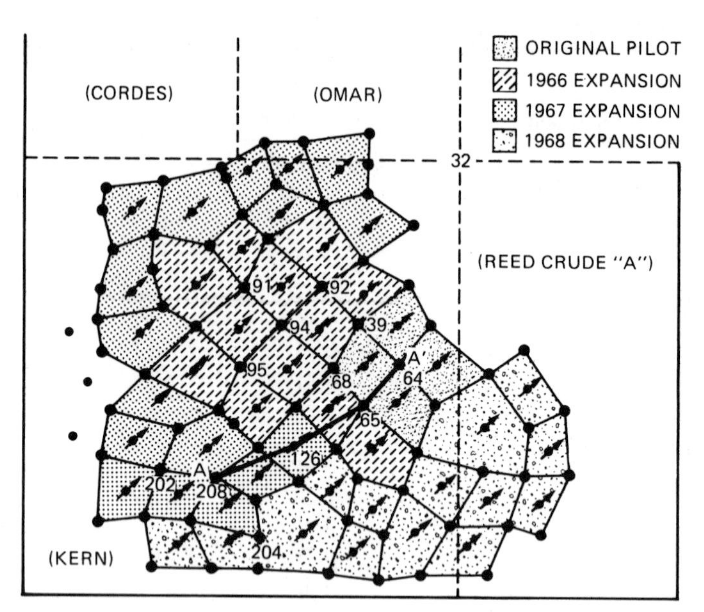

Figure 10.1. Locations of Kern steam displacement pattern expansion.[3] © 1970 SPE-AIME.

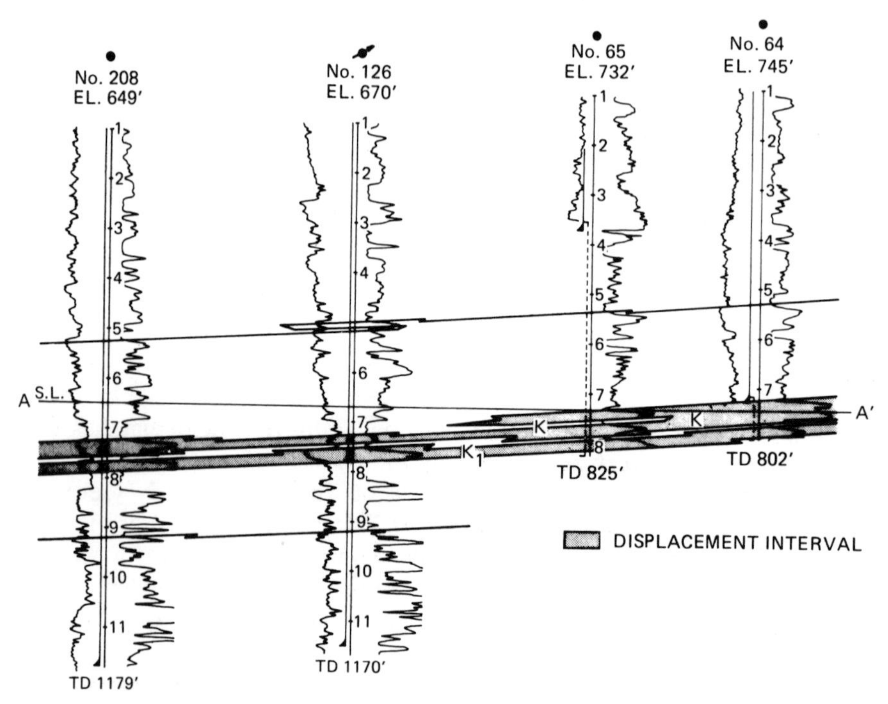

Figure 10.2. Structure section through Kern steam displacement project.[3] © 1970 SPE-AIME.

project is shown in Figure 10.3, which gives the total project steam injection rate, total fluid, and oil production rates.

One project was started in 1971 as an expansion of a small and successful 1969 steamflood pilot. It consisted of 114 inverted 5-spot patterns (85 producers) on 310 acres. The wells were completed in the K1 sand of the reservoir. Table 10.7 gives pertinent reservoir parameters.

Steam was injected at an average quality of 70% into the bottom one-third of the K1 sand. The lower portion of this sand contained a slightly higher water saturation than did the upper portion. Initially most of the injected steam entered the lower part of the sand.

Steam was initially injected in the 114 injection wells at a rate of 280 bbl/d per well. Since July 1974, the steam injection rate was decreased continuously, reaching a value of approximately one-half the original rate in September 1977. After this date, a number of injection wells were shut in and reopened (74 wells) and/or recompleted to a lower zone (26 wells). During this period of operation, the steam injection rate remained approximately 150 bbl/d per well.

After about 0.75 PV of steam had been injected, steam had broken through at most of the producers. Casing failures resulted in 13 of the 85 producers. A reduction in steam injection rates was then made, and several

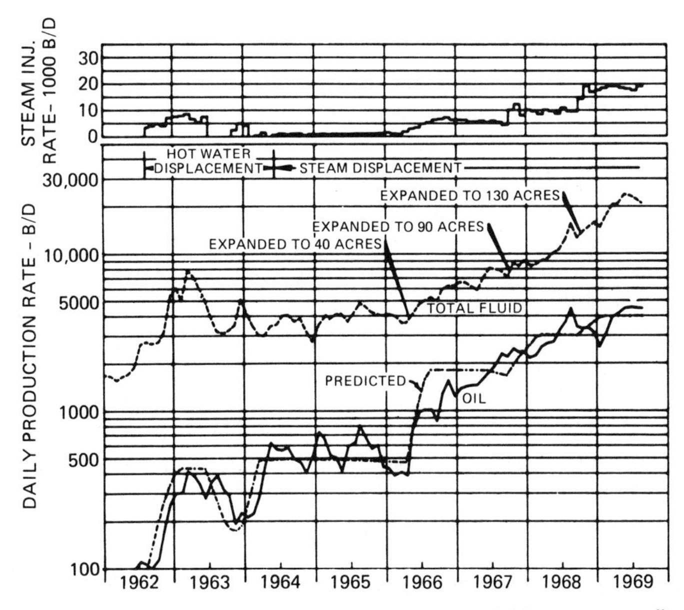

Figure 10.3. Early production and injection history of Kern steam displacement project.[3] © 1970 SPE-AIME.

Table 10.7 Reservoir Data, Green and Whittier K1 Steamflood at Kern River

Project area	310 acres
Depth, subsurface	900–1100 ft
Gross thickness	80 ft
Net thickness	60 ft
Porosity	31%
Permeability	2.4 darcies
Dip	3–6°
Reservoir temperature	90°F
Reservoir pressure	80 psia
Oil gravity	13.5°API
Oil viscosity, reservoir conditions	4500 cp
Oil in place	
At flood start	21.0 MSTB (47% PV)
October 1, 1979	4.7 MSTB (11% PV)
Gas saturation at flood start	3% PV

injectors were recompleted for injection into a deeper zone, not in communication with the K1 sand. After about 1 PV of steam had been injected, 26 of the injection wells were recompleted in a lower sand. The other 88 of the original 114 injection wells in the project were still injecting steam into the original zone as of early 1978.

After injection of about 1.6 PV of steam toward the end of September 1979, some 78% of the oil in place at flood start had been recovered from the project area at a cumulative oil–steam ratio of 0.237 bbl/bbl. An improvement in oil recovery resulted from oil production from adjacent oil sands opened after preheating from the initial steamflood.

Other projects such as the Canfield R-1 sand displacement recovered 57% of the preflood oil in place at an ultimate oil–steam ratio of 0.195 bbl/bbl. Conduction heating from the R-1 sand had raised the shallower R sand temperature from 85 to 165°F by February 1978. Recompletion to this sand was carried out, and project life was estimated to be 4.5 years shorter because of preheating. Numerical simulation results showed that recovery should not be significantly increased, but the OSR would be increased from 0.186 bbl/bbl without preheating to 0.230 bbl/bbl with preheating.

Getty has indicated that steam override problems greatly reduced oil recovery. To alleviate this problem, Getty tested two different techniques to improve vertical sweep efficiencies. In one test, 45 producing wells were recompleted so that the production intervals were open only in the lower half of the sand. The second method consisted of injecting a foam diverter (Corco 180) with the steam into nine injection wells. Corco 180 was injected in 55-gal treatments at intervals of 10 days over a 13-month period.

Both tests were judged successful because a marked increase in the oil production rate was observed. Where foam diverters were used, radioactive tracers indicated more uniform steam injection profiles. However, the steam profiles tended to revert to a gravity override condition during the 10-day intervals between injection of the foam diverter. Work to enhance the stability of the foam is continuing.

Schoonebeek, The Netherlands

The total project steam injection rate and gross fluid and oil production for the Shell/Esso Schoonebeek steam drive pilot are shown in Figure 10.4.[4] Reservoir and project data are given in Table 10.8.

The Schoonebeek pilot project pattern is shown in Figure 10.5. Basically, the pattern is an inverted 5-spot except as modified to an inverted 4-spot to accommodate the reservoir geometry. An east-to-west fault running through the project prevents effective communication of the southern quarter of the

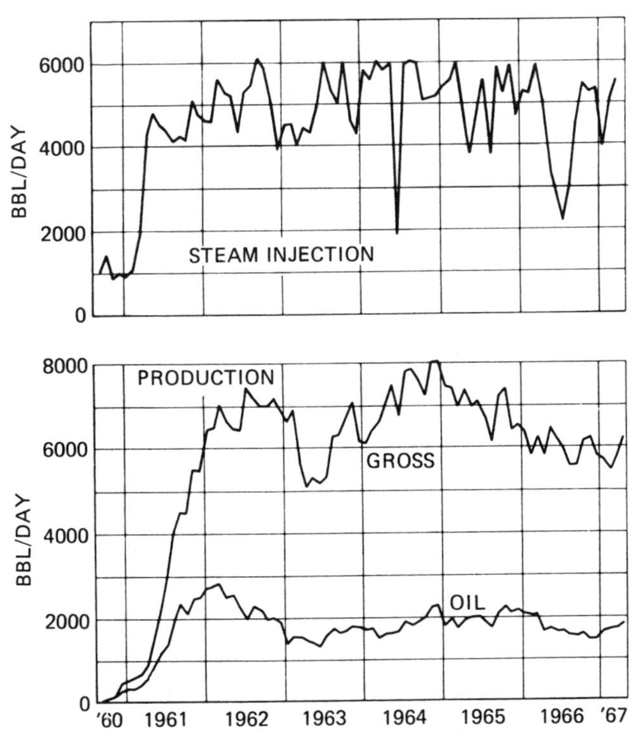

Figure 10.4. Schoonebeek field injection and production rates, steam drive project.[4] © 1968 SPE-AIME.

Table 10.8 Reservoir Data, Schoonebeek Field Pilot Steamflood

Project area	65.5 acres
Depth, subsurface	2500 ft
Gross thickness	80 ft
Net thickness	80 ft
Porosity	30%
Permeability	1000–10,000 md
Dip	6.5°
Reservoir temperature	100°F
Reservoir pressure	120 psig
Oil gravity	25°API
Oil viscosity, reservoir conditions	180 cp
Oil in place	
At flood start	12.1 MSTB (85% PV)
January 1, 1967	8.13 MSTB (57% PV)
Gas saturation at flood start	4% PV

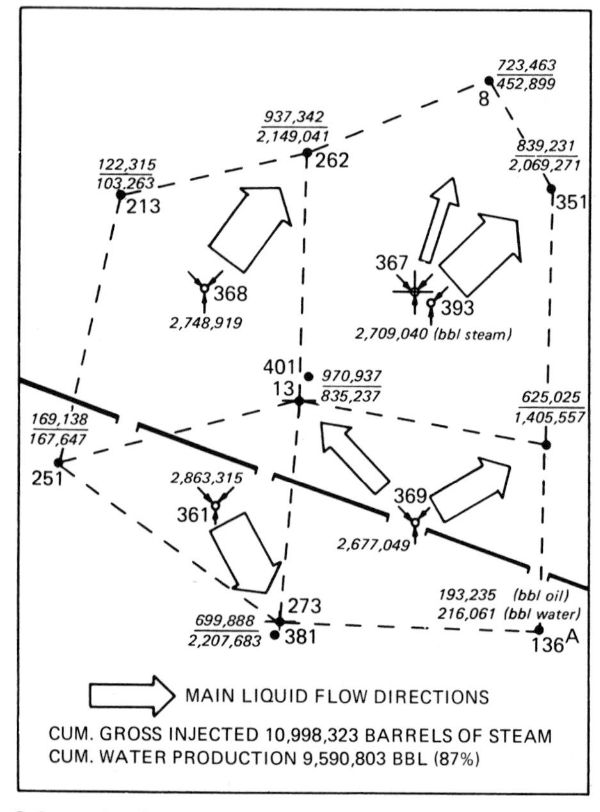

Figure 10.5. Schoonebeek steam drive project cumulative injection and production, April 1, 1967.[4] © 1968 SPE-AIME.

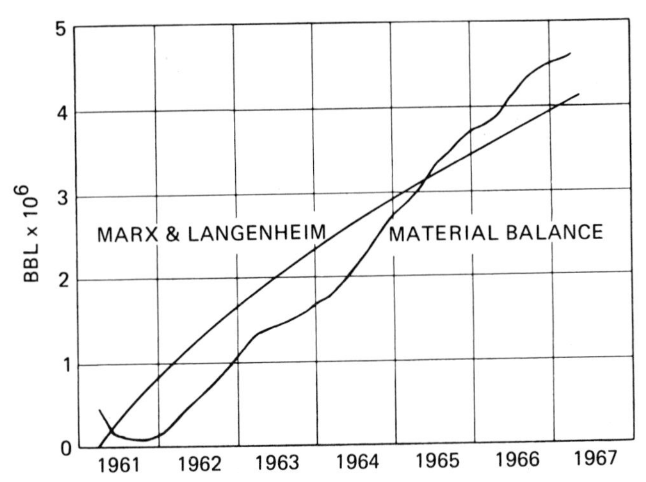

Figure 10.6. Schoonebeek steam drive project pore volume steam zone according to material balance and Marx and Langenheim.[4] © 1968 SPE-AIME.

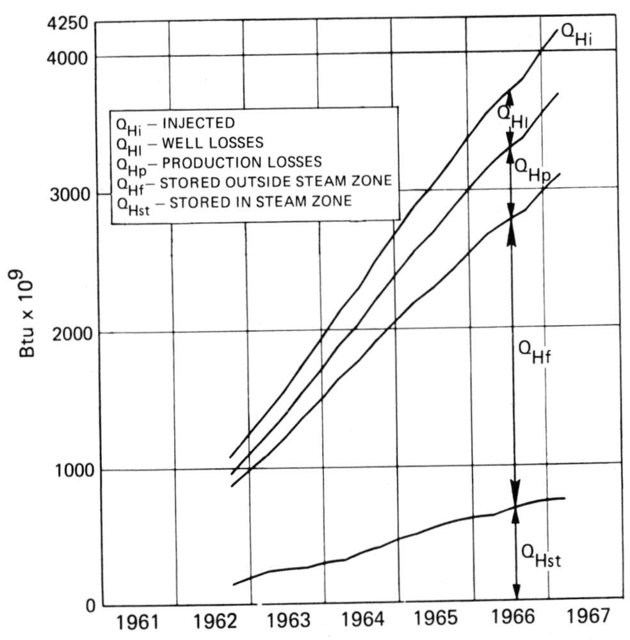

Figure 10.7. Schoonebeek steam drive project, heat distribution.[4] © 1968 SPE-AIME.

project with the northern part. The up-dip part of the sand lies to the southwest, and as the arrows in Figure 10.5 indicate, liquid flow generally has been greater in the down-dip direction to the northeast.

An analysis of cores taken in replacement injection well S-393 showed that steam had entered the upper 36 ft of the formation and had reduced the oil saturation to an average value of 8%. Immediately below this zone, the oil saturation abruptly increases to 35% in a zone displaced by hot-steam condensate. Assuming the steam zone to have an average thickness of 26 ft, the Marx and Langenheim equation (Chapter 3) was employed to obtain an estimate of the size of the steam zone as a function of time. This is compared in Figure 10.6 to an estimate of its volume by material balance calculation.

An analysis of the distribution of the heat injected in the Schoonebeek project is shown in Figure 10.7. Note the analysis shows that the amount of heat stored within the steam zone is much less than that stored outside of it. Further, a significant fraction of the injected heat has been produced.

As of early 1967, oil recovery from the Schoonebeek pilot project had reached 38% of the original oil in place (OOIP), with 33% OOIP produced as a result of the steam drive. At this time, the cumulative OSR was 0.37 bbl/bbl. Ultimate recovery was anticipated to be in excess of 50%.

Mt. Poso, California

Following two small steamflood pilots, Shell initiated Phase I of the Mt. Poso steamflood in November 1971.[21] The flood has been conducted in four phases (field areas) to make the most efficient use of the steam-generating equipment. By 1977, the entire field was being flooded.

The reservoir is an easterly rising homocline with a dip of 6°. The reservoir is closed to the east and north by a normal fault. There are separate producing sands at Mt. Poso. The shallowest and most productive is the only sand that has been flooded to date. The sand has an extremely high (reported 20 darcies) permeability.

Scaled physical model tests indicated that the best strategy would be to arrange injection wells in two rows. One row was placed up dip, along the closure at the Mt. Poso fault. The other row of injectors was placed down dip, along a line just to the east of the original water–oil contact. In the 2100-acre total project area, there are 159 producing wells and 41 injectors. Operating strategy has been to inject approximately 0.2 pore volumes of steam in the down-dip injectors and another 0.7 pore volumes up dip. The purpose of the down-dip injection was to prevent the formation of a cold oil bank low on the structure.

Such an oil bank would be relatively immobile because of the high oil viscosity (280 cp). Several wells located in the central area of the field

Table 10.9 Reservoir Data, Mt. Poso Steamflood

Project area	2100 acres
Depth, subsurface	1800 ft
Gross thickness	75 ft
Net thickness	55 ft
Porosity	33%
Permeability	20 darcies
Dip	6°
Reservoir temperature	110°F
Reservoir pressure	
Original	450 psia
At flood start	100 psia
Oil gravity	16°API
Oil viscosity, reservoir condition	
at flood start	280 cp
Oil in place	
At discovery	266.1 MSTB (90% PV)
At flood start	171.5 MSTB (58% PV)
October 1, 1979	146.2 MSTB (49% PV)

experienced an unexpected loss in productivity. The loss in productivity may have been caused by a buildup of a cold oil bank and/or an emulsion bank near the wellbore. By the end of 1978, down-dip injection had been completed in Phases I and III of the project. Down-dip injection ended in Phase II in July 1979.

Shell also conducted numerical simulations of the Mt. Poso steamflood. These simulations studied the effects of various operating policies. They included history matches of primary depletion and the Phase I steamflood.

After injection of about four-tenths of a pore volume of steam, incremental oil recovery amounted to about 13% of the oil in place at flood start. This is equivalent to an oil–steam ratio of 0.18 bbl/bbl. The water cut declined from its original value of near 100% to about 90% in 1979.

Shell expects to recover over 66 MSTB of incremental oil from the steamflood. This would amount to an increased recovery from 35% at flood start to more than 60% of the OOIP. Additional data may be found in Table 10.9.

Slocum, Anderson County, Texas

The Slocum steamflood was operated by Shell. This was a marginal steamflooding prospect that was rather thin and underlain by a bottom-water zone.[13] The reservoir has a single sand with a gross thickness of 34 ft and a net thickness of 32 ft. Reservoir depth near the crest of the structure is 520 ft. Dips on the flanks reach about 5°. The project contains 49 patterns of inverted 7-, 9-, and 13-spots. There are 258 production wells and 49 injectors; one injector is centrally located in each pattern.

The well spacing is 1 acre per producing well. The sand is generally clean and has a permeability of about 3.5 darcies. The viscosity of the oil at reservoir conditions ranges up to 3000 cp. Since the discovery of the field in 1955, only about 300,000 barrels of oil (about 2% of the OOIP) have been produced. A typical well on primary production produces only 1–2 STB/d. The low natural reservoir pressure discouraged application of steam stimulation.

Steamflooding commenced in 1967. Oil saturation in the reservoir originally was 70% of pore volume. This was reduced to 68% at the beginning of the steamflood. The average oil saturation in June 1979 was about 36%, indicating that almost half of the OOIP had been produced.

During the first year of pattern life, injection rates were 1000–1200 barrels of 70–80% quality steam per day per injector. These rates established an early thermal response in the producing wells. Later the rates were reduced to 500 bbl/d of steam per injector. This lower rate was sufficient to stabilize the steam zone and helped to maintain production in hot wells.

After about 40% of the oil in place from each pattern was produced, injection of produced disposal water was initiated. Disposal water (150°F) was injected at 4000 bbl/d per pattern. This process was continued to depletion.

Table 10.10 Reservoir Data, Slocum Steamflood

Project area	258 acres
Depth, subsurface	520 ft
Gross thickness	34 ft
Net thickness	32 ft
Porosity	34%
Permeability	3.5 darcies
Dip	0°–5°
Reservoir temperature	80°F
Reservoir pressure	
Original	110 psia
At flood start	120 psia
Oil gravity	18–19°API
Oil viscosity, reservoir conditions	1000–3000 cp
Oil in place	
At discovery	15.1 MSTB (70% PV)
At flood start	14.7 MSTB (68% PV)
At termination (September 30, 1979)	7.7 MSTB (36% PV)

Relatively high vertical sweep efficiency was apparently achieved by allowing the steam to enter the sand through the water leg in the sand. As steam penetrated the formation, heat rose into the oil zone, bringing it in contact with a greater portion of the oil than would have been realized by injecting steam into only the oil zone.

In September 1977, steam injection was terminated. However, injection of produced water was continued through June 1979, at which time the flood was concluded.

During the life of the project, 52 Mbbl steam and 25 Mbbl water were injected to produce almost 7 Mbbl oil and 193 Mbbl water. This amounts to a recovery efficiency of 47% of OOIP and a rather marginal overall OSR of about 0.134 bbl/bbl. Further details are given in Table 10.10.

Tia Juana, Venezuela

The Tia Juana M6 steam drive project consists of 19 inverted 7-spot patterns and is operated by Maraven, S. A., in Zulia, Venezuela.[6,22] The project covers 1831 acres and contains a total of 131 producers and 19 injectors. The spacing per project well is 12.2 acres. Additional data are given in Table 10.11.

Primary production began in 1948 and was followed by a period of steam stimulation from 1969 onward. Some 18% of the OOIP is estimated recoverable by primary methods, and another 5–6% is recoverable by cyclic

Table 10.11 Reservoir Data, Tia Juana Steamflood

Project area	1831 acres
Depth, subsurface	1624 ft
Gross thickness	250 ft
Net thickness	125 ft
Porosity	38.1%
Permeability	2.8 darcies
Dip	0.5°
Reservoir temperature	113°F
Reservoir pressure	
Original	780 psia
At flood start	100 psia
Oil gravity	12°API
Oil viscosity, reservoir conditions	
Original	400–4000 cp
At flood start	600–6000 cp
Oil in place	
At discovery	565 MSTB (85% PV)
At flood start	456 MSTB (67% PV)

steam injection techniques. Two primary producing mechanisms operate in this field: solution gas drive and formation compaction. The latter drive mechanism is the more important. The M6 project is designed to evaluate steamflooding on a large scale as a follow-up process after steam stimulation.

Steamflooding began in November 1975 into one of the 19 injectors. Since September 1978, steam has been injected into all of the injection wells, and about 78.7 Mbbl of steam had been injected to July 1, 1982. The project at this time was still relatively early in its flood life. Low water cuts were obtained throughout this portion of the flood, with a cumulative value of 46% by July 1, 1982. The cumulative OSR to this time was 0.26 bbl/bbl.

The injectivity of the steam in the M6 steam drive project varied between 2300 and 2800 bbl/d-well. The lower injection rates were associated with the greater number of operating injection wells and may indicate that additional injectivity is available.

Smackover, Arkansas

Phillips Petroleum Company initiated a steamflood in the Nacatoch sand of the Smackover field in Arkansas in May 1971.[14] The Smackover field was discovered in 1922 and initially produced by solution gas drive. The free gas cap was dissipated rapidly, and the producing mechanism from 1930 until the start of steamflooding was gravity drainage and limited water influx. Table 10.12 gives reservoir data.

By 1964, primary production at Smackover was approaching abandonment rates. The moderately high viscosity of the reservoir oil, the presence of a gas cap, and the poor response to waterflood pilots on the periphery of the reservoir suggested that waterflooding at Smackover would not be economical. A steamflood pilot was initiated in November 1964, and its success led to the commercial steamflood.

By the end of 1978, 10 irregular patterns were in operation and covered an area of 305 acres. There is one injection well per pattern.

The Smackover steamflood is somewhat unique in that fluids other than steam are also being injected. When a pattern was first brought into operation, 80% quality steam was injected at a rate of 50 MBtu/hr (3500 bbl/d) for 5 months. For the next 5 months, steam was injected at the same rate, but air was also injected at a rate of 500 kscf/d. After this second 5-month period, 3000 bbl/d of cold water and 500 kscf/d of air were injected until the flood was terminated.

One reason for injecting air along with steam and water was that the air supplied a large volume of fluid to help displace oil, while requiring less energy than would be required to inject an equivalent reservoir volume of steam. The second reason for injecting air was that at least part of it

Table 10.12 Reservoir Data, Smackover Steamflood

Project area	305 acres
Depth, subsurface	1920 ft
Gross thickness	130 ft
Net thickness	25 ft
Porosity	35%
Permeability	2.0 darcies
Dip	0°–5°
Reservoir temperature	110°F
Reservoir pressure	
Original	1050 psia
At flood start	7 psia
Oil gravity	20°API
Oil viscosity, reservoir conditions	
At discovery	25 cp
At flood start	75 cp
Pore volume	20.73 Mbbl
Oil in place	
At discovery	14.1 MSTB (68% PV)
At flood start	10.69 MSTB (52% PV)
October 1, 1979	5.16 MSTB (25% PV)

migrated into the gas cap and partially maintained gas cap pressure. The total volume of steam injected prior to cold-water injection corresponded to about 0.62 pore volumes.

Performance data from the Smackover project are reported in terms of the total amount of energy injected. To permit comparison with other projects, these data have been converted to an equivalent amount of steam (80% quality, 500 psia). By October 1, 1979, incremental oil recovery by steamflood was about 24% of the oil in place at flood start. This production corresponds to an equivalent basis OSR of about 0.16 bbl/bbl.

Street Ranch, Maverick County, Texas

Conoco conducted steam pilot operations in the San Miguel Tar Sand deposit in South Texas.[23] The formation is about 2300 ft deep. Gross thickness ranges from 20 to 80 ft with the average close to 50 ft. Except for interspersed limestone streaks, the sand is very clean, resulting in porosities of 26–30% and permeabilities of 250–1000 md.

The tar present in the San Miguel-4 sand is one of the most viscous, dense, sulfur-laden hydrocarbons known. It has a gravity of −2° API and is essentially a solid at the 95°F original reservoir temperature. The immobile nature of the tar is illustrated by its 180°F pour point and its estimated

20×10^6 cp viscosity. Additional reservoir and project data are given in Table 10.13.

The process consists of placing a horizontal fracture near the base of the oil sand and injecting initially above fracture pressure. The pilot was performed in a 5-acre inverted 5-spot pilot pattern. During a 31-month period, ending June 1980, the Street Ranch pilot produced 169,040 bbl of −2° API tar with monthly tar production rates occasionally exceeding 300 STB/d. Postpilot core wells indicated residual tar saturations as low as 8% and an average recovery efficiency of better than 50%.

In this process, the production wells were fractured horizontally with cold water and immediately stimulated with high-pressure steam to provide a heated target. Next, a horizontal fracture was propagated from the injection well to the production wells, and a mixture of high-pressure steam and hot water was injected at very high rates and pressures to hold the fracture open, to preheat the formation, and to mobilize the native tar.

It should be noted that this process requires the placement of a horizontal fracture and that the in situ stress state in the Street Ranch area favored this fracture orientation. In most locations, vertical fractures would occur unless steps are taken to alter the in situ stress state of the reservoir.

The daily steam injection rates typically exceeded 100 bbl/ft of sand or 20 bbl/acre-ft. The injection rates and pressures then were reduced to promote injection into the matrix and efficient displacement of the liquified tar to the producing wells. In the final stage, produced water was mixed with fresh water and recycled through the reservoir, and tar production continued to an arbitrary limit.

At the end of the steam injection phase, the cumulative OSR was

Table 10.13 Reservoir Data, Street Ranch Steam Pilot

Project area	5 acres
Depth	1500 ft
Gross thickness	52 ft
Net thickness	40.5 ft
Porosity	27%
Permeability	250–1000 md
Dip	2°
Reservoir temperature	95°F
Reservoir pressure	Not reported
Oil gravity	−2°API
Oil viscosity, original reservoir cond.	20×10^6 cp (estimated)
Oil in place (inside pattern)	
At flood start	204 kSTB (48.2% PV)
At flood termination (June 18, 1980)	94 kSTB (22.2% PV)

0.086 bbl/bbl. After the follow-up water injection or heat-scavenging phase, the OSR was 0.092 bbl/bbl. At these low OSRs, the process would not have acceptable economics if tar were burned as fuel. Conoco was believed to be investigating alternate energy sources such as coal for steam generation.

Peace River, Alberta, Canada

Shell has been piloting a combination of steam stimulation and steam drive in the Peace River oil sands of northwestern Alberta, Canada.[24] Currently, a 10-kSTB/d project is planned. The Peace River oil sands contain about 75×10^9 STB of bitumen in place. The oil sand is found in the Cretaceous Bullhead formation at a depth of 1800 ft. A thin, high-permeability water zone lies at the bottom of the 90-ft-thick sand and provides a path for interwell communication at subfracturing pressure levels. The initial oil saturation averages 77% in a relatively uniform sand having a porosity of 28%. The upper majority of the sand averages 220 md permeability, while the lower portion of the sand averages 1440 md. At the initial formation temperature of 62°F, the 9°API oil has a viscosity of 200,000 cp.

The original pilot consisted of 24 production wells and 7 injection wells arranged in seven 7-acre, 7-spot patterns. In Shell's process, steam is initially injected through the bottom-water zone. A pressure cycle process is conducted where periods of high injection rate alternate with periods of low injection rate. The cycles of pressure buildup and blowdown enhance the ability of injected steam to move upward from the bottom-water zone and tend to increase areal sweep.

To the end of October 1984, the project had produced about 21% of the original bitumen in place since its start-up in November 1979. Actual OSRs are believed to be about 0.25 bbl/bbl.[25] A pressure-monitoring system installed in the injection wells is used to avoid inadvertent fracturing of the formation.

For the 10-kSTB/d project, there will be four clusters of wells, each consisting of 13 7-spots. Each cluster will have two drilling pads from which the wells will be directionally drilled to desired locations downhole. Project start-up was scheduled for late 1986.

REFERENCES

1. Myhill, N. A., and Stegemeier, G. L. "Steam Drive Correlation and Prediction," *J. Pet. Tech.* (February 1978), pp. 173–182.
2. Leonard, J. "Increased Rate of EOR Brightens Outlook," *Oil Gas J.* (April 14, 1986), p. 71.

3. Bursell, C. G. "Steam Displacement—Kern River Field," *J. Pet. Tech.* (October 1970), p. 1225.
4. van Dijk, C. "Steam-Drive Project in the Schoonebeek Field, The Netherlands," *J. Pet. Tech.* (March 1968), p. 295.
5. Pollock, C. B., and Buxton, T. S. "Performance of a Forward Steam Drive Project—Nugget Reservoir, Winkleman Dome Field, Wyoming," *J. Pet. Tech.* (January 1969), p. 35.
6. de Haan, H. J., and Schenk, L. "Performance Analysis of a Major Steam Drive Project in the Tia Juana Field, Western Venezuela," *J. Pet. Tech.* (January 1969), p. 111.
7. Hall, A. L., and Bowman, R. W. "Operation and Performance of Slocum Field Thermal Recovery Project," *J. Pet. Tech.* (April 1973), pp. 402–410.
8. Volek, C. W., and Prior, J. A. "Steam Distillation Drive—Brea Field, California," *J. Pet. Tech.* (August 1972), pp. 899–906.
9. Afoeju, B. I. "Conversion of Steam Injection to Waterflood, East Coalinga Field," *J. Pet. Tech.* (November 1974), pp. 1227–1232.
10. Hearn, C. L. "The El Dorado Steam Drive—A Pilot Tertiary Recovery Test," *J. Pet. Tech.* (November 1972), pp. 1377–1384.
11. Blevins, T. R., Aseltine, R. J., and Kirk, R. S. "Analysis of a Steam Drive Project, Inglewood Field, California," *J. Pet. Tech.* (September 1969), pp. 1141–1150.
12. Bursell, C. G., and Pittman, G. M. "Performance of Steam Displacement in the Kern River Field," *J. Pet. Tech.* (August 1975), pp. 997–1005.
13. Thurber, J. L., and Rahman, K. "Slocum Field," Enhanced Oil Recovery Field Reports, Vol. 5, No. 2 (September 1979), pp. 291–294.
14. Smith, R. V., Bertuzzi, A. F., Templeton, E. E., and Clampit, R. L. "Recovery of Oil by Steam Injection in the Smackover Field, Arkansas," *J. Pet. Tech.* (August 1973), pp. 883–889.
15. French, M. S., and Howard, R. L. "The Steamflood Job, Hefner Sho-Vel-Tum," *Oil Gas J.* (July 17, 1967), pp. 64–66.
16. Blevins, T. R., and Billingsley, R. H. "The Ten-Pattern Steamflood, Kern River Field, California," *J. Pet. Tech.* (December 1975), pp. 1505–1514.
17. Greaser, G. R., and Shore, R. A. "Steamflood Performance in the Kern River Field," SPE 8834, presented at the Joint 1980 SPE/DOE Symposium on Enhanced Oil Recovery, Tulsa, Oklahoma, April 20–23, 1980.
18. Restine, J. L. "Effect of Preheating on Kern River Field Steam Drive," *J. Pet. Tech.* (March 1983) pp. 523–529.
19. Oglesby, K. D., Blevins, T. R., Rogers, E. E., and Johnson, W. M. "Status of the 10-Pattern Steamflood, Kern River Field, California," *J. Pet. Tech.* (October 1982), pp. 2251–2257.
20. Matthews, C. S. "Steamflooding," *J. Pet. Tech.* (March 1983), p. 465.
21. Stokes, D. D. "Steam Drive as a Supplemental Recovery Process in an Intermediate-Viscosity Reservoir, Mount Poso Field, California," *J. Pet. Tech.* (January 1978), pp. 125–131.

22. Armando, H. L., Jr. "The M6 Steam Drive Project, Design and Implementation," Canada-Venezuela Heavy Oil Recovery Symposium, Edmonton, Canada, 1977.
23. Britton, M. W., Martin, W. L., Leibrecht, R. J., and Harmon, R. A. "The Street Ranch Pilot Test of Fracture-Assisted Steamflood Technology," *J. Pet. Tech.* (March 1983), pp. 511–522.
24. Matthews, C. S., and Gorrill, R. G. "Shell Plans Peace River Steamflood," *Petrol. Eng.* (July 1985), pp. 34–42.
25. Cox, B. "Shell to Start Peace River Steam Injection," *Enhanced Recovery Week* (September 22, 1986) p. 1.

Chapter 11

NUMERICAL CALCULATION OF STEAM INJECTION BEHAVIOR

ABSTRACT

Numerical simulators that solve finite-difference equations representing the differential equations describing heat transfer and fluid flow in porous media are the most reliable and accurate methods for predicting the behavior of hot-fluid injection processes. They have been used successfully since the mid-1960s to provide detailed insight into the mechanisms and recovery behavior of hot waterflooding and steamflooding as well as cyclic steam stimulation.

This chapter discusses results calculated by models developed for Exxon by Weinstein and Shutler and co-workers. The Weinstein model calculates flow only in the horizontal direction but accounts for layering (permeability variation with depth) in the reservoir. The Shutler model is two dimensional. Both can be used for either radial or cartesian grids. The Shutler model includes the effects of gravity, capillarity, and vertical cross-flow, which the Weinstein model neglects, and therefore is a more complete representation of the physical processes involved. Both models account for heat conduction two dimensionally, both within the oil formation and outside of it. The assumptions inherent in the two models and a brief description of the solution methods used are discussed.

These models were compared to one- and three-dimensional laboratory physical models, and remarkably close agreement was obtained considering the difficulty in matching true physical model relative permeability characteristics, oil characterization, and other factors. The Weinstein model treats the oil as having a distillable fraction, while the Shutler model assumes the oil to be nondistillable and gas free. For this reason, the Shutler model tends to predict oil recovery conservatively for oils that have a significant distillable fraction. Example results for hot waterflooding and steamflooding with and without follow-up cold-water injection are presented. Effects of reservoir thickness on steamflood behavior and the acceptability of a one-dimensional representation are discussed. While these models are normally used to represent the reservoir in vertical cross sections, the Shutler model can also be used areally.

Later models developed by Coats and others solve the three-dimensional case. Grid orientation strongly affects areal flow distributions calculated using five-point differencing. Realization of this has led to the development of more accurate differencing schemes such as the nine-point method. Computation times become excessive for complex three-dimensional problems requiring large numbers of grid blocks at the present time. Sobocinski has successfully modeled steamflooding of a moderate-viscosity oil in a large multiwell project using a two-dimensional areal steamflood model. More highly stratified or high-viscosity oil systems may require a three-dimensional representation, however.

INTRODUCTION

Because of the high costs associated with injecting steam, it is desirable to have accurate calculations to assist in field analysis and planning. Thus, the oil industry has developed sophisticated computational schemes that simulate the steam injection processes. Analytical techniques must ignore many of the physical processes by invoking simplifying assumptions. Numerical methods incorporate most of the physical processes but suffer from reliability problems and excessive computation time requirements. There has been a need to develop improved calculation procedures. In recent years, the trend has been to develop more stable, faster–running simulators. These have proven capable of efficient calculations for single- and two- to three-well, three-dimensional simulations of pattern elements (such as one-eighth of a 5-spot, one-sixth of a 7-spot). Large, comprehensive, three-dimensional simulations of a whole reservoir under steamflood or even a large multiwell group of patterns are beyond the scope of thermal simulators as of 1986 but in time may become feasible.

The main advantage of numerical simulation over the simple calculations is that complexities of reservoir description, rock and fluid properties, and operational strategies can be accounted for, to obtain a much more comprehensive and reliable prediction of process behavior over long-term operation than is attainable by an analytical or semianalytical method.

This chapter describes two numerical models that simulate energy and mass transport in a multilayered reservoir. The models have been sucessfully applied to a variety of situations. They have accurately matched experimental results and field data from single- and multicycle steam stimulation treatments and have predicted behavior for floods involving cold water, hot water, and steam injection.

Table 11.1 Features of Weinstein and Shutler Steam Injection Models

Characteristic	Weinstein Steam Stimulation Model	Shutler Two-Dimensional, Three Phase Steam Drive Model
1. Fluid flow and heat convection	One-dimensional	Two-dimensional
2. Heat conduction	Two-dimensional	Two-dimensional
3. System	Vertical cross section	Vertical cross section[a]
4. Phases	Oil, water, vapor (steam and/or gas)	Oil, water, vapor (steam and/or gas)
5. Interphase mass transfer	Steam/water Solution gas/free gas	Steam/water Dead oil (no transfer)
6. Injection fluid	Steam, water, gas	Steam or water, no gas
7. Gravity	No	Yes
8. Capillary pressure	No	Yes
9. Rock compressibility	Yes	No
10. Heterogeneity	Layer	Block
11. Grid configurations	Linear Radial Cylindrical Elliptical	Linear Radial Cylindrical
12. Initial conditions	Specified	Capillary, gravity equilibrium or specified
13. Reservoir and fluid properties	Cubic spline functions	Tables

[a]Modified areal version also available.

Numerical models requiring computer solution are more comprehensive than analytical models. They generally include the effects of both fluid and energy flow. Over the years, a number of computer models that solve finite-difference representations of the nonlinear mass and energy balances have been developed. Spillette and Nielsen[1] treated hot waterflooding in two dimensions. Three-phase steam injection models were presented by Weinstein et al.[2] for linear fluid flow, by Shutler for both linear[3] and two-dimensional[4] fluid flow, by Abdalla and Coats[5] for two-dimensional fluid flow, and by Coats et al.[6,7] for three-dimensional fluid flow. A later Coats[8] model and the Weinstein model [2] include the effects of oil distillation and solution gas while the earlier models do not.

Table 11.1 outlines the various features of two of the numerical models, that of Weinstein et al.[2] and the two-dimensional, three-phase steam drive model of Shutler.[3] While Weinstein describes his model as being for steam stimulation and Shutler labels his for steam drive, it is important to realize that both can be used for either process.

The Weinstein one-dimensional model simulates a sequence of alternating thin sands and shales and assumes no vertical communication between the sands. It assumes that the full thickness of each sand is uniformly invaded, but different injection levels per unit thickness of sand can be calculated. The Shutler model accounts for gravity and capillarity, which are neglected in the case of the Weinstein model. The Shutler two-dimensional model, therefore, is able to calculate gravity override and partial sweep of thick sands.

WEINSTEIN ONE-DIMENSIONAL MODEL

In the Weinstein one-dimensional model,[2] the flow of fluids and convective energy is assumed to be linear or radial within a multilayered reservoir while heat conduction is two dimensional across the reservoir and bounding strata. The model considers all three fluid phases with interphase mass transfer between water and vapor and between oil and vapor. It can account for solution gas, distillation, and solvent extraction effects.

To accelerate the solution procedure, the Weinstein model ignores capillary pressure. Each layer can be assigned a different permeability, porosity, thickness, initial water saturation, and initial gas saturation. Rock compressibility can account for formation expansion and compaction. The model can operate with either linear, radial, cylindrical, or elliptical grid geometry. The cylindrical grid orientation is used in modeling laboratory experiments, while the elliptical configuration is used to consider the case of steam injected along a vertical fracture. For boundary conditions, the program can handle either a constant-pressure condition or an implicit rate specification.

Fluid property functions and relative permeabilities are expressed as cubic spline functions. A subsidiary program is needed to generate the four coefficients for each interval of the spline fit. Three-phase relative permeabilities are calculated from two-phase water–oil and gas–oil curves by the Stone method.[9] Fluid properties are represented as products of separate functions of temperature, pressure, and/or solution gas–oil ratio.

SHUTLER TWO-DIMENSIONAL MODEL

The two-dimensional, three-phase model of Shutler[4] describes the simultaneous flow of the three phases—oil, water, and vapor—in two dimensions. Interphase mass transfer between the water and vapor phases is calculated, but it is assumed that the oil is nonvolatile and the hydrocarbon gas insoluble in the liquid phases. The model calculates two-dimensional heat convection within the reservoir and two-dimensional heat conduction in a vertical cross section spanning both the oil sand and adjacent strata (Fig. 11.1).

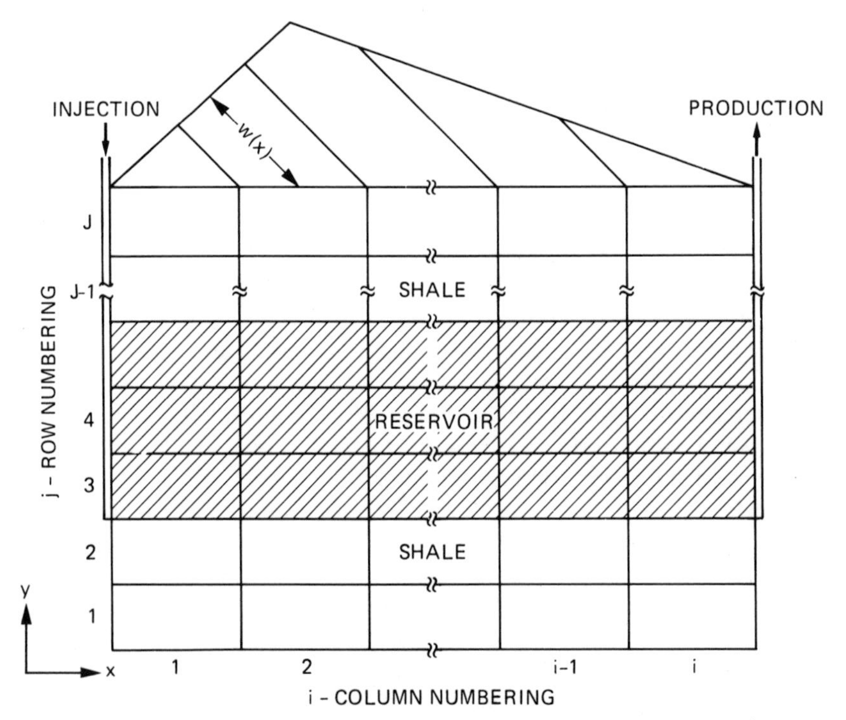

Figure 11.1. Grid orientation for one-eighth 5-spot model (after Shutler[4].) © 1970 SPE-AIME.

Capillary pressure and gravity effects are included. Porosity, horizontal and vertical permeability, and horizontal and vertical thermal conductivity can be input block by block to reflect any desired degree of heterogeneity and anisotropy. The model grid can have a cylindrical orientation or a cross-sectional description of the reservoir in either cartesian or radial coordinates. Initial saturation and pressure distributions may be either specified or calculated by capillary–gravity equilibrium. A specified production schedule or rate, coupled with either a fixed injection or production bottom-hole pressure, can be assumed. An alternative method is to specify the injection rate and production bottom-hole pressure.

Properties of reservoir fluids are input in tabular form as functions of pressure and temperature. Three-phase relative permeabilities and capillary pressures are synthesized from tables or two-phase water–oil and gas–oil curves.

DESCRIPTION OF EQUATIONS SOLVED

The governing fluid flow and energy equations for both the Weinstein and Shutler models are basically the same. Darcy's law provides expressions for velocities of the three phases, oil, water, and vapor. Combination of these relations with the corresponding mass continuity equation yields the partial differential equations governing flow of the three phases within a reservoir sand. Special terms account for interphase mass transfer between water and vapor. The differential equations, which are solved numerically, are given below:

Material Balances (Combined with Darcy's Law)

For the oil phase:

$$\nabla \cdot \frac{kk_{ro}\rho_o}{\mu_o}(\nabla P_o - \rho_o \nabla D) + q_o = \phi \frac{\partial(\rho_o S_o)}{\partial t} \tag{11.1}$$

For the water phase:

$$\nabla \cdot \frac{kk_{rw}\rho_w}{\mu_w}(\nabla P_w - \rho_w \nabla D) + q_w + M_g = \phi \frac{\partial(\rho_w S_w)}{\partial t} \tag{11.2}$$

For the gas phase:

$$\nabla \cdot \frac{kk_{rg}\rho_g}{\mu_g}(\nabla P_g - \rho_g \nabla D) + q_g - M_g = \phi \frac{\partial(\rho_g S_g)}{\partial t} \tag{11.3}$$

where

$$S_o + S_w + S_g = 1 \tag{11.4}$$

and k = absolute permeability
k_r = relative permeability
μ = fluid viscosity
P = phase pressure
ρ = phase density
D = depth
q = production (or injection rate)
ϕ = porosity
S = phase saturation
M_g = interphase mass transfer term (allows for steam condensation)

The subscripts o, w, and g refer to oil, condensed water, and water vapor, respectively. Water and gas phase pressures (P_w, P_g) and oil phase pressure P_o are related to capillary pressures P_c by

$$P_w = P_o - P_{c_{ow}} \qquad P_g = P_o + P_{c_{og}}$$

Energy Balance (Describes Conductive and Convective Heat Transfer)

$$\begin{aligned}\nabla \cdot K_h \nabla T - \nabla \cdot [v_o \rho_o H_o + v_w \rho_w H_w + v_g \rho_g H_g] + Q \\ = \frac{\partial}{\partial t} [\phi(S_o \rho_o H_o + S_w \rho_w H_w + S_g \rho_g H_g) + (1 - \phi)(\rho c)_R T]\end{aligned} \tag{11.5}$$

When applied to adjacent impermeable strata, velocities vanish. In addition, the accumulation term is simplified by setting porosity to zero and replacing $(\rho c)_R$ with $(\rho c)_{R+F}$. In Equation 11.5 v and H are the velocities and specific enthalpies, respectively, of the three flowing phases. Q is a heat source term.

The energy balance assumes instantaneous thermal equilibrium between the fluid and rock matrix. Energy conduction, convection, accumulation in both fluid and rock, and sources or sinks at wells are included in the description of the heat transfer process. In addition, a gas composition balance is written for the two-dimensional model. Details of the difference equations used to approximate the above differential equations are provided in the literature.[2,4,6]

SOLUTION PROCEDURES

For both the Weinstein and Shutler models, the mass and energy balances are rewritten in finite-difference form. The Newton–Raphson method is applied to maintain all nonlinear terms in the solution and thus give a better accounting of the physical phenomena.

For the Weinstein model, the mass balances are combined so as to eliminate saturations at the new time level. The result is an equation in pressure, temperature, and condensation terms, designated as the *pressure equation*, owing to its similarity to the equation of the same name in isothermal simulation. Two pressure equations have been developed, one used when a vapor phase is present, the other when no vapor phase is present.

The iteration method for the solution then proceeds as follows. For the first iteration, the energy equation is solved for the conduction heat losses to the overburden and underburden. This is done once per time step. Since conduction is a slow process compared to the convection occurring within the reservoir, time-lagging conduction leads to negligible error.

A phase-constraint decision is made based on whether or not a vapor phase existed in a particular grid block at the end of the previous time step. The appropriate pressure equation and the energy equation are then solved simultaneously for pressure and temperature. The condensation terms are updated, the phase constraints checked, and the cycle beginning with the pressure–energy solution is repeated until changes in pressure, temperature, and mass transfer terms are less than specified tolerances.

The simultaneous solution of pressure and energy equations avoids convergence problems associated with time and iteration lagging. With both steam and hydrocarbon gas present, the simultaneous solution also facilitates the calculation of vapor phase equilibrium.

Following convergence of the iteration, the individual mass balances are solved for the saturations at the new time level. These saturations may be calculated using either explicit mobilities or semi-implicit mobilities.† Semi-implicit mobilities are required to stabilize the solution when injection or production rates are high or when steam and gas saturations change rapidly near the wellbore.

Solution of the finite-difference equations for the Shutler model is achieved in three stages for each time step by separate, successive solution

†Explicit mobilities are transmissibility terms, including relative permeability and fluid viscosity, that are determined at the old or previous time step. Implicit mobilities are those calculated for the new or current time step. The latter are more accurate to use, but using them slows the calculation because of additional iteration required for their determination. An estimated, or *semi-implicit*, mobility obtained from extrapolation of the values from the previous time steps is used in the Weinstein and Shutler methods.

of the three principal aspects of the problem: fluid flow, heat transfer, and gas phase composition change. In the first stage, the three mass balances are solved simultaneously over the two-dimensional reservoir grid by the alternating-direction implicit procedure[10] or by the strongly implicit procedure.[11] In solving these equations, temperature and gas composition are considered constant. At the end of the first stage, new pressure and saturation distributions are obtained.

In the second stage, the energy balance is solved over a two-dimensional grid extending above the reservoir by application of the noniterative alternating-direction procedure.[12] The velocities and saturations obtained from the first stage are considered constant. At the end of the second stage, a new temperature distribution is obtained.

In the third stage, the finite-difference form of the gas composition equation is solved explicitly, assuming that values of velocity, saturation, and temperature obtained from the first two stages are constant. The resulting values for composition combine with the pressure, saturation, and temperature distributions from the first two stages to provide starting point information for the first stage of the next time step.

EVALUATION OF MODELS

Willman et al.[13] conducted a number of one-dimensional laboratory core oil displacements by hot water and steam. These investigators reported temperature distributions, and Shutler[3] presented corresponding oil recovery curves. These laboratory results have also been simulated with the Weinstein model.[2]

Excellent agreement between calculated and experimental temperature profiles for both hot water and steamfloods is shown in Figure 11.2. Figure 11.3 presents the experimental and calculated oil recovery curves. The hot-water prediction is in good agreement with the experimental data, but the calculated steamflood curve for the dead oil version of the model falls considerably short of the final oil recovery. The discrepancy between the ultimate recovery of 0.54 PV and the calculated value of 0.48 PV is attributed to partial distillation of the Napoleum–Primol mixture used in the experiment.

The same calculation was repeated using the solution gas drive model. In order to simulate the distillation effect, the dissolved "gas" was given the properties of Napoleum. Solution gas–oil ratio data were input as a function of temperature and pressure, and an initial value was assigned. Oil density and viscosity were altered to account for the distillable nature of the oil. This caused no significant changes in the temperature profiles of Figure 11.2 or the hot-water recovery curve of Figure 11.3. However, there was a marked change in the calculated oil recovery by steam injection, as indi-

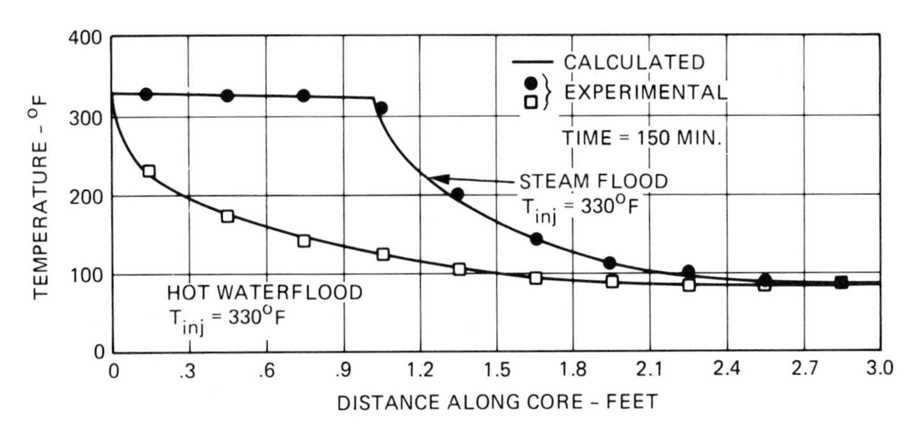

Figure 11.2. Comparison of calculated and experimental temperature profiles for laboratory model (after Weinstein et al.[2]). ©1977 SPE-AIME.

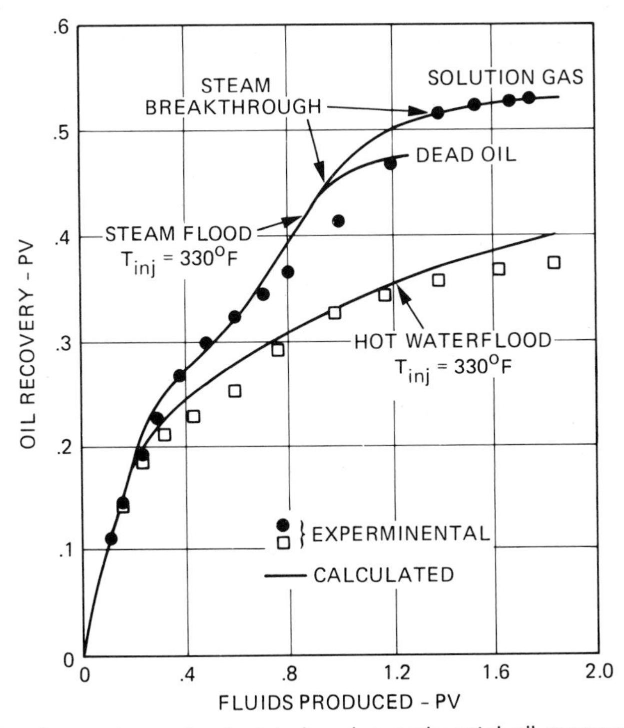

Figure 11.3. Comparison of calculated and experimental oil recovery curves for laboratory model (after Weinstein et al.[2]). © 1977 SPE-AIME.

cated by the uppermost curve in Figure 11.3. The solution gas model closely matched the experimental curves and predicted the experimental ultimate recovery of 0.54 PV.

The mechanism of the Weinstein one-dimensional solution gas model is closely related to actual physical phenomena. Figure 11.4 shows volume fraction Napoleum in the oil versus distance along the core for three times during the flood. The steam distills off the Napoleum and banks it up at the leading edge of the steam zone. The maximum volume fraction and extent of the bank increase with time until the peak reaches the outlet face. Within the bank, the Napoleum aids recovery by solvent extraction of the original oil. The lighter hydrocarbon mixes with the oil, reduces its viscosity, and results in a lower residual saturation than for the original oil. This solvent extraction effect was described by Willman et al.[13] and is also referred to by Pursley[14] as one of the mechanisms responsible for oil displacement by steam injection.

The two-dimensional steam drive model of Shutler was used to simulate laboratory experiments conducted in a 10-in.-thick, homogeneous, scaled model of a one-eighth of a 5-spot flooding pattern.

Comparisons of the calculated and experimental oil recovery curves are shown in Figure 11.5 for both a continuous steamflood and a 0.25-PV steam bank followed by water at the original reservoir temperature. The agreement for both types of flood is good.

In Figure 11.6*A* are shown the 400°F isotherms (which approximately

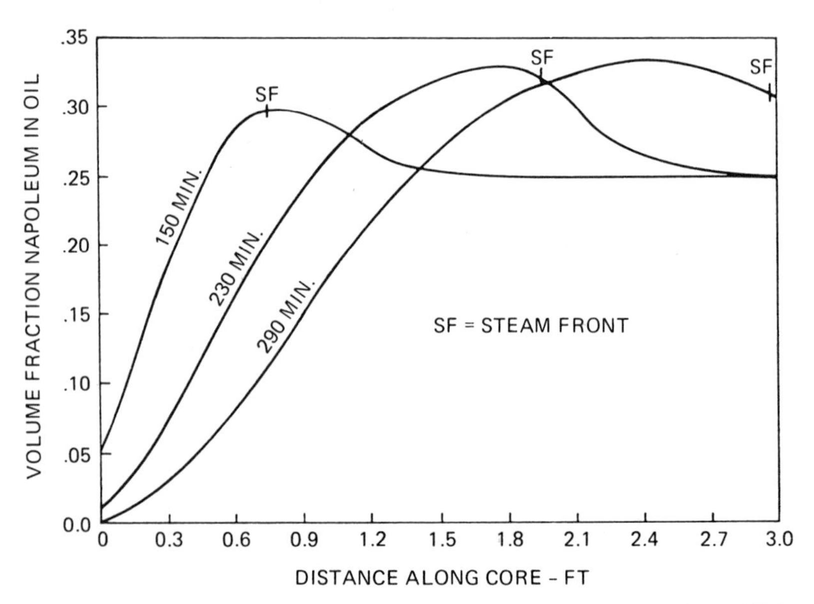

Figure 11.4. Calculated profiles of volume fraction napoleum in oil (after Weinstein et al.[2]). © 1977 SPE-AIME.

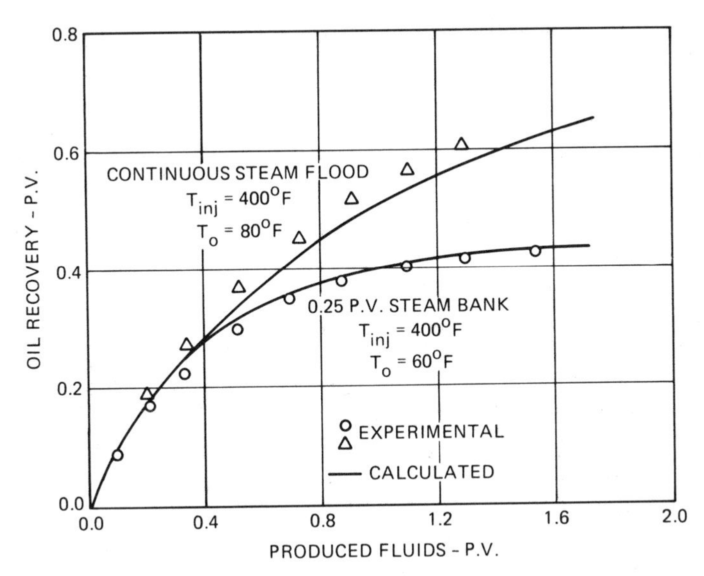

Figure 11.5. Comparison of experimental oil recovery curves with results calculated by the Shutler two-dimensional model.[4] © 1970 SPE-AIME.

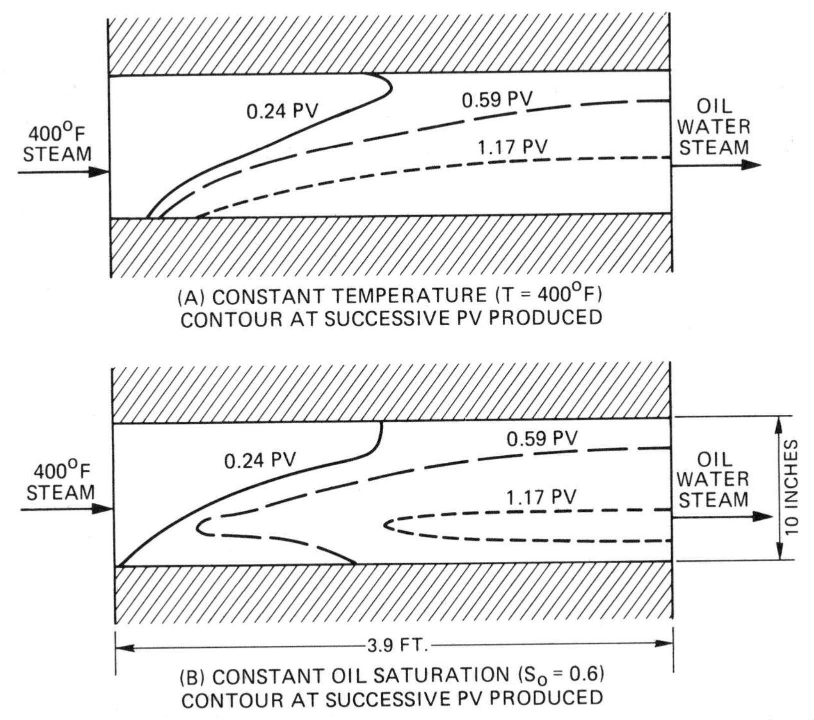

Figure 11.6. Calculated two-dimensional contours of 400°F isotherm and 0.6 oil saturation contours during continuous steamflood (after Shutler[4]). © 1970 SPE-AIME.

represent the leading edge of the steam zone) at successive times during the continuous steamflood. Steam tends to override as a result of gravity segregation.

In Figure 11.6*B* are shown the $S_o = 0.6$ contours at successive times during the continuous steamflood. The earliest contour conforms roughly to the shape of the steam zone. The later contours show that steam continues to sweep across the top and that some underrunning by steam condensate also occurs as a result of gravity segregation from the less dense oil.

The oil recovery predicted by the Shutler two-dimensional, three-phase steam drive model was also compared with that measured for a layered, heterogeneous laboratory model.[15] The case investigated was again one-eighth of a 5-spot, 10-acre pattern with a reservoir description scaled to the Mannville Clearwater sand at Cold Lake in Canada. The model was 8.4 in. thick and extended 2.33 ft between injector and producer. Continuous, impermeable tight streaks constructed of concrete were installed approximately 1.7 in. from the top and bottom of the model. The permeability of the sand was 200 darcies and the porosity was 36%. Initial oil saturation was approximately 93%, with the exception of a 1.3-in. bottom-water zone that was saturated with water.

This experimental study was discussed in more detail by Pursley.[14] Steam was injected at 1000 psia and 70% quality. The first 1.4×10^6 bbl steam (scaled to field conditions) were injected through the water zone only, after which the wells were recompleted to inject into the oil and gas zones only.

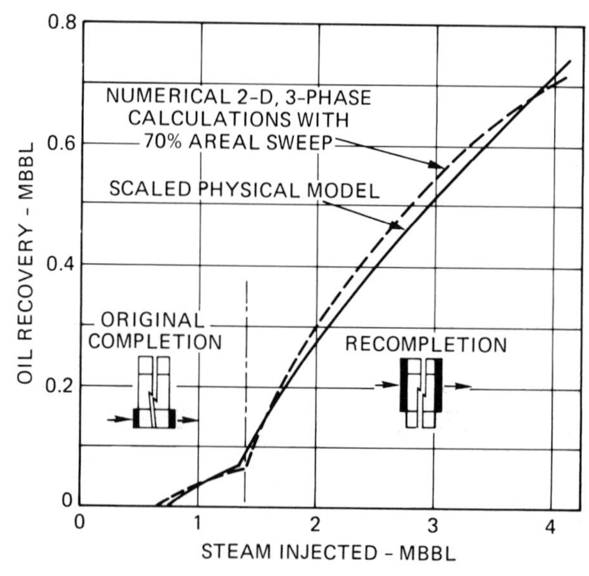

Figure 11.7. Comparison of laboratory model data with two-dimensional, three-phase steam drive calculations (after Pujol and Boberg[15]). © 1972 SPE-AIME.

For the two-dimensional numerical calculations, the areal sweep efficiency was assumed to be 70%.

Figure 11.7 shows oil recovery as a function of steam injected for the model (after adjustment to field conditions) and for the numerical calculation. The numerical predictions indicate slightly higher oil recoveries after recompleting the wells. However, after injection of about 3.8×10^6 bbl steam, oil recovery predicted by the numerical calculations begins to fall slightly below that measured for the laboratory model.

A perfect match between numerical and experimental predictions is impossible to attain, considering the uncertainty existing in some of the parameters used in both of the models. However, over the range of interest, the experimental and numerical predictions are in good general agreement.

EXAMPLE RESULTS USING SHUTLER TWO-DIMENSIONAL MODEL

Hypothetical Steamflood Pilot Studies

The Shutler two-dimensional, three-phase simulator has been used to make a comparative study of cold waterflooding, hot waterflooding and steamflooding for a hypothetical field pilot. The assumed pilot area was 2625 ft long, 1280 ft wide, and 120 ft thick, with a dip of approximately 8.5°. Average porosity was 29%, and initial oil viscosity was 125 cp. Permeability was about 400 md at the top and bottom of the section and graded continuously from above and below to 5 darcies 40 ft from the top. The area was assumed to have been produced for 20 years, mostly by depletion drive, and at the time of the study, the average water and gas saturations were calculated to be 10 and 4%, respectively.

The calculations showed that steam injection followed by cold-water injection gives higher ultimate recovery than a hot-water bank followed by cold-water injection (Fig. 11.8). For a cumulative heat injection of 5.12×10^{12} Btu (0.6 PV steam or 1.4 PV hot water), the steam bank process resulted in 50.9% recovery of the original oil in place compared to 39.5% recovery by the hot-water bank at 2 PV total fluid injected. When the cumulative energy input was 2.2×10^{12} Btu (0.262 PV steam or 0.6 PV hot water), recoveries of 41.7 and 34.6%, respectively, were obtained by the two processes. Both thermal recovery processes are far superior to cold-water injection, which resulted in only 21.2% recovery of original oil in place.

The numerical model results illustrate the considerable advantage of injecting steam rather than hot water. Larger amounts of thermal energy increase the incremental recovery for steam over hot water. Another advantage of steam is that it recovers oil sooner and permits better heat recovery efficiency than does hot water. One disadvantage of steam bank

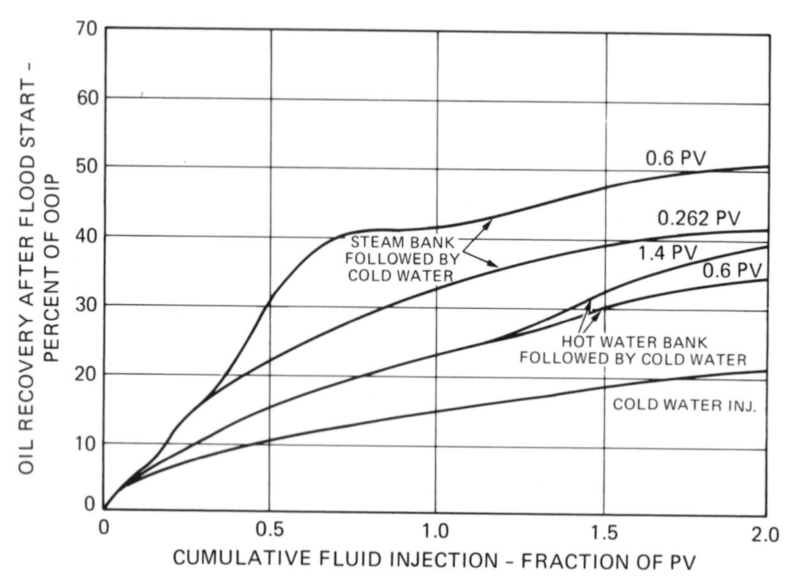

Figure 11.8. Comparison of recoveries for several processes as calculated by Shutler two-dimensional model for hypothetical pilot.

operation is the rapid collapse of the steam zone when cold water is introduced. This causes the flattening of the oil production curves (Fig. 11.8) corresponding to the point where steam injection ends. Steam zone collapse has been observed in the laboratory[14] and can be partially avoided in the field by following the steam with a small bank of hot water at steam temperature prior to cold-water injection.[16]

In all of the cases shown in Figure 11.8, gravity effects were important. There was vertical segregation of the water and steam and preferential movement of the water along the bottom layers. Additional oil recovery by hot-water injection over cold-water injection was mostly due to improved displacement efficiency and, to a lesser degree, to improvement in the vertical sweep efficiency. No reduction in the waterflood residual oil saturation at elevated temperatures was assumed. On the other hand, the additional oil recovery for steam injection compared to hot-water injection is mostly due to an increase in the overall sweep efficiency, the greater volume of reservoir heated with the high-energy input with steam, and the lower residual oil saturation in the steam-invaded zone.

Comparison with One-Dimensional Calculations

Figure 11.6 shows that steam zone advance is not vertically uniform. Yet, a one-dimensional calculation such as the simplified Marx–Langenheim model assumes a vertically uniform steam zone advance (vertical sweep 100%). In Figure 11.9, oil recovery predictions by the Shutler numerical, one-dimensional model[3] are compared with those of the two-dimensional model[4] for two different laboratory systems. The two systems differ significantly in thickness, but the other data are the same (including the same injection rate per unit thickness) as given in Appendix A.

The results of the calculations show the following:

(a) The one-dimensional model predicts optimistically for the thick system ($d/h = 4.7$).

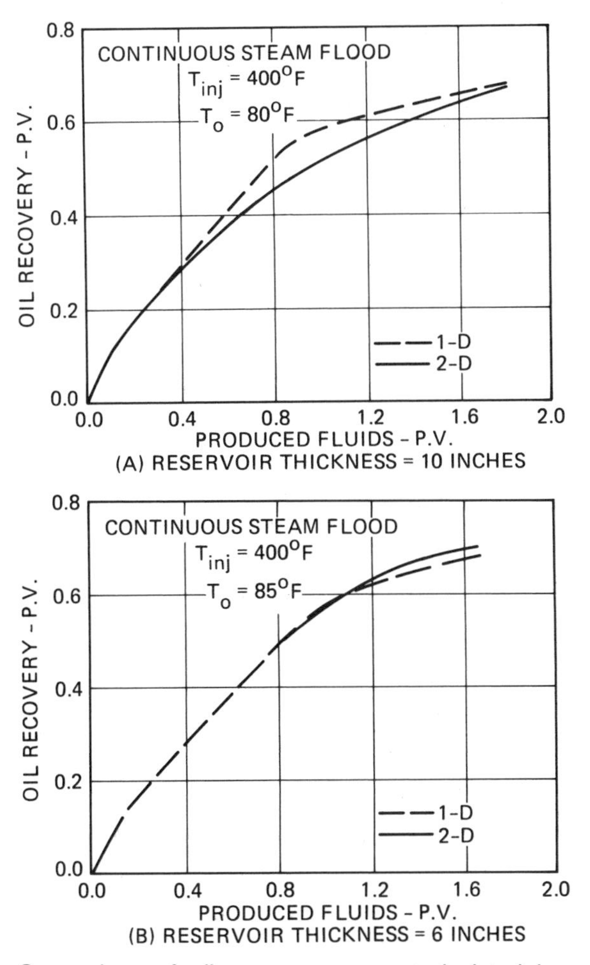

Figure 11.9. Comparison of oil recovery curves calculated by one- and two-dimensional techniques (after Shutler[4]). ©1970 SPE-AIME.

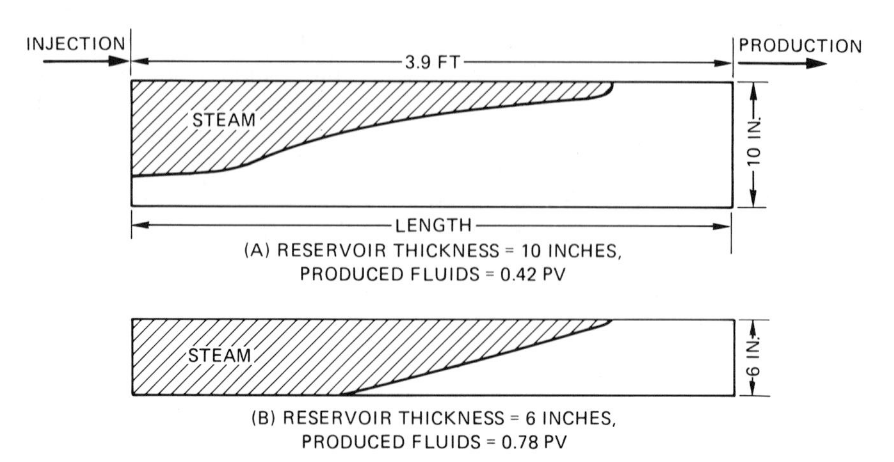

Figure 11.10. Effect of model sand thickness on steam zone distribution (after Shutler[4]). © 1970 SPE-AIME.

(b) The one-dimensional model predicts accurately for the thin system ($d/h = 7.8$) despite some override tendency of the steam zone (see Fig. 11.9). The reason for this can be seen by referring to the shape of the steam zone for the two models depicted in Figure 11.10.

These results led to the rough guideline of a spacing–thickness ratio of 7 or greater being required to justify the use of a one-dimensional Marx–Langenheim model result for steamflooding, discussed in Chapter 9. This guideline is only a very approximate one, however, because severe permeability stratification or high oil viscosity could invalidate the 7:1 rule. (Note that the models are homogeneous, and the original oil viscosity is about 1000 cp for the cases of Figs. 11.9 and 11.10.) Later we shall see that for a 150-cp oil, a one-layer representation can be adequate in a heterogeneous situation at an even lower spacing–thickness ratio than 7:1 if one alters the relative permeability curves to calibrate the simulator to match actually observed field performance. There was no alteration of relative permeability data in the two cases illustrated in Figures 11.9 and 11.10.

Importance of Vertical Permeability

Two calculations made by the Shutler two-dimensional model[4] for a highly viscous (500,000-cp) oil are illustrated in Figure 11.11:

(a) Zero vertical permeability prevented oil bank formation and promoted channelling of steam across the top of the oil sand through an initial gas zone (see Fig. 11.11*A*).

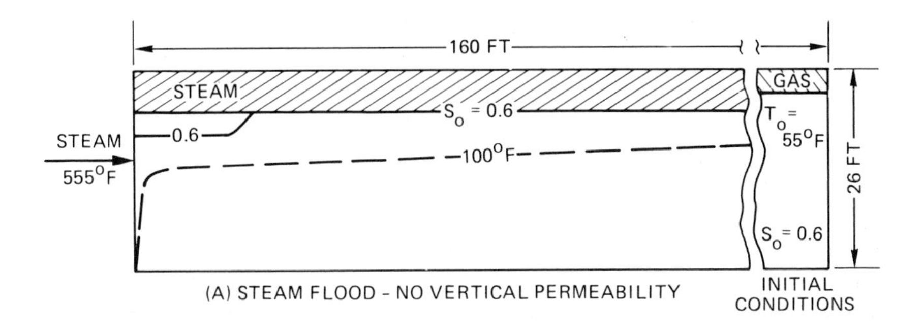

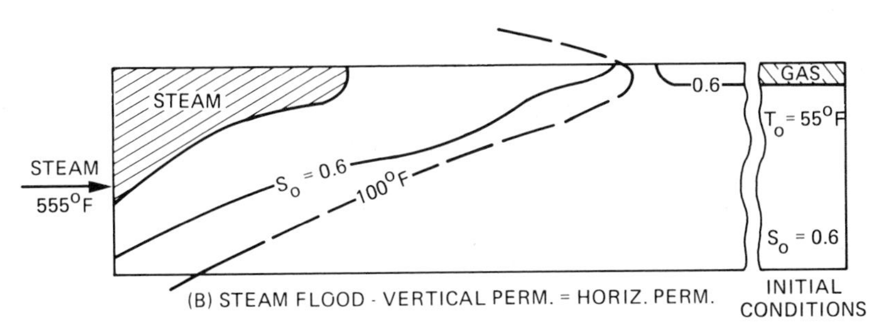

Figure 11.11. Effect of vertical flow on two-dimensional fluid and temperature distributions (after Shutler[4]). ©1970 SPE-AIME.

(b) Vertical permeability set equal to the horizontal permeability permitted efficient displacement (Fig. 11.11*B*), but because the oil bank flowed into the gas zone and cooled, it reduced injectivity to nearly zero for this highly viscous oil. The pressure drop across the model was the same for both cases shown in Figure 11.11.

Areal Effects Using Shutler Model

The Shutler two-dimensional model[4] was basically used as a vertical cross-sectional model in the above examples. However, a modified version has also been used to calculate areal effects.

This version calculates heat conduction in three dimensions but accounts for flow in only two dimensions, in the horizontal (*XY*) plane. Gravity effects are not included in this version, and the steam zone is assumed to be vertically uniform in the oil sand. Figure 11.12 gives calculated positions of the leading edge of the steam zone at three times during a steamflood of one-quarter of a scaled laboratory 5-spot model. The calculation indicates that the sweep efficiency of the steam zone will be remarkably good.

One explanation is that conductive heat transfer controls or at least greatly influences the shape of the steam zone. This prevents steam fingering and promotes good areal sweep. This observation has also been made in conducting laboratory model tests and explains why models of the vertical cross section rather than areal models appear to deserve greater attention in deriving an adequate prediction of the steam drive process.

Coats[6] and others have reported a substantial grid orientation effect on the frontal shape calculated by areal steamflood models employing five-point differencing. When simulating a similar one-quarter of a 5-spot, he observed that use of a grid that is parallel to the diagonal between injector and producer (rather than a grid such as used in Figure 11.12, which he refers to as a diagonal grid) results in a much more pronounced steam finger. The diagonal grid in his model results in a nearly radial steam front growth with cusping occurring only as the front approaches the producer.

The striking difference is illustrated in Figure 11.13. The effect occurs because of the much higher mobility of steam than oil in viscous oil systems. Improved difference equations have been formulated, such as the nine-point difference method, that seek to obtain greater accuracy of frontal movement predictions. The effect of grid type on calculated oil recovery is far less.

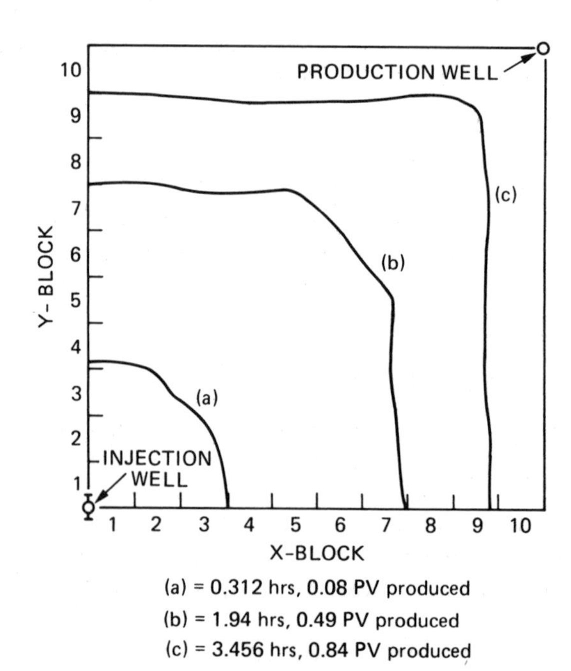

Figure 11.12. Numerical calculation of areal configuration of steam zone, simulation of one-fourth, 5-spot laboratory model.

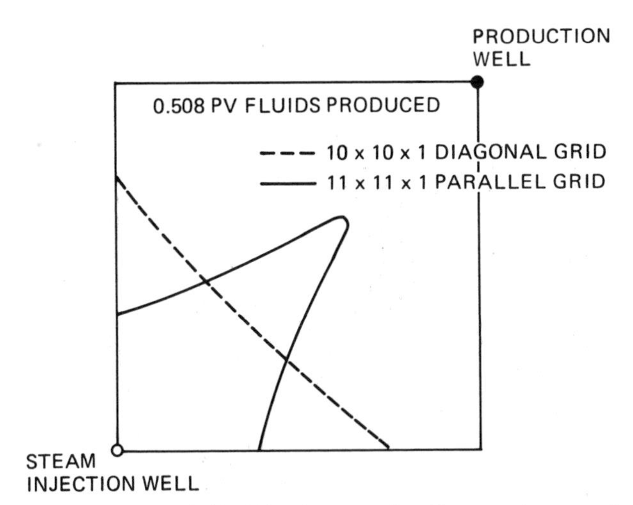

Figure 11.13. Comparison of 20% steam saturation contours calculated using diagonal and parallel grids.[6] ©1974 SPE-AIME.

Coats[6] reports that he has consistently calculated oil recovery curves using parallel grids that are identical in shape to and only slightly lower than those calculated from the diagonal grid despite the large difference in calculated frontal shape or breakthrough time.

Effect of Relative Permeability Curve Variations

Numerical simulations of cold waterfloods, hot waterfloods, and steamfloods were conducted using the two k_w/k_o curves given in Figure 11.14; results of these simulations are given in Figure 11.15. Note that while results of the cold waterfloods and hot waterfloods depend significantly on which curve is used, the steamflood is very little affected by a change in the k_w/k_o curves. This occurs because steam was injected at low pressure in the example shown, and the oil recovery is mostly from the steam zone. In a thinner sand or at a higher pressure, a greater fraction of the oil recovery would be from the hot waterflooded region, and in this case, water–oil relative permeability variation with temperature could more significantly affect overall steamflood oil recovery.

A number of workers have observed shifts in relative permeability curves as temperatures increase. Usually these shifts represent enhanced oil recovery at the elevated temperatures. In general, relative permeability shift effects appear to be only important in the case of hot waterfloods or for steamfloods where significant recovery occurs ahead of the steam zone in the hot-waterflooded region.

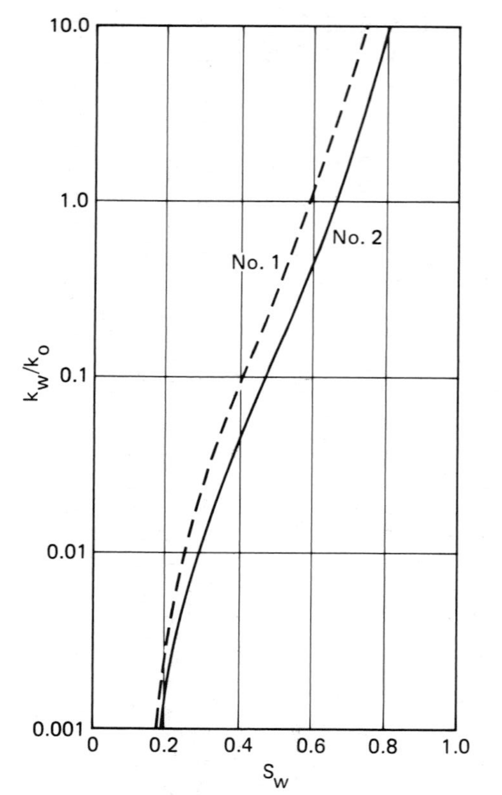

Figure 11.14. Variation in k_w/k_o curves tested in numerical runs.

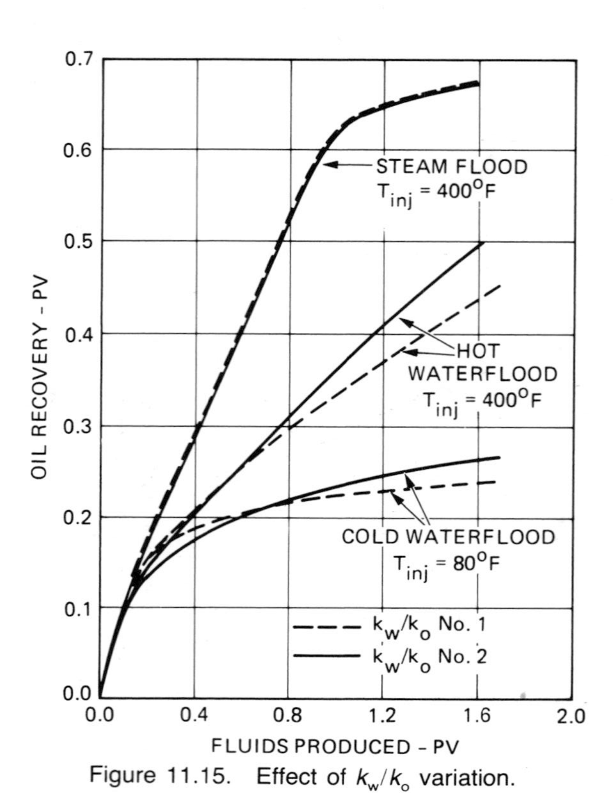

Figure 11.15. Effect of k_w/k_o variation.

Other Applications of Numerical Steamflood Simulators

There exists little published information on specific applications[17–22] of thermal simulation. Chu and Trimble[18] used a three-dimensional, three-phase numerical simulator to history-match five and one-half years of performance of a steam stimulation pattern in the Kern River field, California. The model was subsequently used to optimize operating parameters.

Gomaa, Duerksen, and Woo[19] also used a thermal simulator to model elements of an inverted 5-spot pattern to determine the relative importance of various steamflood and reservoir parameters for the Monarch sand in the Midway-Sunset field, California. Munoz[20] simulated steamflooding in the Tia Juana M-6 project, Venezuela. In this work, a three-dimensional model was prepared for an element of symmetry in a regular inverted 7-spot. These studies addressed the effects of permeability variations, gravity segregation, and positions of completion intervals.

Williams[21] prepared a three-dimensional model of production response of steam stimulation and steam displacement in the steeply dipping North Midway field in California. Another three-dimensional study of a steeply dipping reservoir was reported by Moughamian et al.[22] This work provided information used to design a steamflood in the heavy-oil-bearing Webster sand sequence in the Midway-Sunset field, California.

All of these studies treated a representative segment or an element of a repeated pattern where either steam stimulation or steam displacement was contemplated. None dealt with fieldwide simulation of on-going steamfloods with irregularly spaced wells.

Sobocinski et al.[23] describe the application of an areal two-dimensional thermal model to simulate a multiwell steamflood project in the Georgsdorf field[24] in West Germany. The reservoir consists of a single sand 32–38 m thick containing 120–160-cp oil. Sobocinski states that while simulation studies of the steamflood should ideally be based on three-dimensional models that consider fieldwide aspects of complicated geology and irregularly spaced wells, such models were difficult to justify for Georgsdorf because large amounts of computing time and memory space would be needed. Furthermore, it was not feasible to develop representative cross-sectional models for Georgsdorf because such cross sections were difficult to identify.

Most of the study objectives were related to the areal movement of fluids in a large irregular pattern development, and therefore, an areal model was built. There was concern, however, whether such a model could provide meaningful information because the vertical distribution of fluids and temperature would be modeled only approximately in an areal model. The rough spacing–thickness guideline for a one-layer representation of 7:1 (based in part on the results shown in Fig. 11.9 and 11.10) was not quite satisfied as the spacing–thickness ratio was 5–6 for this case. Nevertheless, it was found that permeability stratification and gravity effects could be adequately accounted for by modifying the relative permeability curves to enable the model to match observed reservoir behavior.

Figure 11.16 shows the modeled portion of the Georgsdorf field where the steamflood projects are being operated. Figure 11.17 shows the 38×12 areal model grid. There were 39 active wells in the modeled area. An excellent match of the cumulative oil production performance of the steamflood that began in 1975 is shown in Figure 11.18. The match was achieved primarily by altering absolute and relative permeability data. The model was then used to predict the effects of several operating alternatives.

At this time, numerical simulation of thermal processes is so expensive that there are almost no models built having more than 1000 grid blocks. These are very small models when considering the complexities of reservoir heterogeneities and the interactions of numerous wells in an actual project. Whole project simulation involving many wells may require a grid so coarse that the model will not adequately represent the physics of the process. Models of this type may be little more than energy/fluids bookkeeping vehicles and are probably of little or no value. Such models may retain correct overall energy and material balances, but the distribution of temperatures and saturations will be inaccurate to such an extent that results will be incorrect. We seem to be in a stage that may require the judicious use of both analytical and numerical models to obtain good answers, especially for those situations where coarse gridding cannot be avoided.

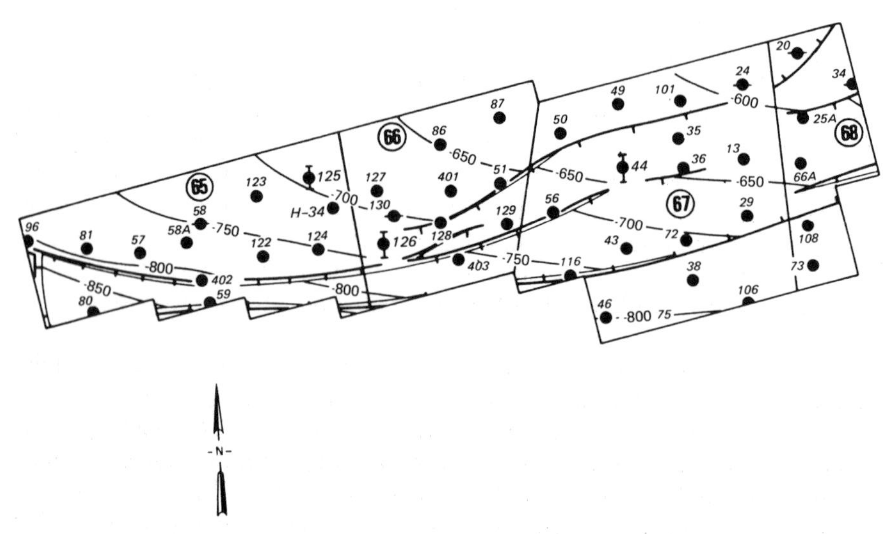

Figure 11.16. Structure contours and well locations for Georgsdorf steamflood I projects.[23] © 1984 SPE-AIME.

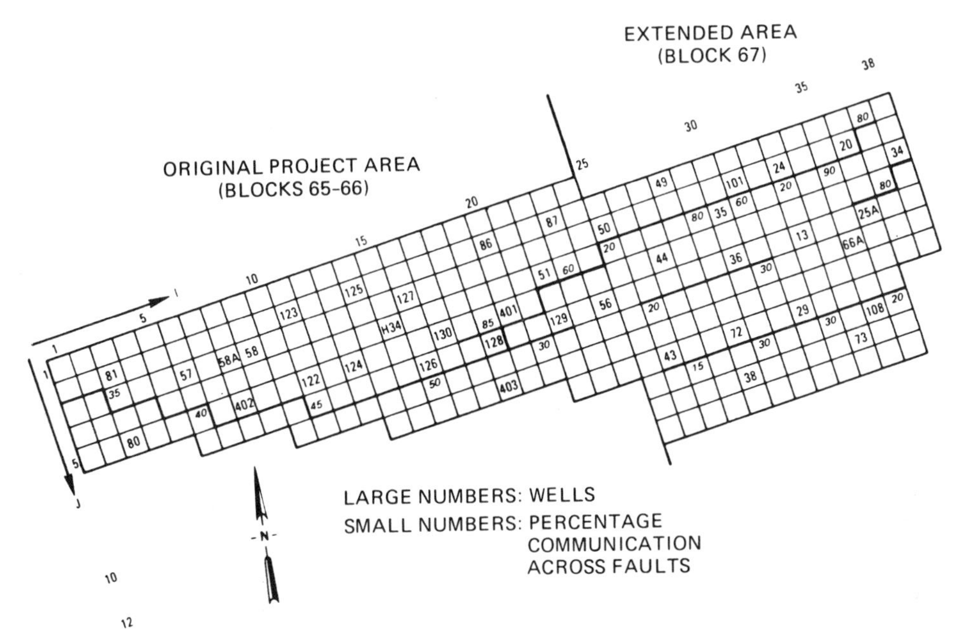

Figure 11.17. Areal model for Georgsdorf steamflood I projects.[23] © 1984 SPE-AIME.

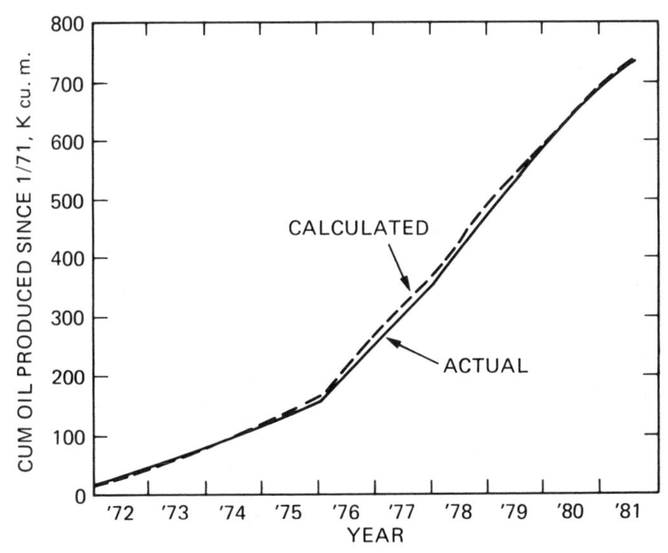

Figure 11.18. Oil recovery vs. time for Georgsdorf blocks 65–67.[23] ©1984 SPE-AIME.

Undoubtedly, further advances in computing hardware and numerical techniques will be required to achieve widespread use of numerical simulators for steamflooding and steam stimulation. The pioneering efforts reported in this chapter will someday appear rudimentary when these advances are achieved.

APPENDIX: CALCULATION DATA USED IN SHUTLER MODEL EXAMPLES

This appendix contains data (see following and Table 11A.1) used in the calculations discussed and indicates values for two different systems modeled by the Shutler two-dimensional model:

(A.1) Laboratory model of one-eighth of a 5-spot: applicable to Figures 11.5, 11.6, 11.9, and 11.10.

(A.2) One-eighth of a $1\frac{1}{4}$-acre 5-spot: applicable to Figure 11.6.

	A.1	A.2
$k(k_x = k_y)$, darcies	132	5
ϕ, fraction	0.375	0.33
L, ft	3.9	160
Sand thickness, ft	0.83	26
K_{hR}, Btu/ft-day-°F	24	35
K_{hob}, Btu/ft-day-°F	12	34
$(\rho c)_R$, Btu/ft^3-°F	36	40
$(\rho c)_{ob}$, Btu/ft^3-°F	30	36
T_i, °F	80	55
S_{wi} (bottom block), fract.	0.1	0.4
S_{gi} (top block), fract.	0	0.4
T_{inj}, °F	400	555
P_{inj}, psi	260	1100
P_p, psi	≥190†	100
μ_o(cp) at T (°F) = 50	5000	500,000
150	126	800
250	20	41
350	7.1	8.8
450	4.0	3.6
550		2.0

†$q_p \approx 1$ bbl/d.

Table 11A.1 Relative Permeability and Capillary Pressure Data

A.1

S_w	k_w	$k_{o_{ow}}$	$P_{c_{ow}}$	S_{o_2} [a]	k_g	$k_{o_{og}}$	$P_{c_{og}}$
0.1	0	1.0	4.111	0.1	0.520	0	4.517
0.2	0.0016	0.875	0.095	0.2	0.410	0.009	0.067
0.3	0.0081	0.735	0.072	0.3	0.310	0.031	0.042
0.4	0.0259	0.590	0.061	0.4	0.220	0.062	0.020
0.5	0.0672	0.420	0.051	0.5	0.140	0.110	−0.001
0.6	0.1000	0.210	0.041	0.6	0.080	0.190	−0.022
0.7	0.1400	0.070	0.031	0.7	0.030	0.335	−0.043
0.8	0.2000	0.016	0.021	0.8	0.005	0.570	−0.064
0.86	0.2500	0	0.011	0.89	0	1.000	−0.085

A.2

S_w	k_w	$k_{o_{ow}}$	$P_{c_{ow}}$	S_{o_2} [a]	k_g	$k_{o_{og}}$	$P_{c_{og}}$
0.24	0	0.916	1.6	0.3	0.19	0	0.38
0.3	0.0066	0.495	0.27	0.4	0.12	0.01	0.28
0.4	0.0190	0.197	0.14	0.5	0.055	0.04	0.21
0.5	0.0420	0.072	0.084	0.6	0.015	0.125	0.16
0.6	0.1060	0.007	0.06	0.7	0.001	0.38	0.12
0.7	0.2300	0	0.04	0.76	0	0.70	0.11

[a] $S_g = 1 - S_{w_{ir}} - S_{o_2}$.

REFERENCES

1. Spillette, A. G., and Nielsen, R. L. "Two-Dimensional Method for Predicting Hot Waterflood Recovery Behavior," *J. Pet. Tech.* (1968), p. 627.
2. Weinstein, H. G., Wheeler, J. A., and Woods, E. G. "Numerical Model for Steam Stimulation," *Soc. Pet. Eng. J.* (February 1977), p. 65.
3. Shutler, N. D. "Numerical Three-Phase Model of the Linear Steamflood Process," *Soc. Pet. Eng. J.* (June 1969), p. 232.
4. Shutler, N. D. "Numerical Three-Phase Model of the Two-Dimensional Steamflood Process, *Soc. Pet. Eng. J.* (December 1970), p. 405.
5. Abdalla, A., and Coats, K. H. "A Three-Phase, Experimental and Numerical Simulation Study of the Steamflood Process," SPE 3600, presented at the Forty-Sixth Annual Fall Meeting of the Society of Petroleum Engineers of AIME, New Orleans, Louisiana, October 3–6, 1971.
6. Coats, K. H., George, W. D., Chu, C., and Marcum, B. E. "Three-Dimensional Simulation of Steamflooding," *Soc. Pet. Eng. J.* (December 1974), p. 573.
7. Coats, K. H. "A Highly Implicit Steamflood Model," *Soc. Pet. Eng. J.* (October 1978), p. 369.
8. Coats, K. H. "Simulation of Steamflooding with Distillation and Solution Gas," *Soc. Pet. Eng. J.* (October 1976), p. 235.

9. Stone, H. L. "Probability Method for Estimating Three-Phase Relative Permeability," *J. Pet. Tech.* (February 1970), p. 214.
10. Douglas, J., Jr., Peaceman, D. W., and Rachford, H. H., Jr. "A Method for Calculating Multidimensional Immiscible Displacement," *Trans. AIME*, **216,** 297 (1959).
11. Weinstein, H. G., Stone, H. L., and Kwan, T. V. "Simultaneous Solution of Multiphase Reservoir Flow Equations," *Soc. Pet. Eng. J.* (June 1970), p. 99.
12. Peaceman, D. W., and Rachford, H. H., Jr. "The Numerical Solution of Parabolic and Elliptic Differential Equations," *J. Soc. Ind. Appl. Math.*, **3,** 28 (1955).
13. Willman, B. T., Valleroy, V. V., Runberg, G. W., Cornelius, A. J., and Powers, L. W. "Laboratory Studies of Oil Recovery by Steam Injection," *J. Pet. Tech.* (July 1961), p. 681.
14. Pursley, S. A. "Experimental Studies of Thermal Recovery Processes," paper presented at Symposium on Heavy Oil Recovery, University of Zulia, Maracaibo, Venezuela, July 1–2, 1974.
15. Pujol, L., and Boberg, T. C. "Scaling Accuracy of Laboratory Steamflooding Models," SPE 4191, presented at the California Regional Meeting of the Society of Petroleum Engineers, Bakersfield, California, November 8–10, 1972.
16. Connally, C. A., Jr., and Marberry, J. E. "Steamflood Pilot, Nacotoch Sand-Troy Field, Arkansas," SPE 4758, presented at SPE Symposium on Improved Oil Recovery, Tulsa, Oklahoma, April 22–24, 1974.
17. Coats, K. H. "Reservoir Simulation: State-of-the-Art," *J. Pet. Tech.* (August 1982), p. 1633–1642.
18. Chu, C., and Trimble, A. F. "Numerical Simulation of Steam Displacement-Field Performance Applications," *J. Pet. Tech.* (June 1975), pp. 765–776.
19. Gomaa, E. E., Duerksen, J. H., and Woo, P. T. "Designing a Steamflood Pilot in the Thick Monarch Sand of the Midway-Sunset Field," *J. Pet. Tech.* (December 1977), pp. 1559–1568.
20. Munoz, J. D. "Numerical Simulation of Steam Drive in the Tia Juana M-6 Project," *The Future of Heavy Crude and Tar Sands*, McGraw-Hill, New York, 1981, Chapter 47.
21. Williams, R. L. "Steamflood Pilot Design for a Massive, Steeply Dipping Reservoir," SPE 10321, presented at the SPE Annual Technical Conference, San Antonio, Texas, October 5–7, 1981.
22. Moughamian, J. M., Woo, P. T., Dakessian, B. A., and Fitzgerald, J. G. "Simulation and Design of Steam Drive in a Vertical Reservoir," *J. Pet. Tech.* (July 1982), pp. 1546–1554.
23. Sobocinski, D. P., Leskova, H. and Greebe, F. "Simulating a Steamflood at Georgsdorf: An Adjunct to Reservoir Management," *J. Pet. Tech.* (November 1984), pp. 1952–1964.
24. Lillie, W. H. E., and Springer, F. P. "Status of the Steam Drive Pilot in the Georgsdorf Field, Federal Republic of Germany," *J. Pet. Tech.* (January 1981), pp. 173–180.

Chapter 12

ADVANCED ONE-DIMENSIONAL ANALYTICAL METHOD FOR STEAMFLOODING AND HOT WATERFLOODING

ABSTRACT

The Marx–Langenheim or Myhill–Stegemeier analytical models for steamflooding described in Chapter 9 require the assumption of a steam zone or steam plus hot-water zone average oil saturation that must be derived independently from laboratory or field data. This chapter presents the Shutler–Boberg graphical techniques utilizing Buckley–Leverett shock theory to calculate saturation distributions in the steam and hot-water zones as well as in the portion of the reservoir still at original reservoir temperature. Since the method is laborious, it is not often used, but it aids in the understanding of how the very low oil saturations observed in the steam zone come about.

The method treats the oil as nondistillable; thus, it underpredicts oil recovery for lighter oils, which have a significant distillable fraction. The Shutler–Boberg method is recommended only for situations discussed earlier, where the assumptions inherent in the Marx–Langenheim heat transfer equation apply. It should be used in preference to the methods of Chapter 9 for oil reservoirs where significant oil depletion prior to steamflooding has occurred, for example, by prior waterflooding. The method accounts for the delay in oil production response when the bank of steam-displaced oil moves through the depleted oil sand before reaching the production well. This effect is neglected in the Marx–Langenheim or Myhill–Stegemeier methods. Use of the Shutler–Boberg method for continuous hot-water injection and for continuous steam injection is illustrated in this chapter.

INTRODUCTION

Mathematical models of the thermal flood processes may be generally categorized as numerical or analytical. Numerical models may be extremely comprehensive and serve as tools for research or advanced reservoir analysis. Analytical models may be much more economical at the expense of accuracy and flexibility and serve as tools for engineering screening of possible reservoir candidates for field testing. The Shutler–Boberg[1] model described in this chapter is analytical, but it is more rigorous in its treatment of fluid flow than the simplified methods discussed in Chapter 9. The Shutler–Boberg model as published in 1972 only applied to steamflooding. In this chapter, further details are given permitting its extension to the case of hot waterflooding as well.

The method is one dimensional and therefore neglects gravity effects. Thus, it is useful only for sands where the spacing–thickness ratio is 7 or greater. The heat transmission is calculated by the Marx and Langenheim equation for the case of steam or the Lauwerier equation for the case of hot waterflooding.

Analytical models of thermal flood processes consist of solutions to simplified heat transfer and one-dimensional fluid flow equations. The Shutler–Boberg[1] model does likewise. However, the earlier analytical methods either neglected heat loss while accurately treating nonisothermal two-phase fluid flow[2,3] or greatly simplified the treatment of multiphase fluid flow while accounting for heat loss.[4–6] The Shutler–Boberg model[1] overcomes these shortcomings by adequately accounting for both heat loss and nonisothermal multiphase fluid flow.

The fluid flow equations, however, do not account for distillation of light ends. The method accounts for oil bank formation and movement that occurs when steam is injected into sands where oil saturation has been reduced by prior waterflooding. In these systems, a delayed oil production response will occur. This behavior is not predicted by the more simplified methods, such as those discussed in Chapter 9, and use of the Shutler–Boberg method is required.

GENERAL APPROACH

The model is based on the physical system illustrated in Figure 12.1. The reservoir is considered in linear cross section and is bounded above and below by impermeable strata. Steam or hot water is injected at one end of the reservoir while oil and water are produced at the other. Heat losses via thermal conduction to adjacent strata are considered.

In the following three sections, we discuss (1) the aspects of heat transfer, which in a general sense apply equally to both steam or hot-water drive; (2) the fluid flow aspects of hot waterflooding and the accompanying considerations in coupling; and (3) the fluid flow aspects of steamflooding and the accompanying considerations in coupling. In the Shutler–Boberg paper,[1] the discussion of steamflooding is further subdivided into three variations in the

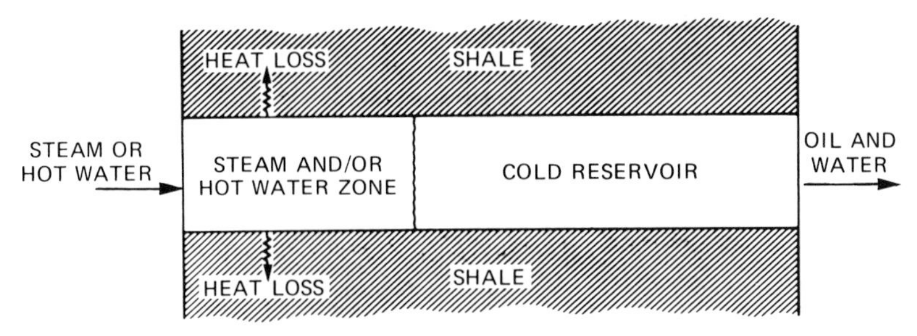

Figure 12.1. Physical system modeled by Shutler–Boberg method.[1] © 1972 SPE-AIME.

calculation procedure, each of which is successively more complex and more accurate. Only the simplest variation is presented here. The reader is referred to the appendices of this chapter for the theoretical bases of concepts and equations. Examples of their applications are presented in Appendix C.

RESERVOIR HEAT TRANSFER

The object of the heat transfer calculation is to obtain a representation of temperature along the length of the reservoir as a function of time during the process (see Fig. 12.2*a*). Any appropriate method for calculating a temperature profile $T(x, t)$ during hot-fluid injection into porous media may be used. Here the methods of Lauwerier[7] and Marx and Langenheim[8] are used for hot-water and steam injection, respectively (see Chapter 3).

In order to make the nonisothermal problem susceptible to application of the isothermal two-phase flow theory of Buckley and Leverett,[9] the Shutler–Boberg method represents $T(x, t)$ by a series of $n + 1$ moving isothermal zones of temperature T_i, $i = 1, \ldots, n + 1$, separated by discontinuities, or "shocks," in temperature (Fig. 12.2*a*). The positions of these shocks are denoted by $x_i(t)$, $i = 1, \ldots, n$, and coincide with the positions of temperatures $T_{i+1/2} = \frac{1}{2}(T_i + T_{i+1})$ (Fig. 12.2). The instantaneous velocities of the shocks are given by slopes of the curves in Figure 12.3,

$$C_i = \frac{dx_i(t)}{dt} \qquad i = 1, \ldots, n \tag{12.1}$$

and the average velocities $\bar{C}_i$ over any period of time t are given by the slopes of the chords from $x(0)$ to $x(t)$ in Figure 12.3. The utility of these velocities will appear later. These shock velocities depend directly on the volumetric heat capacity

$$(\rho c)_{\mathrm{R+F}} = \phi(\rho_{\mathrm{w}} c_{\mathrm{w}} \bar{S}_{\mathrm{w}} + \rho_{\mathrm{o}} c_{\mathrm{o}} \bar{S}_{\mathrm{o}}) + (1 - \phi)(\rho c)_{\mathrm{R}} \tag{12.2}$$

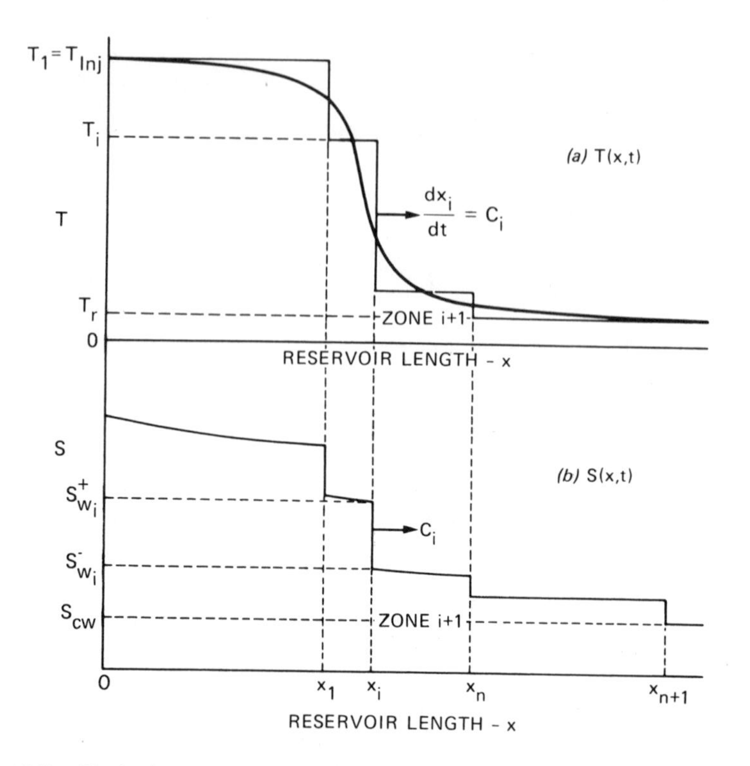

Figure 12.2. Typical temperature $T(x, t)$ and saturation $S(x, t)$ profiles at time t.[1] © 1972 SPE-AIME.

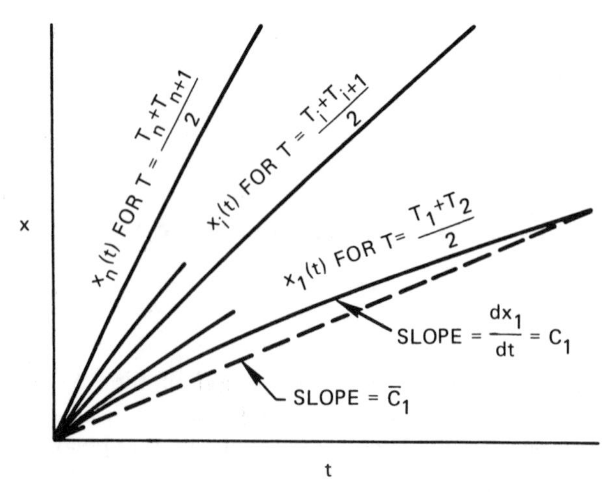

Figure 12.3. Temperature shock positions $x_i(t)$.

in the heated zone, which in turn depends on fluid flow through the appearance of $\bar{S}_w$ and $\bar{S}_o$, the average water and oil saturations in the heated zone.

Experience shows that, depending on the particular problem, the number of zones required for accurate oil recovery prediction varies from 2 to 5. In the case of steam injection, two-zone representations are adequate provided the heated zone ahead of the steam zone is small. If low-quality steam injection or very thin reservoirs are considered, more zones should be used, but a more comprehensive heat transfer calculation than that of Marx and Langenheim (which itself assumes two zones) would be needed to obtain the proper temperature profile.

FLUID FLOW: HOT WATERFLOOD

To describe the fluid flow aspect of the problem, we apply the method of Buckley and Leverett[9] to each of the isothermal zones. The following are the assumptions of this method:

1. Capillary pressure is not important.
2. Constant fluid properties apply within each isothermal zone.
3. Total fluid velocity is constant within each isothermal zone.

According to the Buckley–Leverett theory of simultaneous flow of oil and water, the movement within zone i of planes of constant water saturation S_w is described by

$$\left(\frac{dx}{dt}\right)_{S_w} = \frac{v_i}{\phi} f'_{w_i} \tag{12.3}$$

where f'_{w_i} is the slope of the fractional flow curve and v_i is the fluid velocity for zone i (see nomenclature for definitions). Integration of Eq. (12.3) allows the calculation of the saturation profile within zone i at any time t:

$$x_{S_w}(t) = x_{S_w}(0) + \frac{v_i}{\phi} f'_{w_i} t \tag{12.4}$$

Hence, for the first zone, $i = 1$, given the initial saturation distribution (assumed to be uniform) and the boundary condition S_{wI} (or f_{wI}) and inlet velocity v_I, one can calculate the saturation distribution at any point in time by application of the Buckley–Leverett procedure using the curve for f_{w_1} in Figure 12.4.

When the temperature shock at x_1 is encountered, special consideration is required. Since f_w changes discontinuously across the temperature shock, it

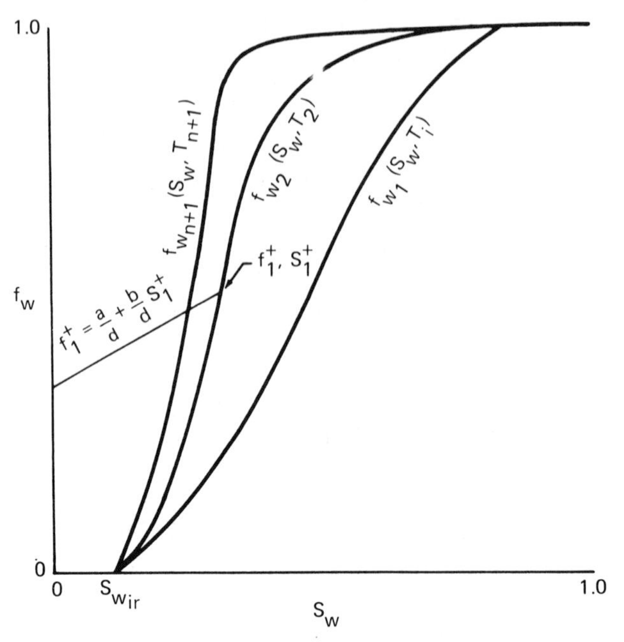

Figure 12.4. *f* vs. *S* for values of *T*.

follows that in any practical situation, a saturation shock will be associated with each temperature shock (Fig. 12.2b). The value of S_w (or f_w) on the downstream side of the first shock, which we denote $S_{w_1}^+$ (and $f_{w_1}^+$), comprises the boundary condition for the second isothermal zone. To obtain this boundary condition and the velocity $v_2 = v_1^+$ and conserve mass at the same time, a mass balance is written on each of the phases (water and oil) across the zone boundary (temperature shock). The resulting equations reduce to the form

$$f_{w_1}^+ = \frac{a}{d} + \frac{b}{d} S_{w_1}^+ \tag{12.5}$$

and

$$v_1^+ = v_1 - d \tag{12.6}$$

where a, b, and d are composed of known quantities, that is, upstream $S_{w_i}^-$ (and $f_{w_i}^-$) and phase densities in both zones (see Appendix A). Equation (12.5) plots as a straight line in Figure 12.4, and its intercept with f_{w_2} yields the values for $f_{w_1}^+$ and $S_{w_1}^+$. With these values known, $S_{w_1}^+$ (and $f_{w_1}^+$) and $v_2 = v_1^+$, the Buckley–Leverett theory may be applied to the second zone.

The procedure just outlined is repeated for each subsequent isothermal zone and accompanying shock. An additional saturation shock at x_{n+1} will be propagated into the unheated zone at a velocity $C_{n+1} > C_n$ (Fig. 12.2). This last saturation shock is not associated with a temperature shock but rather is much the same phenomenon as the Buckley–Leverett shock though its size, shape, and velocity may be different from that obtained from the isothermal flood. Six possible configurations for this shock are described in Appendix B. For this last shock, $S_{w_{n+1}} = S_{cw}$.

In practice, it is usually not necessary to calculate the curved saturation distributions generally associated with Buckley–Leverett saturation profiles. Shortcuts make the calculation proceed more rapidly. These are discussed in Appendix B and are illustrated in the example problem of Appendix C.

To account for heat losses in any analytical description of a thermal process, it is necessary to consider the heat transfer and fluid flow aspects of the problem independently. Thus, we estimate the average fluid saturations $\bar{S}_w$ and $\bar{S}_o$ in the heated region in calculating volumetric heat capacities [Eq. (12.2)] on which temperature shock velocities depend. The shock velocities in turn affect the saturation distributions on which $\bar{S}_w$ and $\bar{S}_o$ depend. More specifically, the shock velocity C_1 affects the average saturation in the first zone because it determines the value $S_{w_1}^-$ at the shock through the requirement that $S_{w_1}^-$ be the saturation that has traveled the same distance as the first shock, that is, $x_{S_{w_1}^-}(t) = \int_0^t C_1 \, dt$.

This velocity C_1 also affects the saturation distribution in the second zone by its inclusion in the parameters of Eqs. (12.5) and (12.6). This explanation applies repeatedly to downstream shock velocities and zonal saturation distributions.

In the case of a hot waterflood, the effect of different fluid saturations on temperature shock velocities is small because the volumetric heat capacities of oil and water differ only by about a factor of 2. Hence, the choice of $\bar{S}_w$ and $\bar{S}_o$ is not critical, the interdependence of shock velocities and fluid saturations is not strong, and the problem of heat transfer–fluid flow coupling is not of great concern. In the case of a steamflood, however, the effect of the average fluid saturations in the steam zone on shock velocity is greater because the volumetric heat capacity of steam is about two orders of magnitude smaller than that of water or oil.

FLUID FLOW: STEAMFLOODING

In the case of steam injection, the Shutler–Boberg model assumes injection of 100% quality steam, assumes it to be incompressible, and neglects any hydrocarbon vaporization. The procedure described in the previous section is applied to all zones downstream of the steam zone. With regard to the steam zone itself, the procedure assumes two-phase fluid flow in the steam zone, and the calculation is analogous to that for hot waterflooding, except that steam and oil flow in zone 1 rather than water and oil.

Actually, steam, liquid water (steam condensate), and oil will be flowing simultaneously, but the simplest version of the Shutler–Boberg model neglects the flow of liquid water. The liquid water saturation in the steam zone is assumed to be at irreducible saturation, $S_{w_1} = S_{wir}$, so the fractional flow of water in the steam zone (f_{w_1}) is zero (see Fig. 12.5). For the two-phase flow of steam and oil,

$$f_s = \frac{k_s(S_s)/\mu_s(T_1)}{k_s(S_s)/\mu_s(T_1) + k_o(S_o)/\mu_o(T_1)} \tag{12.7}$$

where $S_s + S_o = 1 - S_{wir}$. Then, with the Buckley–Leverett relationship,

$$\left(\frac{dx}{dt}\right)_{S_s} = \frac{v_1}{\phi} f_s' \tag{12.8}$$

we can calculate the steam zone saturation profile of Figure 12.5.

Consistent with the assumption of irreducible water saturation in the steam zone is the assumption that all steam condensation occurs at the steam front, that is, no condensation occurs within the steam zone to give rise to a mobile liquid-water phase there. Hence, the total fluid velocity throughout the steam zone is equal to the injection velocity, $v_1 = v_I$.

To obtain the boundary condition $S_{w_1}^+$ and the velocity $v_2 = v_1^+$ for the two-phase second zone during steam injection, shock equations similar to (12.5) and (12.6) are employed (see Appendix A).

The problem of heat transfer–fluid flow coupling is of much greater importance in a steamflood than in a hot waterflood. Because of the great difference between the volumetric heat capacities of steam and those of the liquid phases, the interdependence of temperature shock velocities and

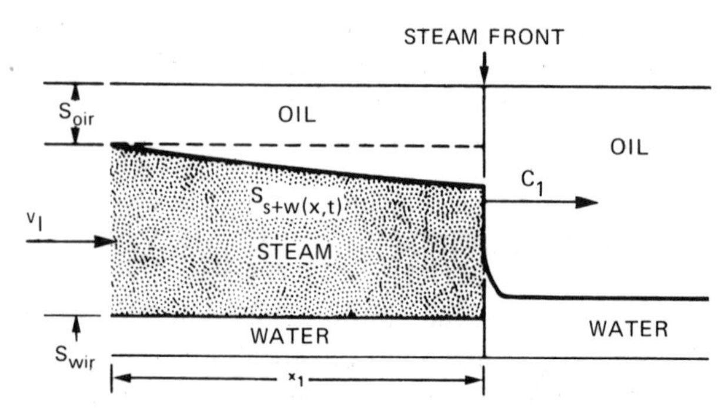

Figure 12.5. Qualitative three-phase, one-dimensional saturation profiles $S(x, t)$ at time t. $S_{w_1}(x, t) = S_{wir}$.[1] © 1972 SPE-AIME.

average fluid saturations in the heated zone is strong. In fact, the volumetric heat capacity of steam is sufficiently small relative to that of oil or water that it may be ignored. Hence, to calculate the volumetric heat capacity $(\rho c)_{R+F}$ of the steam zone, Eq. (12.2) can still be used, but now instead of $\bar{S}_w + \bar{S}_o = 1$, we have $\bar{S}_w + \bar{S}_o = 1 - \bar{S}_s$.

To be consistent with the assumption of $S_{w_1} = S_{wir}$, it follows that $\bar{S}_w = S_{wir}$. Further, to retain the simplicity of this procedure and in light of the error already introduced by the assumption $S_{w_1} = S_{wir}$, the error is accepted in the heat transfer–fluid flow coupling. This is introduced by merely estimating a value of $\bar{S}_o$ to be $\bar{S}_o = S_{oir}$.

In more elaborate variations of the Shutler–Boberg method[1] than the one presented here, accuracy is improved by recalculating $(\rho c)_{R+F}$ as improved estimates are made of S_o and S_w. In addition, the estimate of v_1 is improved by accounting for the steam condensation occurring because of heat losses to the overburden and underburden.

In Figures 12.6–12.12, except where indicated in Figure 12.10, these refinements of the Shutler–Boberg (Analytic) method have been incorporated to improve the accuracy of the technique. In some of the examples, a time–distance averaging approximation has been applied to estimate v_1.

OIL RECOVERY

In this section, we discuss an approach to implement the theory to obtain oil recovery predictions. The approach presented here involves use of average temperature shock velocities. This is much faster than a second variation of the Shutler–Boberg method, which uses instantaneous shock velocities. While the latter may yield more accurate though slightly optimistic results at the price of more computational effort, the method discussed here will yield accurate-to-conservative predictions.

If the shock velocities C_i, $i = 1, \ldots, n$, and the injection velocity v_I are constant, the saturations $S^-_{w_i}$ (or $S^-_{s+w_i}$) and $S^+_{w_i}$ (or $S^+_{s+w_i}$) and velocities v_i will also be constant with time. In this case, these quantities need be calculated only once, and the saturation profile can be estimated at successive times by merely tracking the positions of the shocks.

While the shock velocities will usually decrease with time as a result of heat loss from the reservoir, one may assume a constant average velocity equal to the slope of a chord from the origin to the position x_i of shock i (in Fig. 12.3) at the time of interest. The time of interest will be the period of time over which one wishes to estimate recovery. This will not normally be much greater than the time for breakthrough of zone n, the most advanced thermal zone, since oil recovery rate with respect to produced fluids declines sharply thereafter.

With the shock velocities constant, then, in addition to shock saturations being constant with time, in the case of heavy oils, it appears that most, if not all, zones beyond the first will each have constant saturation distributions with respect to distance, that is, $S_{w_{i+1}} = S_{w_i}$, $i \geq 1$ (see Appendix C). This feature greatly facilitates the oil recovery calculation by the obvious method of integration under the saturation profile. But it makes possible an even faster method of predicting recovery.

If the saturation distributions are constant within each zone, the producing water–oil ratio (which depends on the saturation at the producing end) will be constant during the interval between times of breakthrough of successive zones. Hence, if the producing period is broken into intervals bounded by times of arrival of the shocks at the producing end, the curve of oil recovery versus produced fluids will be composed of straight-line segments whose slopes are determined from the producing water–oil ratio and whose lengths are determined from the associated producing velocities and time intervals. After breakthrough of a zone with a varying saturation, the graphical method of Welge[10] may be used.

At reduced shock velocities, the rate of oil recovery will be reduced with respect to the produced fluids. Since average shock velocities are less than the actual velocities over the first part of the time of interest and greater over the later part, it follows that oil recovery rate (with respect to produced fluids) is conservative over the early time and optimistic over the later time (see Fig. 12.6). If the thermal efficiency is high (relatively low heat loss), however, the shock velocity variation with time is not great, and the average shock velocity approach gives accurate results (see Fig. 12.9).

EXAMPLE RESULTS OF SHUTLER–BOBERG METHOD

Hot-Waterflood Examples

Figures 12.6 and 12.7 present oil recovery curves calculated for the system described in Table 12.1. The broken and the solid lines show the recovery calculated by the method presented here. The curves represented by the points were calculated numerically by the hot-waterflood simulator of Spillette and Nielsen.[11] The dashed line shows recovery calculated by a more simplified method, the method of Jordan, et al.[4,6] The cold-waterflood curve is presented for reference. The early conservatism of the numerical hot-water curve relative to the cold-water curve is an effect of capillary pressure, which is included in the numerical hot-water calculation but not in the other calculations.

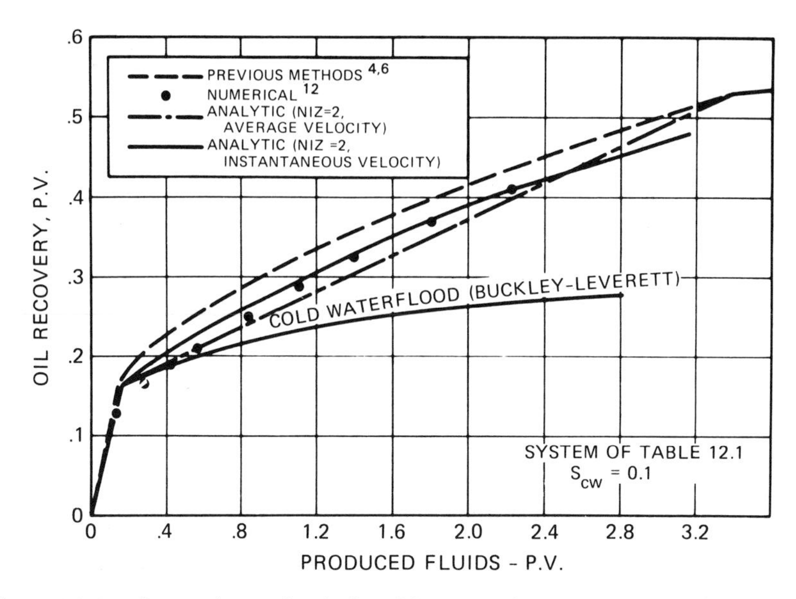

Figure 12.6. Comparison of calculated hot-waterflood recoveries, $S_{cw} = S_{wir}$.

Oil Recovery, No Mobile Water Initially Present. The broken-line curve in Figure 12.6 was calculated by the present method using two isothermal zones (NIZ = 2) and an average shock velocity. This method is extremely simple and rapid. For engineering purposes, it gives sufficiently accurate results. As discussed earlier, this approach usually gives a conservative prediction of oil recovery except at late times.

When only two isothermal zones are selected, the first breakthrough by a heated zone occurs at a later time than would be the case if a greater number of isothermal zones (including zones at lower temperatures) were selected. Since the oil recovery rate is higher before thermal breakthrough than after, any delay in the time of thermal breakthrough causes an overshoot by the recovery curve.

When the thermal front does arrive, however, the ensuing portion of the recovery curve is flatter than it would have been after an earlier breakthrough. The same ultimate recovery is achieved no matter how many zones are used. Since recovery behavior at and immediately after thermal breakthrough is more important than later recovery behavior, it is desirable to concentrate the zones at the leading (low-temperature) end if several zones are used.

The solid line in Figure 12.6 shows the recovery curve calculated by this method for two isothermal zones and instantaneous shock velocities. Since early calculation provides only one point on the recovery curve, this

Table 12.1 Field System, Physical Characteristics for Figures 12.6–12.8

$L = 650$ ft
$h = 60$ ft
$w = 1098$ ft
$\phi = 0.326$
$K_h = 24$ Btu/ft-d-°F
$(\rho c)_{ob} = 36$ Btu/ft^3-°F
$(\rho c)_R = 35$ Btu/ft^3-°F
$(\rho c)_o = 23.5$ Btu/ft^3-°F (average in heated region)
$(\rho c)_w = 63.8$ Btu/ft^3-°F (average in heated region)
$c_w = 1.02$ Btu/lb-°F

Assuming $\bar{S} = 0.5$ in heated region,
$(\rho c)_{R+F} = (1 - \phi)(\rho c)_R + \phi[\bar{S}(\rho c)_w + (1 - \bar{S})(\rho c)_o] = 37.9$
$m_w = 1{,}008{,}000$ lb/d
$T_r = 100$°F
$T_I = 380$°F
$S_{cw} = 0.1$ (or 0.29)
$(\rho_o)_{ST} = 56.1$ lb/ft^3
$(\rho_w)_{ST} = 66.5$ lb/ft^3

T(°F)	ρ_o(lb/ft^3)	ρ_w(lb/ft^3)	μ_o(cp)	μ_w(cp)
100	54.4	66.6	170	0.92
140	52.9	65.9	61	0.61
200	50.4	64.9	21	0.39
380	43.0	59.3	3.5	0.18

S_w	k_w/k_o
0.1	0
0.2	0.008
0.3	0.14
0.4	0.75
0.5	3
0.6	15
0.7	∞

approach is neither as rapid nor as convenient as the approach using an average velocity. However, as we see from the comparison of the recovery curves, in this case, the agreement with the numerical recovery curve is better than in the case in which an average velocity was assumed.

Oil Recovery, Mobile Water Initially Present. Figure 12.7 shows the discrepancy between curves calculated by the present method (and the numerical method) and the simplified method of Jordan et al.[4] early in the flood. This occurs because an oil bank is formed, and its arrival at the

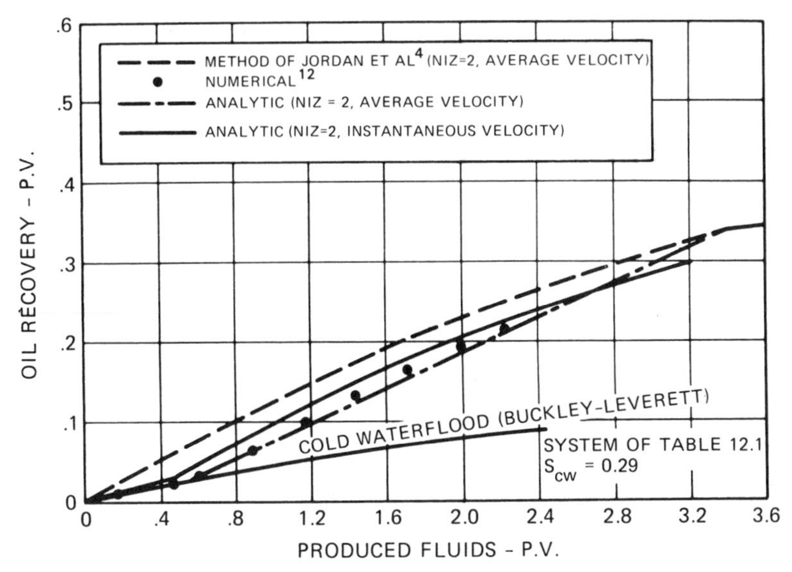

Figure 12.7. Comparison of calculated hot-waterflood recoveries, $S_{cw} > S_{wir}$.

production end (at about 0.5 PV produced fluids) is marked by a sharp increase in oil recovery rate. The previous simplified method[4] did not account for this oil bank formation but rather predicted oil recovery at a fairly uniform rate over the life of the flood. This discrepancy becomes greater with increasing oil viscosity and/or mobile water saturation.

Saturation Profiles. In Figure 12.8, we show comparisons of numerically[11] calculated saturation profiles and those calculated analytically by the Shutler–Boberg method. Figure 12.8*a* shows the saturation profile corresponding to 0.14 PV produced fluids (PVF = 0.14), in Figure 12.6 as calculated by the analytical method, using two isothermal zones and instantaneous shock velocities. The saturations in the heated zone are higher than calculated numerically, and the saturation discontinuity at the temperature shock is more obvious. There then follows the long, flat portion of the saturation profile, which we shall see is a characteristic of a thermal displacement and has been noted by others.[2,3]

Near the point $x/L = 0.7$, the saturation distribution becomes the same as it would be in an isothermal waterflood and declines to the Buckley–Leverett front, near $x/L = 0.9$, which has not yet broken through. The saturation profile at PVF = 1.8 shows how the saturation level downstream of the shock has increased as a result of changing instantaneous shock velocity.

Figure 12.8*b* shows the saturation distribution (at PVF = 1.8 in Fig. 12.6)

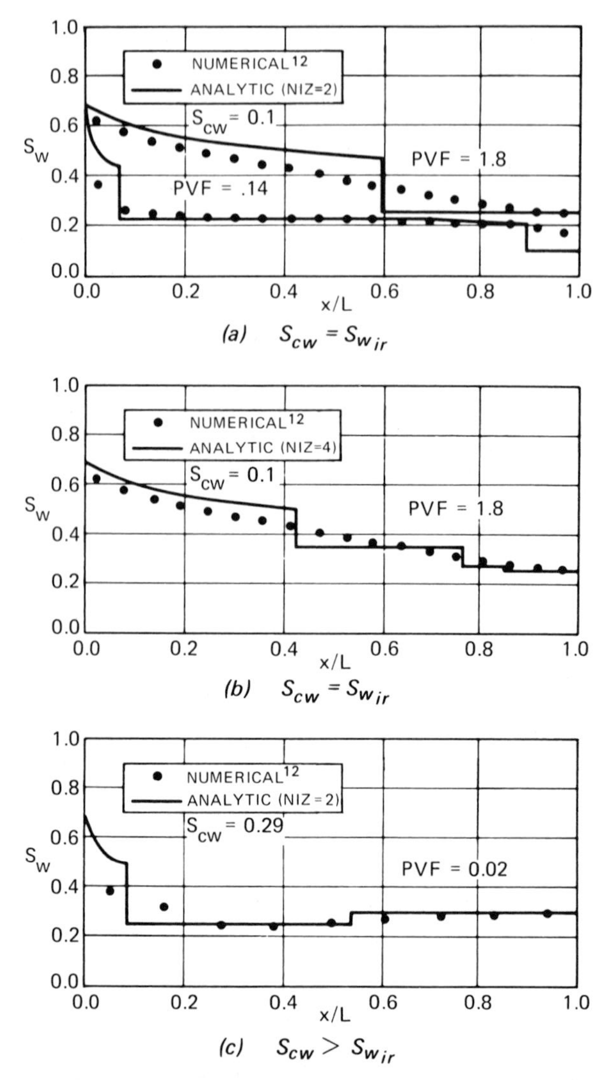

Figure 12.8. Comparison of calculated hot-waterflood saturation profiles.

for the analytical method in which four isothermal zones were used. The agreement of the analytical saturation profile with the numerical profile is much better in this case, but the recovery curve was essentially unchanged (except for time of thermal breakthrough) from that obtained when two zones were used. Experience indicates that there is little need to use more than three isothermal zones if one is merely interested in recovery calculations. However, if one wishes to approximate the saturation distribution in order to estimate pressure drop characteristics, more isothermal zones would be required.

Figure 12.8*c* shows the saturation profile (at PVF = 0.02 in Fig. 12.7) calculated for the hot waterflood of the system initially containing mobile water. The formation of the oil bank is apparent, and the analytical and numerical calculations are essentially in agreement.

Oil Recovery, High Thermal Efficiency. Figure 12.9 shows recovery curves for the system described in Table 12.2. In this case, the thermal efficiency of the process was greater than for the previous system discussed, and as a result, the temperature shock velocities were more nearly constant. Consequently, the difference in calculated recovery between the cases of

Table 12.2 Scaled Laboratory System, Physical Characteristics for Figures 12.9–12.12

$L = 3.88$ ft
$h = 0.83$ ft
$w = 0.68$ ft
$\phi = 0.372$
$K_{hob} = 12$ Btu/ft-d-°F
$K_{hR} = 24$ Btu/ft-d-°F
$(\rho c)_{ob} = 30$ Btu/ft^3-°F
$(\rho c)_R = 36$ Btu/ft^3-°F
$c_w = 1.0$ Btu/lb-°F
$c_o = 0.4$ Btu/lb-°F
$m_s = 182.5$ lb/d
$T_r = 65$°F
$T_I = 400$°F
$H_s = 1168$ Btu/lb
$H_{ws} = 342$ Btu/lb
$H_{wr} = 0$ Btu/lb
$S_{cw} = 0.10$
$S_{wir} = 0.10$
$S_{oir} = 0.14$

T	ρ_o (lb/ft^3)	ρ_w (lb/ft^3)	ρ_s (lb/ft^3)
65	55.3	62.4	
400	48.6	53.7	0.537

T	μ_o(cp)	μ_w(cp)	μ_s(cp)
65	2120	1.01	
400	3.87	0.13	0.0192

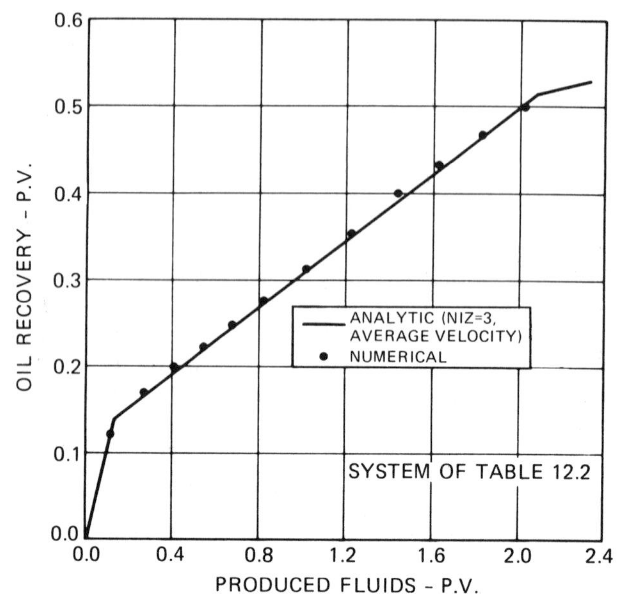

Figure 12.9. Comparison of calculated hot-waterflood recoveries (high thermal efficiency).

average velocity and instantaneous velocities is insignificant. As may be seen from the figure, the analytically calculated recovery using three isothermal zones and an average shock velocity is in good agreement with the recovery curve calculated by the numerical technique.

Steamflood Examples

Figures 12.10 and 12.11 present oil recovery curves for the system described in Table 12.2 calculated using the analytical method, by a more rigorous numerical technique,[12] and by simplified methods developed earlier by Landrum[5] and Willman.[6] The numerical results are represented by the points and are again accepted as the standard of reference. As mentioned earlier, only two isothermal zones are considered in all cases, and only average shock velocities are used. Appendix C gives an example calculation employing the Shutler–Boberg method using average velocities and no iterations to refine the fractional flow curves as improved estimates of fluid saturations are obtained.

Oil Recovery, No Mobile Water Initially Present. The solid lines of Figure 12.10 were calculated by the two variations of the Shutler–Boberg method

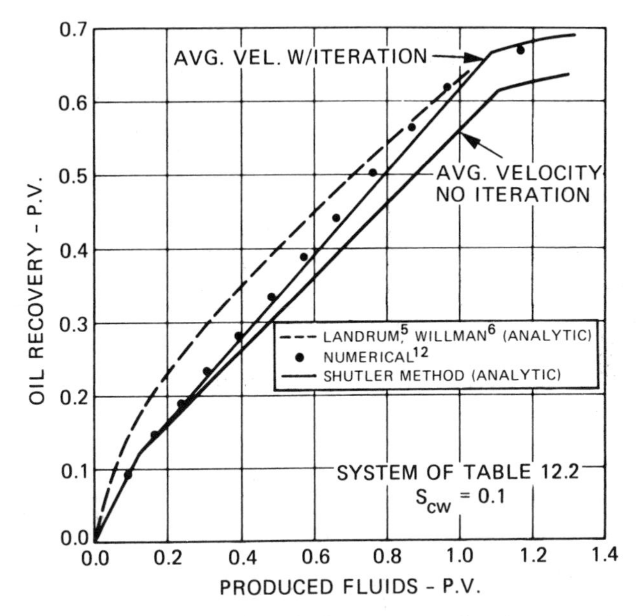

Figure 12.10. Comparison of calculated steamflood recoveries, $S_{cw} = S_{wir}$.[1] © 1972 SPE-AIME.

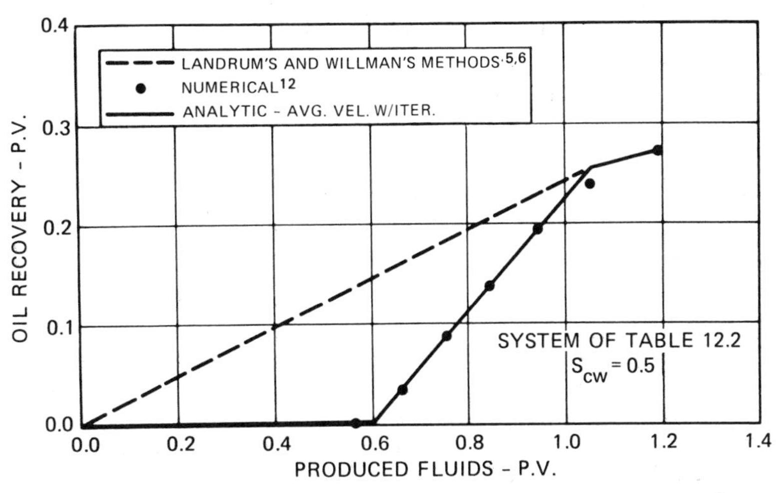

Figure 12.11. Comparison of calculated steamflood recoveries, $S_{cw} > S_{wir}$.[1] © 1972 SPE-AIME.

previously mentioned; the agreement with the numerical curve is seen to be good for the method employing iteration and conservative using the more approximate method. The dashed line represents the recovery calculated by methods developed by Landrum[5] and Willman.[6] The basis for comparison of these methods with the Shutler–Boberg analytical method requires some discussion.

If Landrum's and Willman's methods were independently applied to a given system, their predicted recovery curves would not agree because of differences in (a) the way recovery from the steam zone is calculated and (b) the fluid saturations assumed to determine steam zone heat capacity and, hence, steam front velocity. With regard to steam zone recovery, in Willman's method, the oil saturation in the steam zone is ordinarily assumed to be instantaneously reduced to a residual saturation as determined in the laboratory.

This approach ignores the throughput-dependent aspects of both the steam–oil displacement and distillation; the approach will yield an optimistic prediction of oil recovery as a function of fluids produced, but the ultimate oil recovery will be correct. While the mechanism of distillation is neglected in the Shutler–Boberg method, its effect can be approximated by a shifting of the gas–oil fractional flow curve to allow a lower residual oil saturation.

To avoid the effects of these various approximations within the steam zone, Landrum's and Willman's methods have been treated as if the steam zone oil recovery and steam front velocity were the same as in the Shutler–Boberg method using iteration. Under these circumstances, Landrum's and Willman's methods give essentially the same recovery curve, which is still somewhat optimistic with respect to the numerical curve over most of the life of the flood. The reason for the optimism in each of the methods is basically the same, neglect of the effect of steam–oil displacement on downstream saturation profiles.

Oil Recovery, Mobile Water Initially Present. Figure 12.11 compares recovery curves calculated for a system in which the initial water saturation was high and mobile. In such cases, as in the case of the hot waterflood, an oil bank is formed; even more markedly for this example than for the hot-waterflood example (as a result of a more viscous cold oil and higher water saturation), oil production prior to breakthrough of the oil bank is essentially nil.

The recovery curves calculated by the numerical program and by the analytical method are in excellent agreement. Because of the inability of Landrum's and Willman's methods to account for oil bank buildup, the recovery prediction by those methods is optimistic during the early part of the flood. Other simplified methods such as those discussed in Chapter 9 also suffer from this deficiency.

Saturation Profiles. Figure 12.12 shows three-phase saturation profiles as calculated by the Shutler–Boberg method, the numerical method,[12] and Landrum's and Willman's methods.[5,6] Figure 12.12*a* shows profiles corresponding to PVF = 0.092 in Figure 12.10. Again, the analytical and numerical methods are in essential agreement. The characteristic constant-saturation distribution downstream of the steam zone is present, and the saturation shock in the cold zone is smaller and travels more slowly than would the corresponding Buckley–Leverett shock in an isothermal waterflood. In fact, the isothermal waterflood Buckley–Leverett shock is the one shown by the

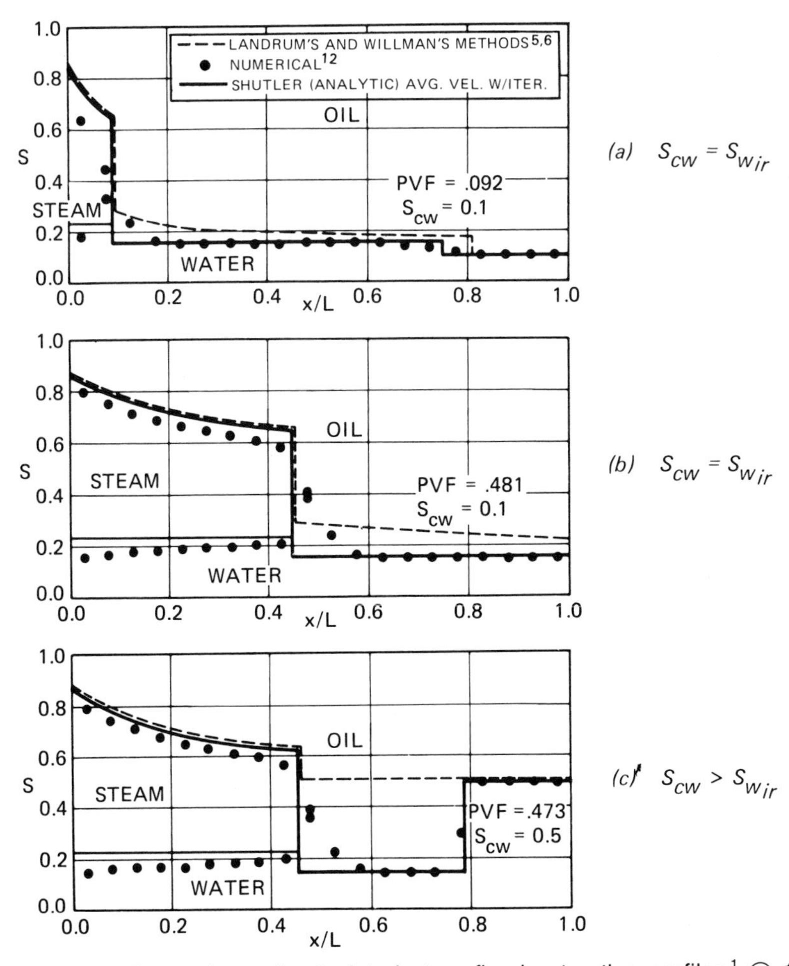

Figure 12.12. Comparison of calculated steamflood saturation profiles.[1] © 1972 SPE-AIME.

dashed line and is the one that is assumed by the Landrum and Willman methods.

This assumption neglects the effect of the hot-zone displacement in downstream saturations. In Figure 12.12*b*, the saturation profiles are shown at a later time after breakthrough of the cold saturation shock. The reason for the optimism of Landrum's and Willman's methods becomes more obvious now as we see the increasing difference between saturation profiles calculated by the different methods.

Figure 12.12*c* shows the saturation profiles (at PVF = 0.437 in Fig. 12.11) for the steamflood of a system in which the initial water saturation was mobile. Here the analytical method and numerical method demonstrate dramatically the oil bank formation. Landrum's and Willman's methods, not being able to account for this oil bank, predict a constant saturation profile downstream from the steam zone.

APPENDIX A: DERIVATION OF SHOCK EQUATIONS

To obtain the mass balances across the discontinuities ("shock" conditions), we consider the interval $x_a \rightarrow x_b$ in Figure 12A.1. A mass balance on the water phase in this region yields

$$(vf\rho_w)_{x_a} - (vf\rho_w)_{x_b} = \phi \frac{d}{dt}\left[\int_{x_a}^{x_i} (S\rho_w)\, dx + \int_{x_i}^{x_b} (S\rho_w)\, dx\right] \quad (12A.1)$$

Here we have broken the region into its two continuous parts for the purpose of writing the accumulation term. If one considers x_a and x_b to be stationary but $x_i = x_i(t)$, carrying out the indicated differentiation in Eq. (12A.1) yields

$$(vf\rho_w)_{x_a} - (vf\rho_w)_{x_b} = \phi\left[(S\rho_w)_i^- \frac{dx_i}{dt} + \int_{x_a}^{x_i} \frac{\partial(S\rho_w)}{\partial t}\, dx - (S\rho_w)_i^+ \frac{dx_i}{dt} + \int_{x_i}^{x_b} \frac{\partial(S\rho_w)}{\partial t}\, dx\right] \quad (12A.2)$$

If one takes the limit of Eq. (12A.2) as x_a and x_b approach x_i, one obtains the shock condition for the water phase:

$$(vf\rho_w)_i^- - (vf\rho_w)_i^+ = \phi[(S\rho_w)_i^- - (S\rho_w)_i^+] \frac{dx_i}{dt} \quad (12A.3)$$

where $dx_i/dt = C_i$ is the velocity of the shock at x_i (− and + superscripts indicate upstream and downstream sides of the shock, respectively.)

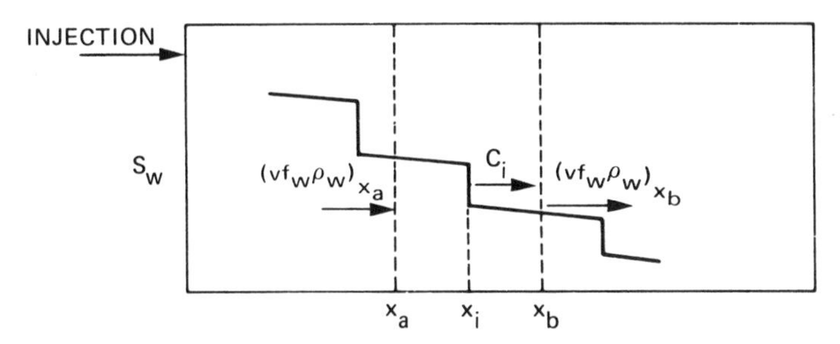

Figure 12A.1. Discontinuous saturation distribution.

More generally, if steam is considered flowing in zone i in the presence of oil and mobile water, the shock condition for steam and water becomes

$$[\bar{\bar{v}}(f_s\rho_s + f_w\rho_w)]_i^- - (vf\rho_w)_i^+ = \phi[(S_s\rho_s + S_w\rho_w)_i^- - (S\rho_w)_i^+]C_i \tag{12A.4}$$

and for oil,

$$[\bar{\bar{v}}(1 - f_s - f_w)\rho_o]_i^- - [v(1-f)\rho_o]_i^+ = \phi\{[(1 - S_s - S_w)\rho_o]_i^- - [(1-S)\rho_o]_i^+\}C_i \tag{12A.5}$$

(For convenience, the w subscript on f and S in the two-phase zone have been omitted.) Here $\bar{\bar{v}}_i$ is the total fluid velocity in zone i averaged over both distance through the zone and time.

The working shock equations for shock i now become [on elimination of v_i^+ between Eqs. (12A.4) and (12A.5)]

$$f_i^+ = \frac{a}{d} + \frac{b}{d}S_i^+ \tag{12A.6}$$

where

$$b = \frac{C_i}{\bar{\bar{v}}_i/\phi}$$

$$r_{w_i} = \frac{\rho_{w_i}}{\rho_{w_{i+1}}} \qquad r_{s_i} = \frac{\rho_{s_i}}{\rho_{w_{i+1}}} \qquad r_{o_i} = \frac{\rho_{o_i}}{\rho_{o_{i+1}}}$$

$$a = a_s + a_w \qquad a_s = r_{s_i}[f_{s_i}^- - bS_{s_i}^-] \qquad a_w = r_{w_i}[f_{w_i}^- - bS_{w_i}^-]$$

$$d = r_{o_i} + b(1 - r_{o_i}) + a_s\left(1 - \frac{r_{o_i}}{r_{s_i}}\right) + a_w\left(1 - \frac{r_{o_i}}{r_{w_i}}\right)$$

and

$$\bar{\bar{v}}_{i+1} = \bar{\bar{v}}_i d \tag{12A.7}$$

After finding f_i^+ and S_i^+ from the intersection of (12A.6) with $f_{i+1}(S)$ (Fig. 12.4), the remainder of the system behaves as in the case of hot-water injection. When steam is not present in zone i, Eqs. (12A.6) and (12A.7) are still applicable with $f_s^- = S_s^- = 0$.

APPENDIX B: CHARACTERISTICS OF TWO-PHASE SATURATION DISTRIBUTIONS

Hot Zone

In the usual case, the zone boundary water saturation S_i^+ will have a velocity greater than the velocity C_i of the shock. This may be seen by considering the shock condition (12A.3) for hot water and neglecting thermal expansion for simplification. Equation (12A.3) then may be written in the form

$$\frac{f_i^+ - f_i^-}{S_i^+ - S_i^-} = \frac{C_i}{v_i/\phi} = b \tag{12B.1}$$

This is the equation for the line between $f_i^+ S_i^+$ and $f_i^- S_i^-$ having a slope of b (see Fig. 12B.1). If the slope $f_i'^+$ is greater than the slope b, Eq. (12.3)

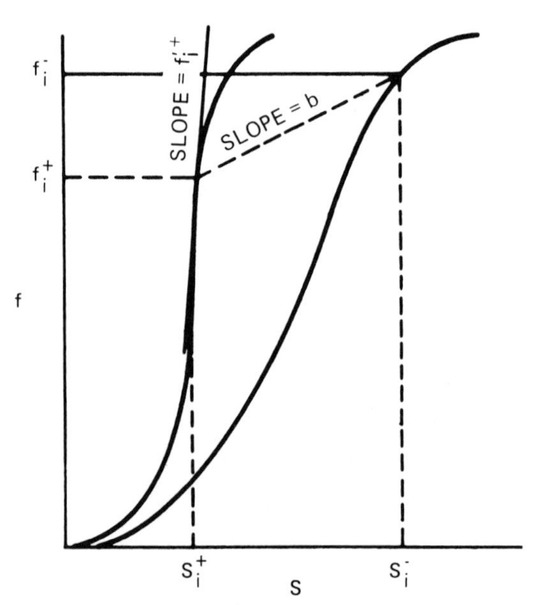

Figure 12B.1. *f* versus *S*.

tells us that the saturation S_i^+ will travel at a rate greater than C_i. Since $b > 0$, inspection of Figure 12B.1 leads to the conclusion that such will be the usual case for a hot waterflood.

If one considers the case of constant C_i and corresponding constant S_i^+, the condition just discussed, that $(dx/dt)_{s_i^+} > C_i$, will cause the saturation in zone i to be constant and equal to S_i^+ from x_i to $x_{s_i^+}$ whereas in Eq. (12.4),

$$x_{S_i}{}^+ = \int_0^t \frac{v_i}{\phi} f_i{}'^+ \, dt \tag{12B.2}$$

assuming $x_{S_i^+} = 0$ at $t = 0$ (see Fig. 12B.2). If $x_{S_i^+} < x_{i+1}$ that is, $(dx/dt)_{S_i^+} < C_{i+1}$, the saturation distribution throughout the remainder of the zone will vary according to Eq. (12.4) and is obtained by the usual Buckley–Leverett method. If $x_{S_i^+} > x_{i+1}$, that is, $(dx/dt)_{S_i^+} > C_{i+1}$, the saturation distribution will be constant throughout zone i from x_i to x_{i+1} (see Fig. 12B.3).

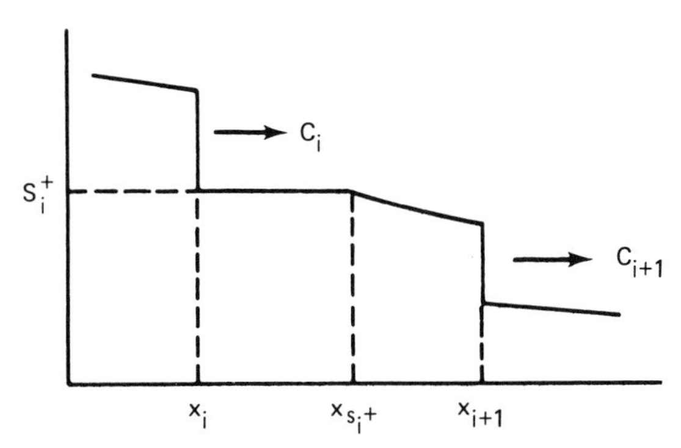

Figure 12B.2. Saturation profile: $C_i < (dx/dt)_{S_i}{}^+ < C_{i+1}$.

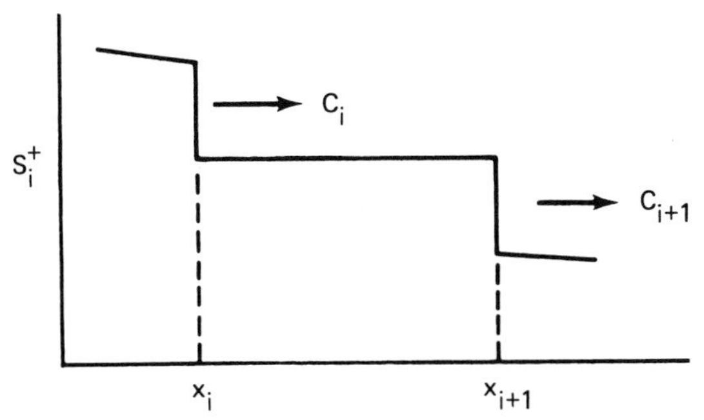

Figure 12B.3. Saturation profile: $C_i < C_{i+1} < (dx/dt)_{S_i}{}^+$.

As a practical matter, the larger the step changes in temperature, the greater the amount by which $(dx/dt)_{S_i^+}$ exceeds C_i and the greater the likelihood that $(dx/dt)_{S_i^+}$ will exceed C_{i+1}. Since the saturation distribution is much easier to work with if it is constant in each zone, and since the number of zones (beyond 3) has little effect on the recovery prediction, it appears that only a few zones and correspondingly large temperature shocks (at least at the hottest end) should be considered.

If one considers the case of varying C_i and correspondingly varying S_i^+, the condition $(dx/dt)_{S_i^+} > C_i$ will not yield a constant saturation distribution because S_i^+ itself is varying with time. To simplify the problem of calculating a saturation distribution at a point in time (corresponding to instantaneous values of x_i and C_i), one may assume that the condition $(dx/dt)_{S_i^+} > C_i$ does yield a constant saturation distribution. This amounts to the assumption that changing zone boundary conditions are instantaneously propagated downstream. The greater the amount by which the saturation velocities exceed the shock velocities, the smaller the error due to this assumption of instantaneous propagation.

Cold Zone

The saturation distributions that may result in the cold zone $(n+1)$ are varied. To simplify matters, again we assume C_n and S_n^+ to be constant and, further, that the initial saturation distribution S_{cw} is constant. It is also convenient to refer to the *Buckley–Leverett* saturation S_{BL} and shock x_{BL}, which are defined as the saturation that satisfies

$$\frac{f_{BL} - f_{cw}}{S_{BL} - S_{cw}} = \left(\frac{\partial f}{\partial S}\right)_{S_{BL}} \tag{12B.3}$$

and the shock that travels with velocity

$$C_{BL} = \frac{v_{n+1}}{\phi}\left(\frac{f_{BL} - f_{cw}}{S_{BL} - S_{cw}}\right) \tag{12B.4}$$

Here S_{BL} is determined by the construction shown in Figure 12B.4*a*. The saturations between S_{cw} and S_{BL} are contained in the shock at x_{BL}. To allow otherwise would give rise to physically meaningless results, that is, triple-valued saturations. If S_{cw} is greater than the saturation corresponding to the maximum value of f' (i.e., $S_{f'_{max}}$), no Buckley–Leverett shock will occur in an isothermal waterflood.

There are six possible saturation distributions that may arise in zone $n+1$ $(x > x_n)$. These are illustrated in Figures 12B.4 and 12B.5. The heights of the saturation profiles correspond quantitatively with the accompanying

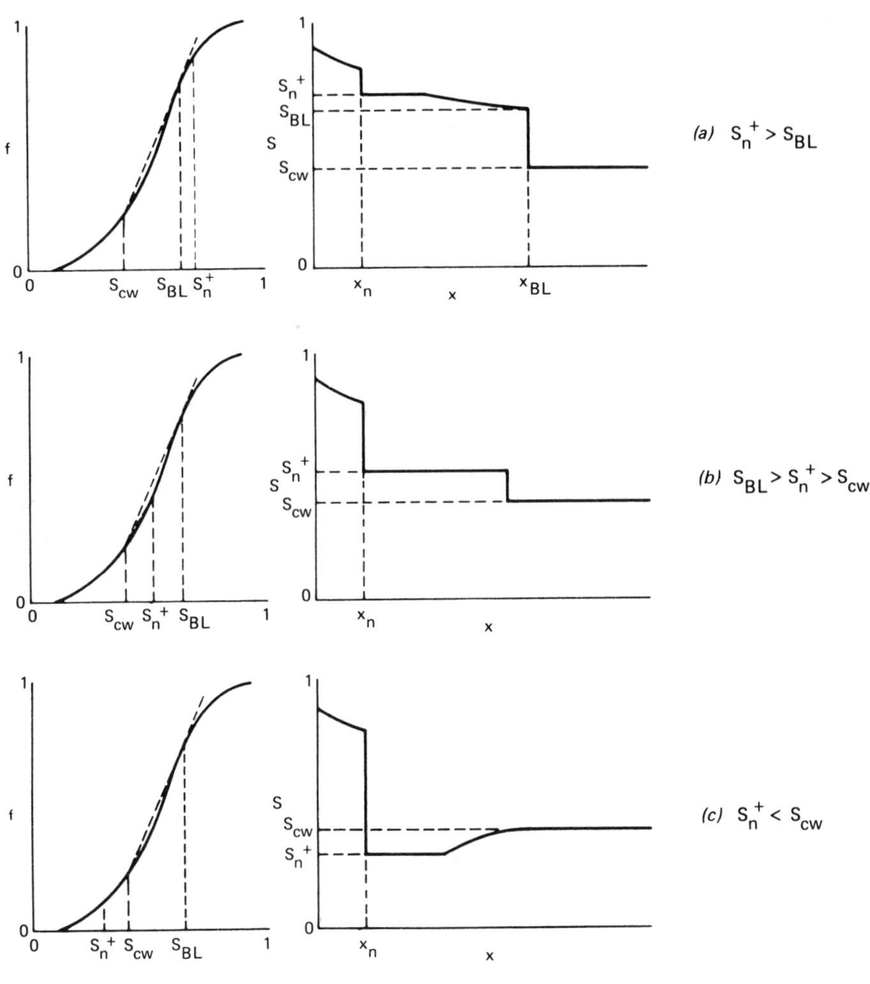

Figure 12B.4. $S_{cw} < S_{f'_{max}}$, [1] © 1972 SPE-AIME.

f-versus-S graphs, which show the construction that yields the various distributions. The x-coordinate positions of the saturations are only qualitative. In each case, it will help to realize that since there are no internal sources or sinks, no saturation may occur downstream of x_n that does not lie between S_n^+ and S_{cw}.

Figure 12B.4 considers the cases when the initial saturation distribution is such that a Buckley–Leverett front would form under isothermal conditions, that is, $S_{cw} < S_{f'_{max}}$.

Figure 12B.4*a* shows the case when $S_n^+ > S_{BL}$, which is accompanied by the consequence that $C_n < (dx/dt)_{S_i^+} < C_{BL}$. Hence, the saturation is constant at $S = S_n^+$ out to the farthest point to which it has traveled at its velocity as given by Eq. (12.3), that is, to the point at which it meets the normal Buckley–Leverett distribution. This distribution is similar to Figure 12B.2 and also appears in Figure 12.8*a*.

Figure 12B.4*b* shows the results when S_n^+ falls intermediate between S_{cw} and S_{BL}. Here, all saturations between S_n^+ and S_{cw} are contained in a shock, since to allow otherwise would give rise to triple-valued saturations. The velocity of the shock,

$$C_{n+1} = \frac{v_{n+1}}{\phi}\left(\frac{f_n^+ - f_{cw}}{S_n^+ - S_{cw}}\right) \tag{12B.5}$$

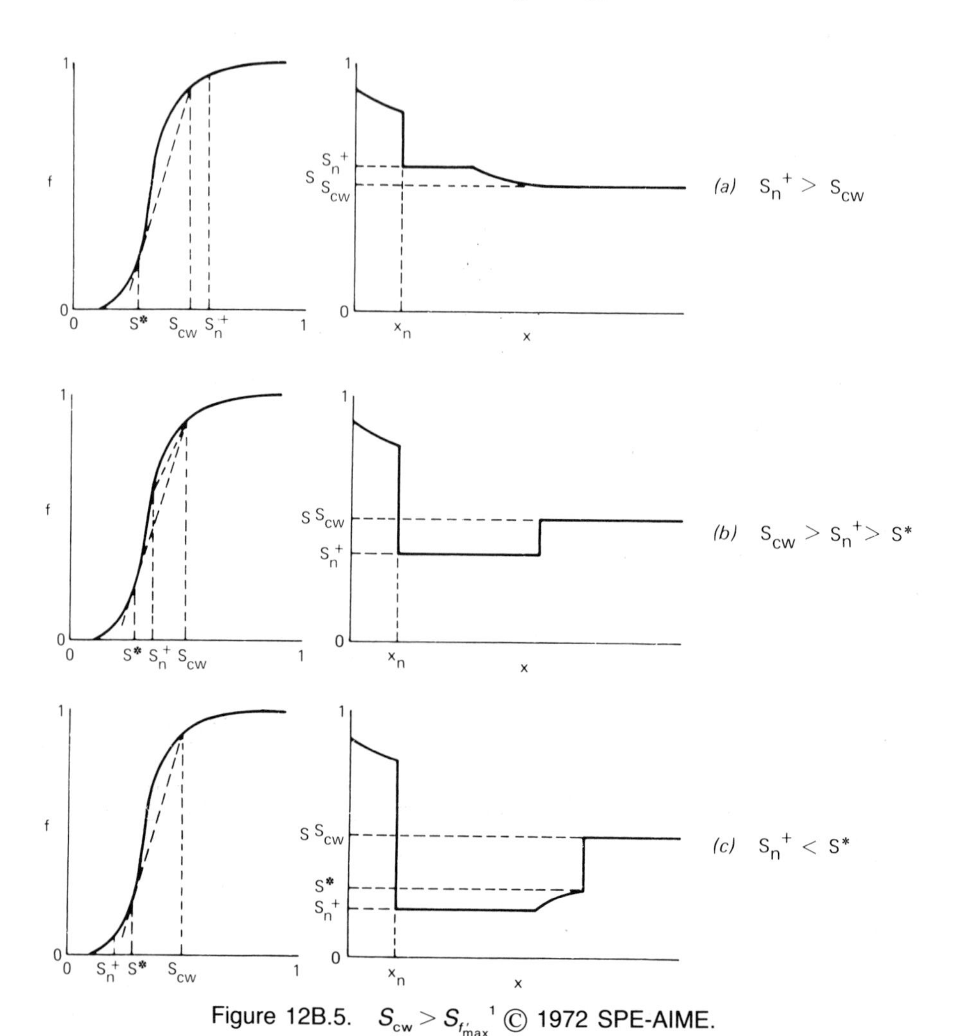

Figure 12B.5. $S_{cw} > S_{f'_{max}}$ [1] © 1972 SPE-AIME.

and its height are less than those for the normal Buckley–Leverett shock that would have occurred in an isothermal displacement. The saturation is constant at S_n^+ from x_n to the position of the shock. This distribution is similar to Figure 12B.3 and also appears in Figure 12.12*a*.

In Figure 12B.4*c*, S_n^+ is less than S_{cw}, and no shock forms. The saturation is constant at S_n^+ from x_n to the point given by Eq. (12B.2) for $i - n$, from which the distribution varies according to Eq. (12.4) for the saturations between S_n^+ and S_{cw}.

In Figure 12B.5, we consider the case for which no Buckley–Leverett front would form in an isothermal displacement, that is, $S_{cw} > S_{f'_{max}}$.

Fig. 12B.5*a* shows that, for $S_n^+ > S_{cw}$, no shock forms. The distribution is constant at S_n^+ from x_n to the point at which it meets the profile that would exist for an isothermal displacement.

Fig. 12B.5*b* represents a case in which a new reference saturation S^* is considered. This is the counterpart of S_{BL} for oil displacing water. That is, it is the saturation that would be associated with a normal Buckley–Leverett front for isothermal displacement of water by oil. If S_n^+ falls between S_{cw} and S^*, all saturations between S_n^+ and S_{cw} are contained in a shock that moves with a velocity proportional to the slope of the chord from S_n^+ to S_{cw}. The saturation is therefore constant at S_n^+ from x_n to the front. This distribution also appears in Figures 12.8*c* and 12.12*c*.

Figure 12B.5*c* shows the case for which $S_n^+ < S^*$. In this case, the profile (which is a counterpart of Fig. 12B.4*a* for oil displacing water) is constant at S_n^+ to the point at which it meets the normal distribution that would arise in an isothermal oil–water displacement. All saturations between S^* and S_{cw} are contained in a shock that moves with a velocity proportional to the slope of the chord from S^* to S_{cw}.

APPENDIX C: EXAMPLE STEAMFLOOD CALCULATION

A calculation technique using the Shutler–Boberg method is demonstrated for the reservoir described in Table 12C.1. The pattern is a 10-acre, 2:1 line drive pattern with 660 ft between injector and producer and 330 ft between adjacent like wells. It is assumed that 70% of the pattern area will be swept by steam. Hence, $w = 231$ ft.

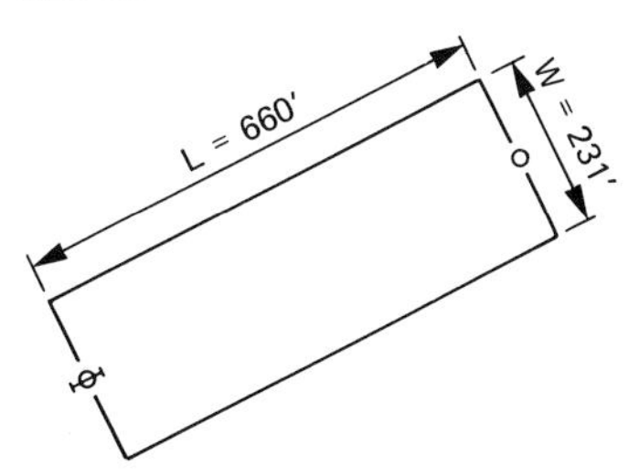

Table 12C.1 Example Problem, Reservoir and Fluid Data

$L = 660$ ft

$h = 100$ ft

$w = 231$ ft

$D = 1000$ ft

$\phi = 0.30$

$K_{hob} = 34$ Btu/ft-d-°F

$K_{hR} = 34$ Btu/ft-d-°F

$(\rho c)_{ob} = 38$ Btu/ft^3-°F

$(\rho c)_R = 36$ Btu/ft^3-°F

$c_w = 1.0$ Btu/lb-°F

$c_o = 0.4$ Btu/lb-°F

$T_r = 65$°F

$P_I = 700$ psia

$X_{surf} = 0.7$ (steam quality injected at surface)

$H_{wv} = 710$ Btu/lb

$H_{ws} = 492$ Btu/lb

$H_{wr} = 33$ Btu/lb

$P_w = 50$ psia (pumped-off producing well pressure)

$S_{cw} = 0.40$

$S_{wir} = 0.10$

$S_{oir} = 0.14$

$r_w = 0.3$ ft (effective radius of producing and injection wells)

$k = 2$ darcies

T	ρ_o(lb/ft^3)	ρ_w(lb/ft^3)	ρ_s(lb/ft^3)
65	55.3	62.4	
503	45.2	48.8	1.526

T	μ_o(cp)	μ_w(cp)	μ_s(cp)
65	212.0	1.01	
503	0.80	0.102	0.01796

S_w	k_w	k_{owo}
0.1	0	1.0
0.2	0.0016	0.868
0.3	0.0081	0.733
0.4	0.025	0.591
0.5	0.062	0.420
0.6	0.10	0.213
0.7	0.14	0.070
0.8	0.20	0.016
0.86	0.25	0

S_s	k_s	k_{ogo}
0.02	0	1.0
0.1	0.002	0.570
0.2	0.030	0.335
0.3	0.082	0.188
0.4	0.150	0.110
0.5	0.225	0.061
0.6	0.315	0.030
0.7	0.420	0.010
0.76	0.480	0

Steam Injection Rate

Using Muskat's equation for a direct line drive,[13] the initial water injectivity into the sand may be estimated:

$$q_{\mathrm{wi}} = \frac{0.002254kk_{\mathrm{w}}\mathrm{h}\,\Delta P/(\mu_{\mathrm{w}}B_{\mathrm{w}})}{(L/a) - 1.17 + (2/\pi)\ln(a/r_{\mathrm{w}})}$$

where $\Delta P = 700 - 50 = 650$ psi
$k_{\mathrm{w}} = 0.025$ (at $S_{\mathrm{w}} = 0.4$)
$\mu_{\mathrm{w}} = 1.01$ cp
$B_{\mathrm{w}} = 1.0$
$a = 330$ ft (distance between like wells)
$L = 660$ ft (distance between unlike wells)

$$q_{\mathrm{wi}} = \frac{0.002254(2000)(0.025)(100)(700-50)/(1.01)(1.0)}{(660/330) - 1.17 + (2/3.14)\ln(330/0.3)}$$
$$= 1371 \text{ bbl/d}$$

The actual steam injection rate on a condensed barrel basis will be somewhat less than the waterflood injection rate because of

1. the flow resistance of the oil bank, which will increase as the flood proceeds, and
2. the volume occupied by gaseous steam (we shall estimate based on the above considerations that an average injection rate of 1000 bbl/d of steam as condensate at 60°F is injected throughout the flood).

Wellbore Heat Loss

Only an approximate estimate of wellbore heat loss can be made since this changes with time and the duration of the flood is not known accurately. However, if wellbore heat loss is a small fraction of the total injected, only an approximate estimate is necessary. If the flood were to last 8 years, the average quality might be based on the wellbore heat loss rate at 4 years. Assuming the use of 7-in. casing and $3\frac{1}{2}$-in. tubing from Figure 2.9,

$$Q_{\mathrm{WHL}} = \text{wellbore heat loss rate}$$
$$= 410 \text{ Btu/hr-ft}$$

The steam injection rate is given by

$$m_{\mathrm{s}} = 1000(350)/24 = 14{,}580 \text{ lb/hr}$$

The average steam quality at the bottom hole is given by Eq. (2.5):

$$\bar{X}_{\mathrm{I}} = 0.7 - \frac{410(1000)}{(14580)(710)} = 0.66 \text{ lb vapor/lb total injected}$$

Reservoir Heat Transfer

1. The two-zone model of Marx and Langenheim is assumed; hence $T_1 = T_{\mathrm{I}} = 503°\mathrm{F}$ and $T_2 = T_{\mathrm{r}} = 65°\mathrm{F}$.
2. For simplicity, the average velocity $\bar{C}_1$ of the temperature shock is used. To obtain $\bar{C}_1$ from $t = 0$ to t_{TBT} (the time to reach breakthrough at the producing well) we note that $x_1 = L$ at t_{TBT}; hence,

$$\bar{C}_1 = L/t_{\mathrm{TBT}} \tag{12C.1}$$

Thus, what is needed is t_{TBT}.

Constants in Eqs. (3.1a), (3.3d), (3.3e), and (3.6a) are combined in a manner to facilitate subsequent calculations. From Eq. (3.1a),

$$\begin{aligned}(\rho c)_{\mathrm{R+F}} &= \phi[c_{\mathrm{o}}\rho_{\mathrm{o}}\bar{S}_{\mathrm{o}} + c_{\mathrm{w}}\rho_{\mathrm{w}}\bar{S}_{\mathrm{w}}] + (1-\phi)(\rho c)_{\mathrm{R}} \\ &= 0.30[0.4(45.2)\bar{S}_{\mathrm{o}} + 1.0(48.8)\bar{S}_{\mathrm{w}}] + (0.70)(36) \\ &= 5.42\bar{S}_{\mathrm{o}} + 14.6\bar{S}_{\mathrm{w}} + 25.2\end{aligned} \tag{12C.2}$$

From Eq. (3.6a), where $x_{\mathrm{s}} = x_1(t_{\mathrm{TBT}}) = L$,

$$\begin{aligned}\frac{\xi_{\mathrm{s}}}{\theta} &= \frac{4K_{\mathrm{hob}}w(T_{\mathrm{I}} - T_{\mathrm{r}})(\rho c)_{\mathrm{ob}}L}{h(\bar{X}_{\mathrm{I}}H_{\mathrm{wv}} + H_{\mathrm{ws}} - H_{\mathrm{wr}})m_{\mathrm{s}}(\rho c)_{\mathrm{R+F}}} \\ &= \frac{4(34)(231)(503-65)(38)(660)}{100(0.66(710) + 492 - 33)(7292)(24)(\rho c)_{\mathrm{R+F}}}\end{aligned}$$

(*Note*: $m_{\mathrm{s}} = 7292$ lb/hr because only half the injected steam enters the line drive element described above.) Hence,

$$\xi_{\mathrm{s}}/\theta = 21.26/(\rho c)_{\mathrm{R+F}} \tag{12C.3}$$

From Eqs (3.3d) and (3.3e),

$$\begin{aligned}\frac{\tau}{\theta} &= \frac{4K_{\mathrm{hob}}(\rho c)_{\mathrm{ob}}t}{h^2(\rho c)^2_{\mathrm{R+F}}} = \frac{4(34)(38)t}{(100)^2(\rho c)^2_{\mathrm{R+F}}} \\ &= \frac{0.5168t}{(\rho c)^2_{\mathrm{R+F}}}\end{aligned} \tag{12C.4}$$

4. To obtain a value for $\bar{C}_1$, assume $\bar{S}_w = S_{wir} = 0.1$ and $S_o = S_{oir} = 0.14$ in the steam zone in Eq. (12C.2):

$$(\rho c)_{R+F} = 5.42(0.14) + 14.6(0.1) + 25.2$$
$$= 27.42 \text{ Btu/ft}^3\text{-°F}$$

From Eq (12C.3),

$$\xi_s/\theta = 21.26/27.42 = 0.7754$$

From the table in Appendix A of Chapter 3,

$$\tau/\theta = 1.56$$

Hence, from Eq. (12C.4) at $t = t_{TBT}$,

$$t_{TBT} = \frac{(27.42)^2(1.56)}{0.5168} = 2270 \text{ days} \quad (6.2 \text{ yr})$$

From Eq. (12C.1),

$$\bar{C}_1 = (660)/2270 = 0.2907 \text{ ft/d}$$

Reservoir Fluid Flow

1. (a) Using the relative permeability data for k_s and k_{ogo} and viscosity data $\mu_s = 0.01796$ cp and $\mu_o = 0.80$ cp at 503°F, one can calculate f_s for the steam zone by Eq. (12.7):

$$f_s = \frac{k_s(S_s)/\mu_s(T_1)}{k_s(S_s)/\mu_s(T_1) + k_o(S_o)/\mu_o(T_1)}$$

In this simplified version of the Shutler–Boberg method, it is assumed that $S_{w_1} = S_{wir}$, and hence $f_{w_1} = 0$. Thus, $f_s = f_{s+w}$. Further, subsequent graphical construction is easier if one plots f_{s+w} versus S_{s+w} so this is done here (see Fig. 12C.1b).

(b) Using relative permeability data for k_w and k_{owo} and the viscosity data $\mu_w = 1.01$ cp and $\mu_o = 212.0$ cp at 65°F, one can calculate f_w for the cold zone. The results are plotted in Figure 12C.1*a*.

2. To obtain the fluid velocity v_1 in the steam zone, where $v_1 = v_I$, the steam injection rate, m_s, is converted to linear velocity:

$$v_1 = \frac{m_s}{\rho_s wh} = \frac{7292(24)}{1.526(231)(100)}$$
$$= 4.96 \text{ ft/d}$$

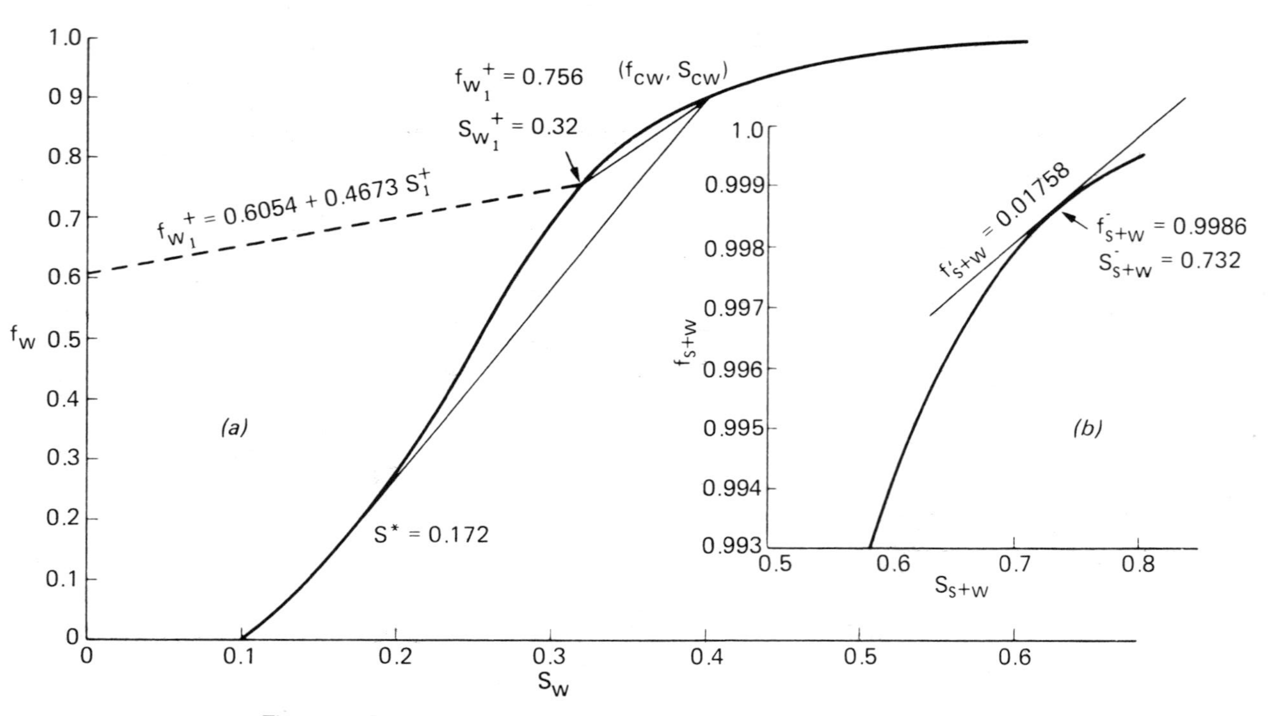

Figure 12C.1. Graphical constructions for example problem.

3. The next quantities of interest are the fluid flow parameters at the upstream side of the shock x_1, that is, f_s^-, $f_{w_1}^-$, S_s^-, and $S_{w_1}^-$. It is assumed that $S_{w_1}^- = S_{wir} = 0.1$ and $f_{w_1}^- = 0$. Here S_s^- is that saturation that moves with a velocity $(dx/dt)_{S_s^-}$ given by Eq. (12.8) and is at a position $x_1(t)$ at time t. Having assumed a constant shock velocity $\bar{C}_1$, and since v_1 is constant, it follows that $\bar{S}_s$ will not vary with time and will be the saturation whose velocity, from Eq. (12.8), is equal to $\bar{C}_1$:

$$\left(\frac{dx}{dt}\right)_{S_s} = \frac{v_1}{\phi} f'_{s+w} = \bar{C}_1 \tag{12C.5}$$

In Figure 12C.1b, the values f_{s+w}^- and S_{s+w}^- are those values of f_{s+w} and S_{s+w} that correspond to the point on the curve at which the slope is f'_{s+w}. Hence, rewriting Eq. (12C.5) and substituting values from above yields

$$f'_{s+w} = \frac{\bar{C}_1}{v_1/\phi} = \frac{0.2907}{4.96/0.3} = 0.01758$$

From the point of tangency of the line of slope 0.01758 in Fig. 12C.1b,

$$f_{s+w}^- = 0.9986 \qquad S_{s+w}^- = 0.732$$

from which it follows that

$$S_{s_1}^- = S_{s+w}^- - S_{w_1}^- = 0.732 - 0.1 = 0.632$$

$$f_{s_1}^- = 0.9986$$

4. To obtain the constants in the straight-line Eq. (12.5),

$$f_{w_1}^+ = \frac{a}{d} + \frac{b}{d} S_{w_1}^+$$

Equations (12A.6) and (12A.7) and associated equations are used:

$$b = \frac{\bar{C}_i}{v_i/\phi} = 0.01758$$

$$r_{s_i} = \frac{\rho_{s_i}}{\rho_{w_{i+1}}} = \frac{1.526}{62.4} = 0.024455$$

$$r_{w_i} = \frac{\rho_{w_i}}{\rho_{w_{i+1}}} = \frac{48.8}{62.4} = 0.78205$$

$$r_{o_i} = \frac{\rho_{o_i}}{\rho_{o_{i+1}}} = \frac{45.2}{55.3} = 0.81736$$

$$a_s = r_{s_i}[f_{s_i}^- - bS_{s_i}^-]$$
$$= 0.024455[0.9986 - 0.01758(0.632)]$$
$$= 0.02415$$
$$a_w = r_{w_i}[f_{w_i}^- - bS_{w_i}^-]$$
$$= 0.78205[0 - 0.01758(0.1)] = -0.001375$$
$$d = r_{o_i} + b(1 - r_{o_i}) + a_s\left(1 - \frac{r_{o_i}}{r_{s_i}}\right) + a_w\left(1 - \frac{r_{o_i}}{r_{w_i}}\right)$$
$$= 0.81736 + 0.01758(1 - 0.81736) + 0.02415 \times \left(1 - \frac{0.81736}{0.024455}\right) - 0.001375\left(1 - \frac{0.81736}{0.78205}\right)$$
$$= 0.03762$$
$$a = a_s + a_w = 0.022775$$
$$f_{w_1}^+ = \frac{0.022775}{0.03762} + \frac{0.01758}{0.03762} S_1^+$$
$$= 0.6054 + 0.4673 S_1^+ \tag{12C.6}$$

5. From the intersept of the line of Eq. (12C.6) with $f_w(S_w)$ in Figure (12C.1), $f_1^+ = 0.756$, $S_1^+ = 0.32$.

From Eq. (12A.7), it also follows that

$$v_2 = v_1 d = 4.96(0.03762) = 0.1866 \text{ ft/d}$$

6. Next the shape of the saturation profile in the cold zone must be determined by referring to Appendix B. In Appendix B, the first classification is with regard to S_{cw} and $S_{f'_{max}}$. From Table 12C.1, $S_{cw} = 0.4$; by inspection of Figure 12C.1*a*, $S_{f'_{max}} \cong 0.26$. Hence, $S_{cw} > S_{f'_{max}}$, and we look to Figure 12B.5.

A comparison between S_1^+, S_{cw}, and S^* is now needed. From the chord construction in Figure 12C.1*a*, which is illustrated in Figure 12B.5, $S^* = 0.172$; hence,

$$(S_{cw} = 0.4) > (S_1^+ = 0.32) > (S^* = 0.172)$$

Thus, the saturation profile will be of the shape of Figure 12B.5*b*; that is, the saturation S_2 in the cold zone will be constant at $S_2 = S_1^+ = 0.32$ out to some distance x_2 and then will increase to S_{cw}.

7. To calculate $t_2(L)$, the time to breakthrough of the saturation $S_2 = S_1^+$ at $x = L$, Eq. (12B.5) is employed. The velocity of the saturation shock C_2 is

$$C_2 = \frac{v_2}{\phi}\left[\frac{f_1^+ - f_{cw}}{S_1^+ - S_{cw}}\right] = \frac{v_2}{\phi} f_2' \tag{12C.7}$$

Constructing a chord between f_{cw}, S_{cw} and $f_{w_1}^+$, $S_{w_1}^+$, we obtain $f_2' = 1.788$, and hence, $C_2 = 1.112$ ft/d. Then

$$t_2(L) = \frac{L}{C_2} = \frac{660}{1.112} = 594 \text{ days}$$

8. In calculating oil recovery, it is convenient to consider two time periods:

A. $0 < t < t_2$

Here t_2 is the breakthrough time for the constant saturation S_2 at $x = L$. From step 7, $t_2 = 594$ days. Hence,

$$\text{PVF}_{t_2} = \frac{v_2 t_2}{\phi L} = \frac{(0.1866)(594)}{(0.3)(660)} = 0.560$$

Note that v_2 applies because the fluid being produced is at T_2, the original reservoir temperature. During this time period from 0 to t_2, the fractional flow of water is constant at $f_{cw} = 0.899$. Hence, the incremental oil produced to time t_2 is

$$\begin{aligned}\text{PVO}_{t_2} &= \frac{v_2 t_2}{\phi L}(1 - f_{cw}) \\ &= \frac{(0.1866)(594)}{(0.3)(660)}(0.101) \\ &= 0.0565\end{aligned}$$

B. $t_2 < t < t_1$

Here, t_1 is the breakthrough time for the shock at x_1, that is, the steam front; from step 4 of the section on reservoir heat transfer,

$$t_1(L) = t_{\text{TBT}} = 2270 \text{ days}$$

The total fluid produced in time $t_1(L)$ is given by

$$\text{PVF}_{t_1} = \frac{v_2 t_1}{\phi L} = \frac{(0.1866)(2270)}{(0.3)(660)} = 2.139$$

Note that v_2 still applies because the fluid being produced is still cold. During this time period from t_2 to t_1, the fractional flow of water at the outflow face is constant by virtue of the saturation profile of the type in Figure 12B.5*b* at $f_L = f_1^+ = 0.756$. Hence, the oil produced to time t_1 is given by

$$\begin{aligned} \text{PVO}_{t_1} &= \text{PVO}_{t_2} + v_2 \frac{t_1 - t_2}{\phi L} (1 - f_L) \\ &= 0.0565 + \frac{(0.1866)(2270 - 594)}{(0.3)(660)} (0.244) \\ &= 0.442 \end{aligned}$$

C. $t_1 < t$

This final period after steam breakthrough usually need not be considered in a piston-type flood modeled by this calculation because of economic considerations. During this period, f_L continuously changes, and the associated portion of the recovery curve is not straight. However, the recovery at any time may be easily calculated by the Welge technique.[10]

Calculated Oil–Steam Ratio, R_{os}

The pore volume in barrels, N_{PVF}, is calculated as

$$\begin{aligned} N_{PVF} &= \frac{\phi hwL}{5.61} = \frac{(0.3)(100)(231)(660)}{5.61} \\ &= 815{,}294 \text{ bbl} \end{aligned}$$

The barrels (condensed water at 60°F) injected as steam is

$$N_s = \frac{24 m_s t}{350} = 500\,t$$

Thus, the oil–steam ratio at 594 and 2270 days is tabulated:

t(days)	PVO	N_s(bbl)	N_o(bbl)	R_{os}(bbl/bbl)
594	0.0565	297000	46100	0.155
2270	0.442	1135100	360000	0.317

Incremental oil recovery in pore volumes as a function of the pore volumes of produced fluids is given in Figure 12C.2.

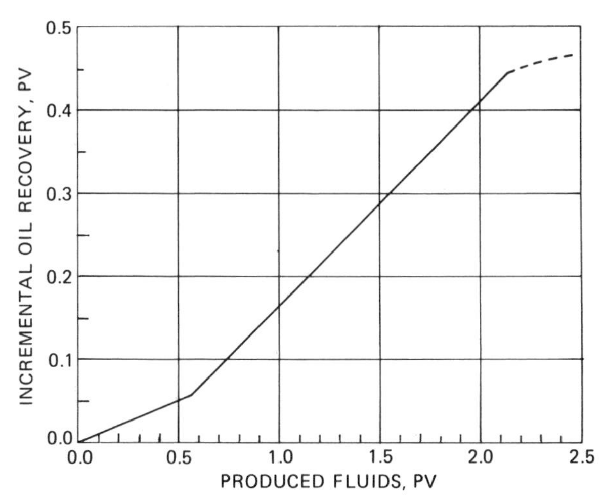

Figure 12C.2. Oil recovery versus fluid production for example problem.

NOMENCLATURE

C	shock velocity, ft/d
c	heat capacity, Btu/lb-°F
D	depth, ft
f	fractional flow (of water if no phase subscript),
	$= \dfrac{k_{\mathrm{w}}(S_{\mathrm{w}})/\mu_{\mathrm{w}}(T)}{k_{\mathrm{w}}(S_{\mathrm{w}})/\mu_{\mathrm{w}}(T) + k_{\mathrm{o}}(S_{\mathrm{o}})\mu_{\mathrm{o}}(T)}$
f'	slope of f vs. S $(\partial f/\partial S)_t$
h	net sand thickness, ft
H_{wv}	latent heat of vaporization at T_{I}, Btu/lb
H_{ws}	enthalpy of water at T_{I}, Btu/lb
H_{wr}	enthalpy of water at T_{r}, Btu/lb
K_{h}	thermal conductivity, Btu/d-ft-°F
k	permeability, ft^4-cp/lb-d (relative permeability if subscripted)
L	length, ft
m	mass injection rate into linear element, lb/d
NIZ	number of isothermal zones
n	number of zones at $T > T_{\mathrm{r}}$
PVF	pore volumes of fluids produced
PVO	pore volumes of oil produced
r	density ratio (see Appendix A)

S saturation (of water if no phase subscript)
T temperature, °F
t time, days
Q_{WHL} wellbore heat loss per ft of depth, Btu/hr-ft
v velocity, ft/d
w width, ft
x coordinate position, ft
X_i steam quality injected at sandface, dimensions
θ $(\rho c)_{R+F}/(\rho c)_{ob}$
μ viscosity, cp
ρ density, lb/ft^3
ϕ porosity

Subscripts

BL result of isothermal Buckley–Leverett analysis
cw connate (initial) water
est estimated
F fluids
I injection
i zone number
ir irreducible
L outlet end
o oil
ob overburden
R rock
r initial reservoir conditions
S constant saturation
s steam
ST stock tank conditions
T constant temperature
t time
TBT to thermal breakthrough
w water

Superscripts

− upstream side of shock
+ downstream side of shock

Overbars

$\bar{}$ average
$\bar{\bar{}}$ averaged with respect to time and distance

REFERENCES

1. Shutler, N. D., and Boberg, T. C. "A One-Dimensional, Analytic Technique for Predicting Oil Recovery By Hot Water or Steamflooding," *Soc. Pet. Eng. J.* (December 1972), pp. 489–498.
2. Fayers, F. J. "Some Theoretical Results Concerning the Displacement of a Viscous Oil by a Hot Fluid in a Porous Medium," *J. Fluid Mech.* (Part 1), **13,** 65 (1965).
3. Dougherty, E. L., and Sheldon, J. W. "The Use of Fluid–Fluid Interfaces to Predict Behavior of Oil Recovery Processes," *Soc. Pet. Eng. J.* (June 1964), p. 171.
4. Jordan, J. K., Rayne, J. R., and Marshall, S. W. III. "A Calculation Procedure for Estimating the Production History During Hot Water Injection in Linear Reservoirs," presented at the Twentieth Technical Conference on Petroleum Production, The Pennsylvania State University, University Park, May 9–10, 1957.
5. Landrum, B. L., Smith, J. E., and Crawford, P. B. "Calculation of Crude Oil Recoveries by Steam Injection," Texas Petroleum Research Committee-Oil Recovery Conference Proceedings, Paper 1389-G, October 29–31, 1959.
6. Willman, B. T., Valleroy, V. V., Runberg, G. W., Cornelius, A. J., and Powers, L. W. "Laboratory Studies of Oil Recovery by Steam Injection," *Trans. AIME*, **222,** 1–681 (1961).
7. Lauwerier, H. A. "The Transport of Heat in an Oil Layer Caused by the Injection of Hot Fluid," *Appl. Sci. Res., Sec. A*, **5,** 145 (1955).
8. Marx, J. W., and Langenheim, R. H. "Reservoir Heating by Hot Fluid Injection," *Trans. AIME* **216,** 312 (1959).
9. Buckley, S. E., and Leverett, M. C. "Mechanism of Fluid Displacement in Sands," *Trans. AIME*, **146,** 107 (1942).
10. Welge, H. J. "A Simplified Method for Computing Oil Recoveries by Gas or Water Drive," *Trans. AIME*, **195,** 91 (1952).
11. Spillette, A. G., and Nielsen, R. L. "Two-Dimensional Method for Predicting Hot Waterflood Recovery Behavior," *J. Pet. Tech.* (June 1968), p. 627.
12. Shutler, N. D. "Numerical Three-Phase Model of Linear Steamflood Process," *Soc. Pet. Eng. J.* (June 1969), pp. 232–246.
13. Muskat, M. *Physical Principles of Oil Production*, McGraw-Hill, New York, 1949, p. 653.

Chapter 13

INTRODUCTION TO DRY IN SITU COMBUSTION PRINCIPLES

ABSTRACT

In situ combustion for the case of air injection without water injection is discussed in this chapter. Because of the high expense of investment and operation and the troublesome nature of this process, it is not used nearly as much as steamflooding, although many projects have been conducted. Nevertheless, it is applicable to deeper and thinner reservoirs than steamflooding and can be operated on much wider well spacings.

The major initial investment in an in situ combustion project is for air compression, which is normally quite a large requirement for this process. A method is given for calculating the ratio of air injected to oil produced for a well-behaved project.

Gravity override and severe channeling have been observed in many projects. When heat arrives at a producing well, the well normally must be shut in to avoid downhole destruction of liners and tubulars. This problem has caused reduced sweep efficiency and low oil recovery for many projects.

Calculation of the air–oil ratio is based on the quantity of air required to burn off the coked residuum of oil left as the hot combustion products and nitrogen pass through the reservoir. A method for estimating the coked residuum, or "fuel," concentration is given to permit making estimates of the air–oil ratio.

The combustion front cannot advance if the air flux passing through the front is too low because frontal temperatures will drop below the minimum level at which combustion can occur. A method for estimating the maximum feasible well spacing to permit this minimum air flux to be exceeded at all times during the course of the fire flood is also presented. Methods of ignition to start a fire flood and other operating characteristics are also discussed.

INTRODUCTION

A great deal has been written about in situ combustion, possibly more than any other thermal recovery technique. Despite the theoretically efficient heat utilization of in situ burning, steam drive is producing much more incremental oil commercially. The thermal efficiency and lower surface fuel requirement of in situ burning compared to steam is outweighed by three factors:

1. higher initial investment, owing to the greater cost of air compressors compared to steam generators;
2. damage to well completions, owing to the severity of downhole conditions especially after hot fluids begin to be produced; and
3. lack of process control as evidenced by poor sweep efficiency, incomplete combustion, and well productivity impairment because of high gas–oil ratios and attendant high producing-well bottom-hole pressures.

The second factor may become critical if the combustion front overrides or channels because of areal or vertical permeability heterogeneities and reaches the vicinity of the producing wells early in the life of the project. In fact, the infeasibility of producing a hot well often leads to early project termination and attendant poor sweep. Damage of liners in injection wells has also been reported shortly after ignition.[1]

FACTORS AFFECTING SHAPE OF COMBUSTION FRONT

Recognition of reservoir situations favoring a more uniform vertical sweep of the combustion front (as opposed to a severely channeling overburn) is most important in the design of a combustion drive project. For convenience, we shall refer to the more uniform combustion front as a *centerburn* and the severely channeling type as an *overburn* (see Fig. 13.1). It should be recognized, however, that gravity forces predominate when a gas is injected into an oil sand, so there is always a tendency for injected air to override the oil column. The following factors tend to counter the override tendency and permit a reasonably uniform displacement:

1. Thin Sand

As Figure 13.2 indicates, even though there is an initial tendency to override, if the sand is thin (less than 15 ft), heat transfers to the base of the

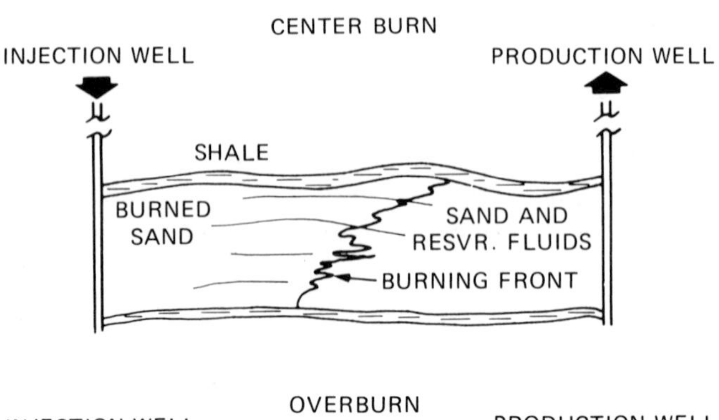

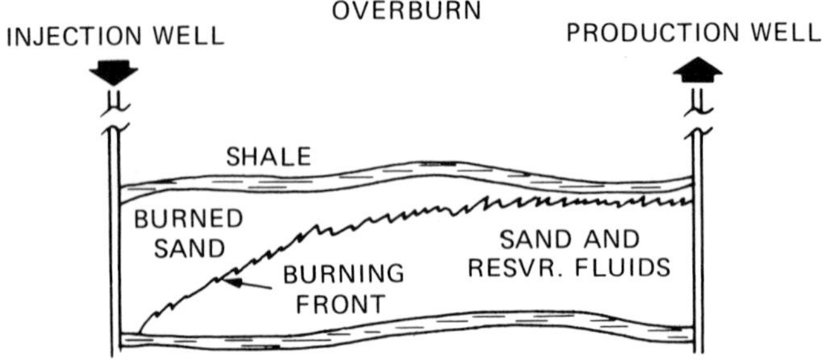

Figure 13.1. Two types of underground burning.

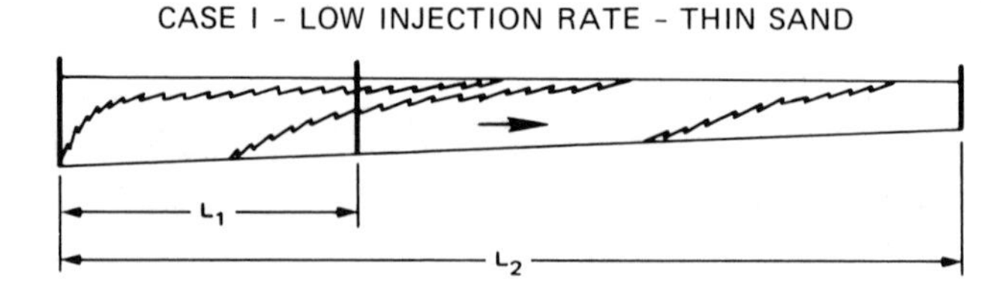

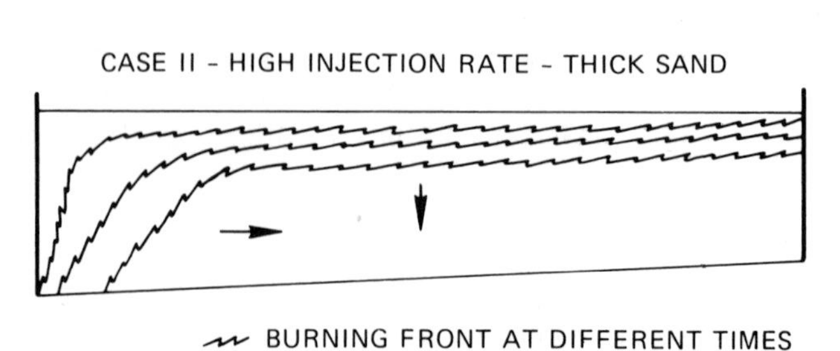

Figure 13.2. Burning front at different times.

sand rapidly enough that oil viscosity can be reduced there. This permits the combustion front to advance at the bottom more rapidly in a thin sand than would be possible in a thick sand. Conduction heating of a point beneath an overriding combustion front is greatly affected by the distance through which the heat must travel.

The time to obtain a given elevation in temperature by conduction of heat is proportional to the square of the distance from the heat source. Sands less than 15 ft thick can often experience centerburns while sands thicker than 40 ft will generally experience overburns. In between lies a gray area; pilot testing is usually the only way to answer the question as to what type of burn will be experienced by sands 15–40 ft thick.

2. Lower Injection Rate

Even in thick sands where overburning is pronounced, the use of a reduced injection rate can permit a more uniformly advancing combustion front. In Figure 13.2, the use of a high injection rate (case II) may cause a rapidly channeling front with early breakthrough at the producing well. Lowering the injection rate permits more time for downward heat transfer, reducing the severity of conditions at producing wells. One operator[2] recommends not exceeding gas production rates of 700 kscf/d per producing well in unconsolidated sands (to avoid sand cutting of liners).

If air injection rates are too low, the air flux (scf/hr-ft^2) and resulting temperatures at the combustion front will drop to levels below which combustion cannot be sustained. This is an important consideration in thin, lower permeability sands where air injectivity is low and where the cooling effect of heat transfer to overburden and underburden is greater than for a thick sand. This aspect will be discussed further in a later section.

3. Low Oil Viscosity

As in the case of steam drive, a high oil viscosity will prevent rapid displacement until the oil sand has been heated. In sands containing a high-viscosity oil, air will seek the path of least resistance; unless lithology prevents it, a severe overburn will usually occur because of gravity effects. For this reason, industry during the 1970s intensified investigation of the use of in situ burning as a tertiary oil recovery process for light- to medium-gravity oils. However, recently industry interest in this application has subsided somewhat.

4. Low Vertical Permeability

Shale barriers and laminae may aid in obtaining a centerburn by preventing or reducing upward migration of injected air. However, if a gas channel already exists at the top of the reservoir, a low vertical permeability may have a detrimental effect, since oil cannot migrate upward to partially block the channel. This effect was discussed in Chapter 11 for steam drive.

5. Absence of Gas Cap

A gas cap, or thin layer of high gas saturation at the top of the sand, will promote overburning since this will provide a path of least resistance to injected air. This will be particularly detrimental in the case of a sand containing a highly viscous oil.

6. Higher Permeability at Base of Sand or Bottom-Water Leg

A high permeability at the top of the sand will promote an overburn. Conversely, if high permeability or bottom water exists at the base of the sand, a more uniform advance of the combustion front can be expected, since these preferential flow paths will in part compensate for gravitational effects.

7. Steep Dip

Injected air and combustion front movement will be more rapid toward up-dip wells than toward wells low on the structure. It is generally advisable to locate air injection wells high on the structure to obtain better control of combustion front movement. Although a centerburn may not be obtained, it may be possible to obtain a controlled overburn. Gates and Sklar[1] discuss a project where this was successfully done.

8. Use of Wide Spacing

In permeable sands where air mobility is not restricted severely, it is generally best to use a wide well spacing. Wells on close spacing may encounter the corrosive conditions and high temperatures associated with the combustion front early in the life of the project. With wells more widely spaced, longer times are permitted for the combustion front to stabilize as a "tilted" interface and move on to obtain good vertical sweep.

At this time, only qualitative guidance can be offered on the subject of vertical sweep of the combustion front. Field pilots reported in the literature have demonstrated that vertical sweep can be nearly 100% in thin sands (less than 15ft thick). Vertical sweep was much lower than this in the South Belridge pilot conducted in the late 1950s in a sand 30 ft thick where the well spacing was very close. Table 13.1 lists the well spacing–sand thickness ratio

Table 13.1 Vertical Sweep of Several Forward Combustion Pilots

Field	Operator	Spacing–Thickness Ratio,[a] d/h	Thickness, h (ft)	Vertical Sweep
South Belridge, CA	General Petroleum	8	30	Poor
Schoonebeek, Holland	NAM	6	44	Poor
Melones, Venezuela	Mene Grande	15	22	Good
Fry, IL	Marathon	4–6	40	Good
North Government Wells, TX	Mobil	37	18	Good
Delaware-Childers, OK	Sinclair	6	45	Excellent
Delhi Field, LA	Sun	154	8	Excellent[b]

[a]Approximate ratio of distance between injection well and first producing well to net sand thickness.
[b]Reportedly 100% vertical sweep efficiency obtained.

for several projects and indicates whether these projects experienced good or poor vertical sweep efficiency of the combustion front.

Most of the early pilots demonstrated that the areal sweep of the combustion front is remarkably uniform. As in the case of steam drive, heat transfer away from viscous fingers of injected air causes temperatures to drop below the combustion point and thus retard the progress of the front where these fingers occur.

Two- or three-dimensional calculation methods are now available for predicting the shape of the combustion front as a function of time.[3–6] However, the difficulty and high cost of these calculations greatly restricts their use. Furthermore, scaling of the complex phenomena associated with in situ burning in a physical, three-dimensional model is at best qualitative because complete scaling of all variables is impossible. Binder et al.[7] discuss a series of experiments where they approximately scaled the overburn process for a thick sand.

Simplified calculation methods have resorted to either of the following two assumptions regarding the shape of the front.

1. The front is *vertical* and advances in pistonlike fashion either radially or linearly away from the injection well (*thin-sand approximation*; see Refs. 8–15).
2. The front is *horizontal* and overlays the entire distance between injection and producing wells (*thick-sand approximation*; see Refs. 16 and 17). A model of this type would be similar to the Vogel model described for steam drive in Chapter 9.

Probably more studies have been made for the first case, and some of these will be discussed in the next section.

TEMPERATURE AND SATURATION DISTRIBUTIONS IN VICINITY OF COMBUSTION FRONT

Numerous laboratory combustion tube runs have been performed that give a clear picture of the temperatures around the combustion front. One such study[14] produced the sequence of temperatures shown in Figure 13.3 as a function of axial distance in a 36-in. tube oriented vertically with the burning occurring downward.

An air heater was used to obtain ignition, and the heater was left on for the first 2.46 hr of the run. The temperature distribution at 5.06 hr is more nearly characteristic of the type of distribution that might be obtained in the field. After a long period of injection, the temperature at the injection point will be close to ambient reservoir temperature. Moving toward the combus-

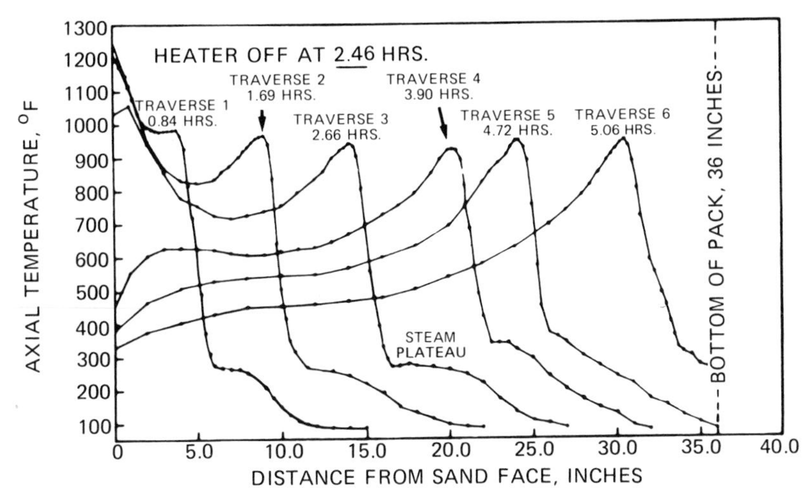

Figure 13.3. Axial temperatures as functions of distance from sandface and time.[14] © 1966 SPE-AIME.

tion front, temperatures generally rise to the peak temperature, which occurs at the combustion front itself.

Because of the low heat capacity of the flue gases and superheated steam, temperatures ahead of the combustion front drop rapidly until saturated steam temperature at the pressure of the model is reached. Until most of the water vapor is condensed, the temperature will drop only slightly for some distance to form a steam plateau. Further ahead, temperatures drop gradually to original sand pack temperature as vaporized light ends condense and move with the hot-water bank ahead of the steam plateau.

Because the injected air has a low heat capacity, it cannot move the bulk of the heat forward. Hence, dry combustion, which involves only air injection, will produce a steam plateau that is relatively small in comparison with wet combustion, which combines air and water injection. With wet combustion, the high heat capacity and high latent heat of vaporization of water are utilized to recover the heat behind the combustion front to produce a large steam plateau that is highly effective in aiding oil displacement.

In the case of a highly viscous oil, the water and gas moving ahead of the steam are relatively less effective in displacing oil than are the condensing steam and hot gases immediately ahead of the combustion front. Figure 13.4*a* shows the temperature distribution (top curve) and saturation distribution (center curve) for displacement of a highly viscous oil at a high initial oil saturation (virgin reservoir case).

For dry combustion, the steam zone is small relative to the size of the

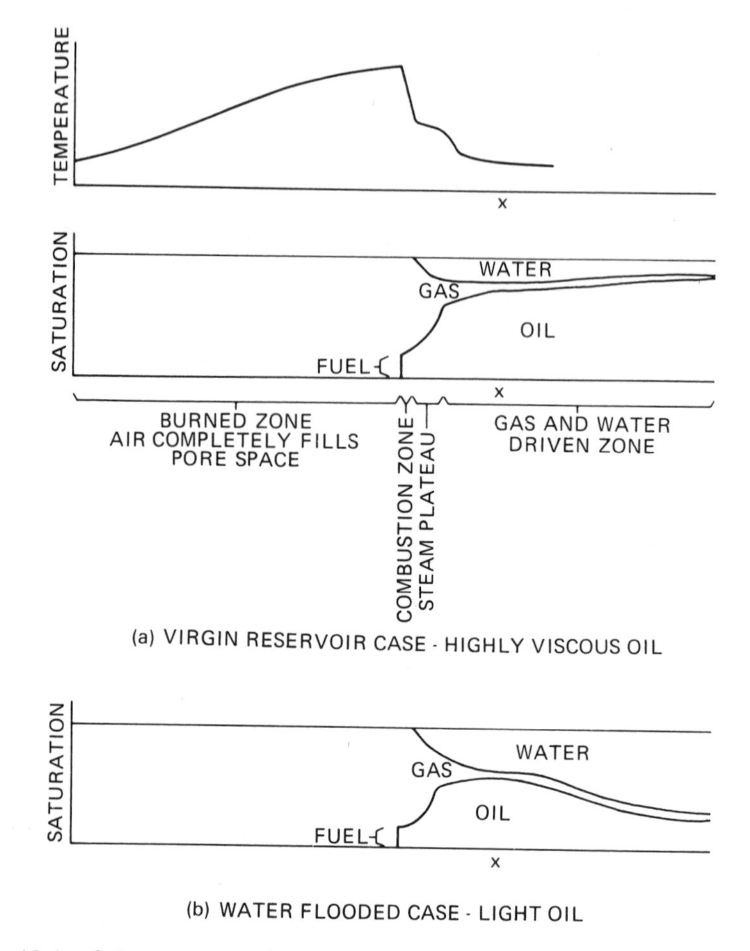

Figure 13.4. Schematic temperature and saturation distribution forward dry combustion, viscous oil.

burned zone (which contains only air, the residual oil left by the steam and hot flue gases having been consumed as fuel). In addition, the connate water will have been vaporized and displaced ahead. With a highly viscous oil, the gas- and water-driven zone downstream of the steam zone will be nearly at original oil saturation, because these fluids cannot displace a high-viscosity oil effectively. For lower viscosity oils, gas and water drive ahead of the heated zone become increasingly important and must be taken into account.

When applied as a tertiary process, it must be recognized that, as in the case of steam (see Fig. 11.6, Chapter 11), an oil bank will form ahead of the heated zone, and oil production will be delayed until this bank arrives at producing wells. Figure 13.4*b* shows schematically the saturation distribution that would exist for displacement of a light oil that had been waterflooded before initiating combustion. Note the buildup of the oil bank.

ESTIMATING DRY-COMBUSTION AIR REQUIREMENTS

One might conclude, and rightly so, that for the case of a pistonlike, advancing combustion front, the process will be essentially completed when the steam zone reaches the producing well. At least, one might as well shut the well in and avoid troublesome hot-well operation since little oil will remain behind the steam front to be recovered. When the combustion front reaches the producing well, we may write the following material balance equation representing the terminal conditions in the reservoir:

$$N_p = b_o V_b \phi (S_{oi} - S_{of})/5.61 \tag{13.1}$$

where N_p = oil produced (STB)
b_o = reciprocal formation volume factor ~1 for heavy oils)(STB/RB)
V_b = bulk volume of burned zone (ft^3)
ϕ = porosity
S_{oi} = original oil saturation
S_{of} = oil saturation consumed as fuel

The volume of injected air required (N_a) is a function of the size of the burned volume and the fuel saturation:

$$N_a = C_A V_b \phi S_{of} \rho_f \tag{13.2}$$

where N_a = cumulative air injected (scf)
C_A = air required to consume a unit mass of fuel (scf/lb)
ρ_f = density of fuel (lb/ft^3)

Combination of Eqs. (13.1) and (13.2) gives a relation for the terminal cumulative air–oil ratio for a pistonlike, advancing combustion front:

$$R_{Ao} = \frac{5.61 C_A \rho_f S_{of}}{b_o (S_{oi} - S_{of})} \tag{13.3}$$

The quantity C_A is obtained by considering the stoichiometry of the combustion reaction for complete combustion:

$$(1 + \tfrac{1}{4}n)O_2 + CH_n \rightarrow CO_2 + \tfrac{1}{2}n\, H_2O$$

Note that n is the atomic ratio of hydrogen to carbon in the fuel. In the absence of experimental data, a value of $n = 1.6$ is frequently used. The moles of oxygen required to burn 1 mole of fuel is $1 + n/4$. Since air contains 21% oxygen, the number of moles of air required is $(1 + n/4)/0.21$. One mole of fuel contains $12 + n$ pounds of fuel, since the atomic weight of carbon is 12 and the atomic weight of hydrogen is 1. If all injected oxygen were consumed, then according to the above chemical equation, the standard cubic feet of air required per pound of fuel is

$$C_A' = \frac{379(1 + n/4)}{0.21(n + 12)} = \frac{1805(1 + n/4)}{n + 12} \quad \text{(scf/lb)} \tag{13.4}$$

To obtain C_A, we must account for the fact that not all of the injected air enters into the above reaction. Some of the air is held up behind the combustion front in the burned zone. Some of the air may completely bypass the combustion zone. This can happen at the cooler top and bottom edges of the sand, where air fluxes might not be high enough to sustain combustion. Some of this unused oxygen passes on through the oil sand (particularly in sands cooler than 100°F) and is produced as free oxygen and/or carbon monoxide representing a partial utilization.

On the other hand, some oxygen will be utilized in the reservoir in low-temperature oxidation reactions. These reactions evolve considerably less heat than the combustion reaction and are generally undesirable in that they can produce corrosive acids and additional coke or fuel that must be consumed. The latter effect is essentially the same as "asphalt blowing," a refinery process in which air is injected to produce asphalt from less viscous heavy-oil residuum. Customarily, these complex effects are combined into an estimate of the overall oxygen utilization efficiency ϵ defined by

$$C_A = C_A'/\epsilon \tag{13.5}$$

Equations (13.3)–(13.5) combine to give

$$R_{Ao} = \frac{10125(1 + n/4)\rho_f}{\epsilon(n + 12)b_o} \frac{S_{of}}{S_{oi} - S_{of}} \tag{13.6}$$

For heavy oils, an approximate form of Eq. (13.6) is obtained by assuming

$$n = 1.6$$
$$b_o = 1.0$$
$$\rho_f = 62.4 \text{ lb/ft}^3$$

which is

$$R_{Ao} = \frac{65036 S_{of}}{\epsilon(S_{oi} - S_{of})} \tag{13.6a}$$

Obviously, to estimate the air–oil ratio, a great deal hinges on accurate knowledge of

1. the initial oil saturation and
2. the fuel concentration.

The air–oil ratio is directly proportional to the fuel concentration and inversely proportional to the difference between the initial and fuel oil saturations.

ESTIMATING FUEL CONCENTRATION

Combustion Tube Test

The conventional approach to estimating the fuel concentration is to conduct a combustion tube test in the laboratory. Tests of this type are discussed by Penberthy[14] and Wilson and Reed.[18] Combustion tubes vary as to diameter and length. Tubes as large as 8 in. in diameter and 15 ft in length have been reported.[19] Most have been operated vertically downward, but horizontal operation has also been employed. Wilson and Reed[18] rotated their horizontal tube to minimize gravitational effects.

In general, the tubes are constructed of thin-walled stainless steel and contain a sand pack. The tube is mounted within a pressure jacket. It is usually insulated, but some investigators have further minimized heat transfer at the wall by employing compensating heaters along the tube to simulate adiabatic conditions as nearly as possible. For small-diameter tubes that use only insulation, higher air fluxes must be employed to obtain a sustained burn. To simulate the lower fluxes and more nearly adiabatic conditions in the reservoir, larger tubes and/or temperature compensation must be employed.

Fuel consumption is calculated only for that part of the run in which burning proceeds as uniformly as possible and after the ignition heater has been turned off. The amount of hydrocarbon burned is calculated from the volume of CO and CO_2 produced during the steady-state period and the volume of sand traversed by the combustion front during this period. This assumes that all combustion products are CO_2, CO, and H_2O. Showalter[20] presented the following relationship for the apparent carbon–hydrogen ratio derived from stoichiometric considerations:

$$R = (A + B) \Big/ \left(\frac{21 - (A + B + C)}{2.37} + \frac{B}{6} \right) \qquad (13.7)$$

where R = weight of carbon/weight of hydrogen in fuel (lb/lb)
A = mole percentage of CO_2 in exhaust gas
B = mole percentage of CO in exhaust gas
C = mole percentage of O_2 in exhaust gas

Note that $n = 12/R$, the atomic hydrogen–carbon ratio.

The volume of air required to burn 1 lb of fuel is given by

$$v_a = \left[\left(\frac{R}{R+1}\right)\left(\frac{2.667(A + 0.5B + C)}{A+B}\right) + \left(\frac{8}{R+1}\right)\right] \Big/ 0.01873 \tag{13.8}$$

If complete combustion is obtained, $B = C = 0$, and Eq. (13.8) simplifies to

$$v_a = \left[2.667\left(\frac{R}{R+1}\right) + \frac{8}{R+1}\right] \Big/ 0.01873 \tag{13.8a}$$

If $n = 1.6$, $R = 7.5$; hence,

$$v_{a1.6} = 176 \text{ scf/lb}$$

The fuel concentration, F_c (in lb/ft^3), is then calculated as

$$F_c = \frac{V_a}{v_a V_b} \tag{13.9}$$

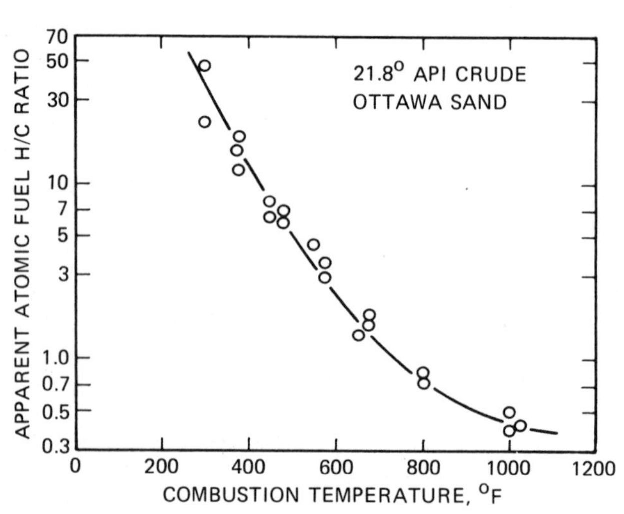

Figure 13.5. Effect of temperature on apparent hydrogen–carbon ratio of fuel.[21] © 1962 SPE-AIME.

where V_a = volume of air injected during steady portion of run (scf)
V_b = volume of sand pack traversed by combustion front (as indicated by temperature measurements) during steady portion of run (ft^3)

Alexander et al.[21] pointed out that Eq. (13.8) is not rigorous because it assumes that all oxygen reacts to form H_2O, CO_2, and CO and neglects the formation of low-temperature oxidation products. In short-core tests, they observed that the apparent value of n was a function of the combustion temperature, the value $n = 1.6$ corresponding to a combustion temperature of about 650°F (see Fig. 13.5).

Correlations of Laboratory Combustion Tube Data

A. *Showalter Correlation.* Figure 13.6 shows Showalter's[20] correlation with crude gravity of fuel concentration F_c and V_a/V_b, the standard cubic feet of air per cubic feet of burned sand obtained from a series of long tube tests. The air requirement and fuel consumption increase exponentially for very heavy oils. Under the test conditions,

1. A 16°API oil had an $F_c = 1.4\ lb/ft^3$, $V_a/V_b = 260\ scf/ft^3$, and
2. A 40°API oil had an $F_c = 0.9\ lb/ft^3$, $V_a/V_b = 140\ scf/ft^3$.

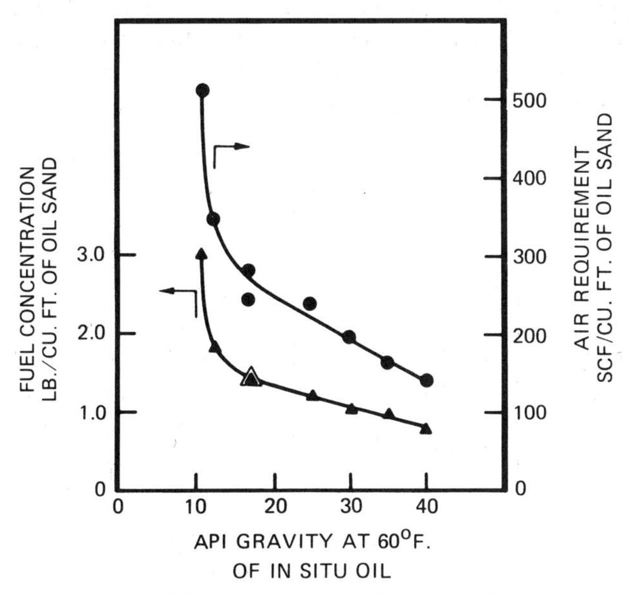

Figure 13.6. Combustion drive fuel concentration and air requirement versus oil gravity.[20] © 1963 SPE-AIME.

Showalter's data also show a slight effect of increasing the F_c and V_a/V_b as test pressure increased (see Fig. 13.7).

B. *Alexander Correlation.* The short-core data of Alexander et al.[21] show a similar correlation with oil gravity but with greater scatter of data. They reported their data as $\bar{F}_c$, the fuel concentration per 100 lb rock, which is related to F_c by

$$\bar{F}_c = 100F_c/(1-\phi)\rho_r = F_c/1.635(1-\phi) \tag{13.10}$$

If $\phi = 0.35$, 1 ft^3 of rock weighs 106 lb. The short-core laboratory results have been correlated by the relation

$$\bar{F}_c = 0.487 + 0.0974C_c + 0.000412\mu_o - 0.458 \times 10^{-7}(\mu_o)^2 \tag{13.11}$$

where C_c = Conradson carbon of oil[2]
μ_o = oil viscosity (kinematic) at 130°F (cs)

C. *Couch Correlation.* Couch et al.[15] used a one-dimensional, three-phase numerical simulator of a combustion tube test developed by Gottfried[12] to study the variation of fuel concentration with porosity and permeability. The objective of this work was to develop a correlation that would permit adjusting laboratory tube run measurements of $\bar{F}_c$ as a function of porosity and permeability for a 15, 25, and 36°API oil (see Figs. 13.8–13.10). The calculated values are higher than those determined experimentally, and their use as absolute values will likely lead to overly pessimistic estimates of air requirements.

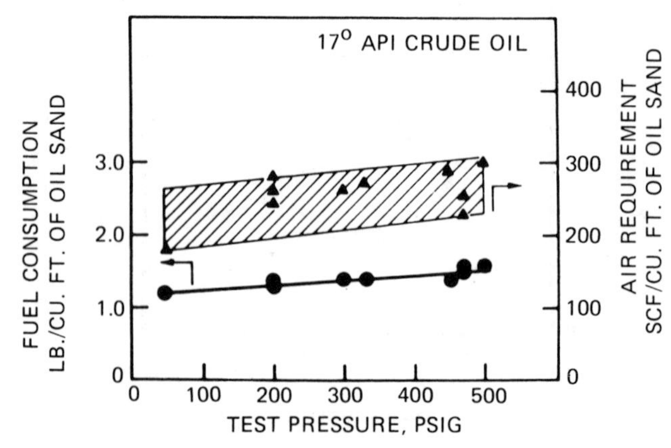

Figure 13.7. Combustion drive fuel consumption and air requirement versus test pressure.[20] © 1963 SPE-AIME.

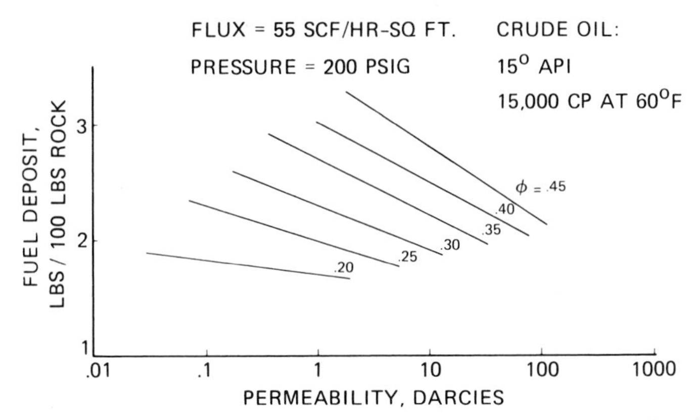

Figure 13.8. Effect of porosity and permeability on in situ combustion fuel.[15] © 1970 SPE-AIME.

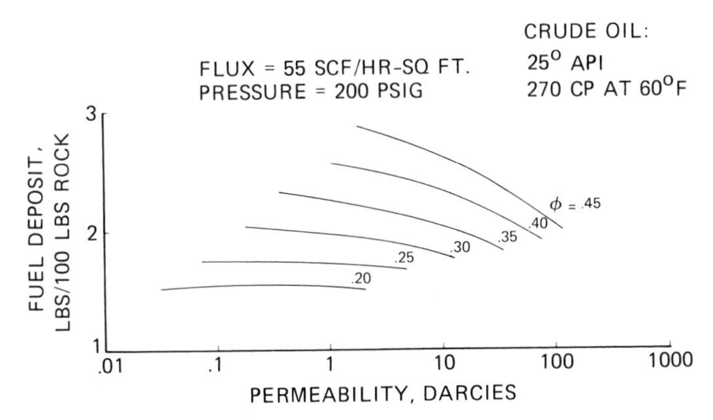

Figure 13.9. Effect of porosity and permeability on in situ combustion fuel.[15] © 1970 SPE-AIME.

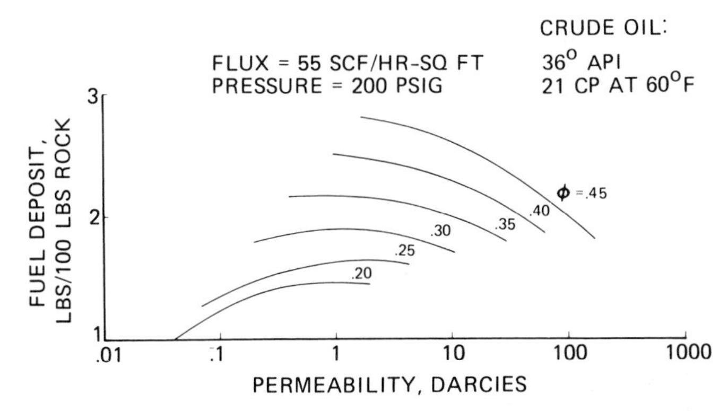

Figure 13.10. Effect of porosity and permeability on in situ combustion fuel.[15] © 1970 SPE-AIME.

Correlations of Apparent Field Fuel Deposition

The relation between the fuel concentration predicted by tube run tests and the reservoir fuel concentration may be good or poor depending on many factors. Factors that can cause disparity between laboratory and field results are the following:

1. *Differences in Initial Temperature and Pressure of the System.* A low formation temperature may cause a large fraction of injected air to be utilized in low-temperature oxidation reactions. A low pressure will enlarge the steam zone preceding the combustion front and permit remaining hydrocarbon to be reduced to a lower residual oil saturation prior to passage of the combustion front.
2. *Differences in Clay Content.* Some workers have noted that the presence of clays reduced the fuel concentration. Possibly, this occurs because of catalytic effects of clays on the cracking reaction, which would tend to reduce the residual fuel. For this reason, tube runs should be conducted with sand containing the natural clays in the reservoir.
3. *Differences in Peak Temperatures.* Differences in air flux and heat loss between the laboratory tube and the field can cause differences in peak temperatures. These factors can cause differences in the amount of fuel buildup resulting from low-temperature oxidation (hydrogen stripping and polymerization). It is generally advisable to operate combustion tubes so that peak temperatures are comparable to those expected under field conditions (600–800°F).

Orkiszewski Correlation. In analyzing the results of several dry combustion field pilots and numerous tube runs, Orkiszewski[23] found that apparent field values of $\bar{F}_c$ correlated strongly with the °API crude gravity and the initial reservoir temperature. He split $\bar{F}_c$ into two components: F_1, which he termed the contribution of the natural residuum, and F_2, termed the contribution of residuum formed in situ by auto-oxidation. Here F_1 and F_2 are plotted as functions of °API and T_r, respectively, in Figure 13.11; $\bar{F}_c$ is obtained by

$$\bar{F}_c = F_1 + F_2 \tag{13.12}$$

and

$$S_{of} = \frac{\bar{F}_c(1-\phi)\rho_r}{100\phi\rho_f} \tag{13.13}$$

where ρ_f = fuel density (lb/ft^3)
ρ_r = rock density (lb/ft^3)

Equation (13.6) is used to compute the air–oil ratio. Note that for

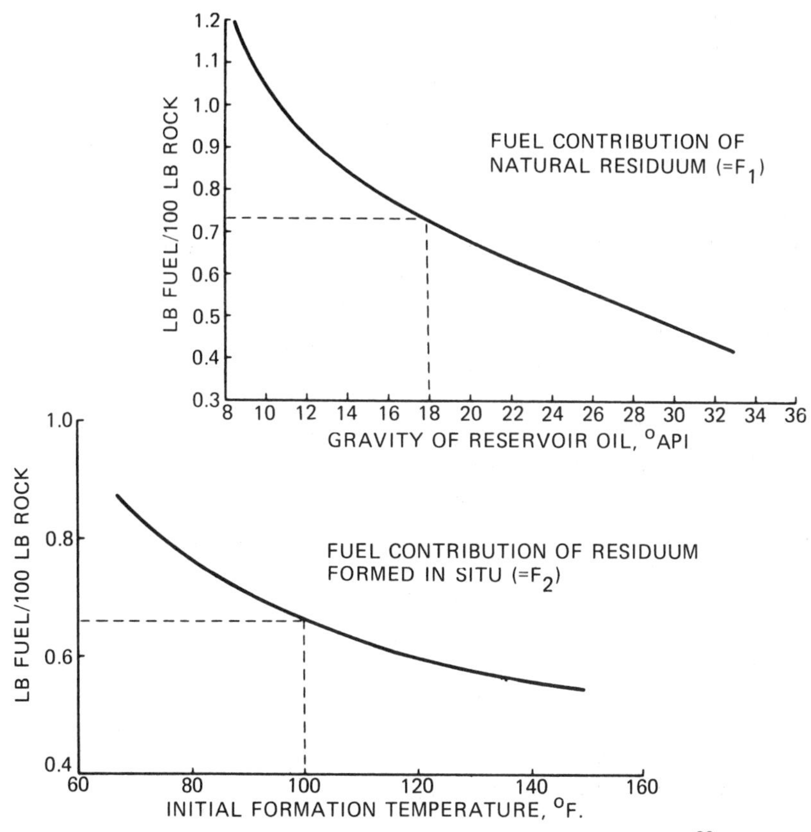

Figure 13.11. Orkiszewski correlation of fuel deposition.[23]

unconfined projects where fluid capture is incomplete, the air–oil ratio computed by Eq. (13.6) must be increased to account for this:

$$R_{Ao}' = R_{Ao}/f_{cap}$$

where R_{Ao}' = air–oil ratio corrected for unconfinement (scf/bbl)
R_{Ao} = air–oil ratio for confined project (scf/bbl)
f_{cap} = fraction of injected nitrogen produced (captured)

ESTIMATION OF MINIMUM AIR FLUX REQUIRED TO SUSTAIN COMBUSTION

In thin, tight sands, the minimum air flux to sustain combustion may become an important consideration. Air flux will normally be sufficiently high to sustain combustion when the front is near the injection well. However, as it moves radially outward, its area expands; for a given air injection rate, the flux decreases. As this happens, temperatures at the front decrease.

Chu's Solution

Analysis of temperatures at the combustion front has seen extensive mathematical treatment for the thin-sand approximation of a moving vertical combustion front.[8–10,13] This theory is not applicable to the case of an overriding combustion front. The most complete analyses to date have been made by Chu[8,13] for a radially expanding vertical front of infinitesimal thickness. His model assumes the following:

1. Sand and impermeable rock are homogeneous and isotropic.
2. Fuel concentration F_c is constant at all points.
3. The flow of fluids is only horizontal.
4. Heat transfer by vaporization and condensation of water is negligible. This assumption precludes prediction of the small steam plateau. Later work where Chu took this into account showed only a minor effect on computed temperatures at the front.[13]
5. The thermal conductivities of the sand and impermeable rock are identical (ρc may differ, however).

Chu numerically solved the following equation describing heat flow in the reservoir under the above assumptions:

$$\frac{\partial^2 T}{\partial r^2} + \left(1 - \frac{V\rho_g c_g}{2\pi K_h}\right)\frac{1}{r}\frac{\partial T}{\partial r} + \frac{\partial^2 T}{\partial y^2} + \frac{V\rho_g N_o g}{2\pi K_h}\frac{1}{r}\delta(r - r_f) = \frac{1}{\alpha}\frac{\partial T}{\partial t} \quad (13.14)$$

Where (using Chu's original nomenclature)

V = air rate per unit thickness of sand (scf/hr-ft)
$\rho_g c_g$ = average air–flue gas product of density and heat capacity (Btu/ft^3-°F)
g = heat liberated per mass of oxygen consumed (Btu/lb)
N_o = mass concentration of O_2 in injected air (lb/lb)
Z = fuel concentration (lb/ft^3)
r_f = combustion front radial distance from injection well (ft)
$\alpha = K_h/(\rho c)_{R+F}$ (ft^2/hr)
δ = Dirac delta distribution (1/ft)

In the impermeable rock, $V = 0$.

The following initial and boundary conditions apply:

$t = 0$ $\qquad T = T_0(r, y)$
$r = r_w$ $\qquad T = T_I$ (all y)
$r = r_e$ $\qquad \dfrac{\partial T}{\partial r} = 0$

$$y = 0 \text{ (center plane of sand)} \qquad \frac{\partial T}{\partial y} = 0$$

$$y = \infty \qquad \frac{\partial T}{\partial y} = 0$$

Typical results obtained by Chu[8] giving temperatures as a function of position around the combustion front are shown in Figure 13.12. One can see clearly the preponderance of heat that is left behind the combustion front by studying the temperature distributions that are shown as isotherms (r vs. y) and as T-versus-r and T-versus-y plots.

Chu's computations of the peak temperature at the front give an indication of the maximum distance the front can penetrate. When peak temperatures have declined to the order of 500–700°F, combustion for practical purposes ceases. A more accurate determination of the minimum temperature at which ignition occurs might be made by studying combustion tube results. Use of the peak temperature to determine the minimum flux may be somewhat optimistic since all other temperatures will be below the peak.

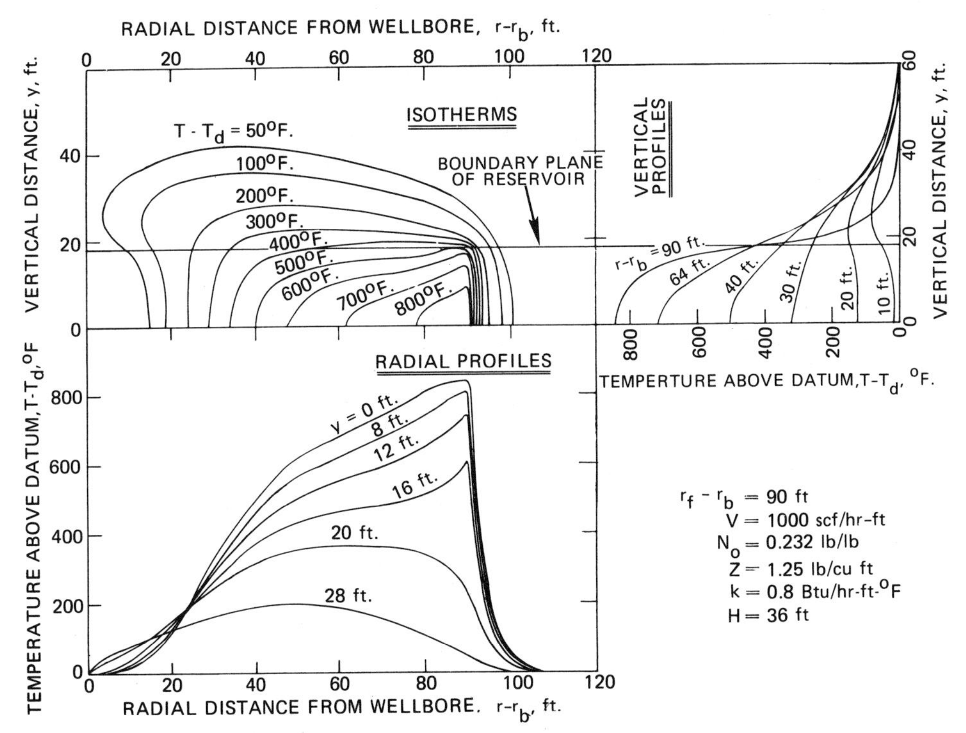

Figure 13.12. Typical isotherms and corresponding radial and vertical temperature profiles.[8] © 1963 SPE-AIME.

Chu discussed the alternative use of the minimum front temperature (at the edge of the sand). He presents numerous plots of the radial location of the point at which the sand edge or minimum combustion front temperature, $T_{f,min} - T_r = 650°F$, where he presumed combustion would cease. This criterion is probably too conservative, since for a formation at 100°F, it would presume that combustion ceases below 750°F. Indeed, numerous field projects have been operated on wider spacing and at lower rates than Chu's calculations would predict using this criterion.

Figure 13.13 shows Chu's results for the peak temperatures as a function of $r - r_b$ and y, the distance from the midplane of the sand. For the case shown, a sand having a thickness $H = 36$ ft and other conditions cited, the midplane peak temperature stabilized at about 220°F higher than the temperatures at the edge of the sand. This factor must be borne in mind when designing on the basis of the midplane peak temperature considerations, since combustion will at best be incomplete (some oxygen will be produced). The combustion may in fact cease, since the assumption that the front is ignited over its entire thickness no longer will be valid.

In view of these considerations, the following guidelines are suggested:

1. Design for an average combustion front temperature estimated to be approximately 100°F below the peak temperature.
2. Determine the minimum temperature required for sustained combustion in the laboratory ($T_{f,min}$). If these data are unavailable, assume $T_{f,min} = 600°F$.
3. Calculate the design minimum peak combustion front temperature,

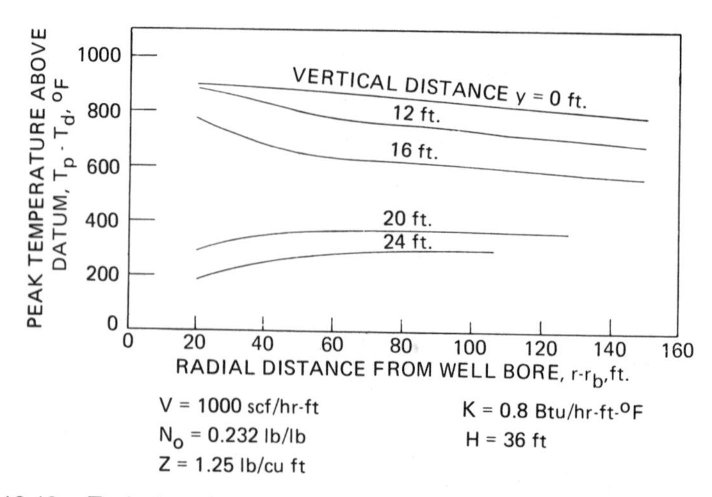

Figure 13.13. Typical peak temperature profiles at various vertical levels.[8] © 1963 SPE-AIME.

T_{fp}. Note that

$$T_{fp} - T_r = T_{f,min} + 100 - T_r$$

where T_{fp} = center-plane peak combustion front temperature, T_p(°F)
T_r = original reservoir temperature, T_d(°F)

4. Use Figures 13.14–13.18, calculated by Chu to determine the maximum radial advance of the combustion front as a function of†:
V = air rate per foot of sand (scf/ft-hr)
N_o = oxygen concentration (0.232 lb/lb for air)(lb/lb)
F_c = fuel concentration [Z] (lb/ft^3)
K_h = thermal conductivity [K] (Btu/ft-hr-°F)
h = net sand thickness [H] (ft)
r_w = well radius [r_b] (ft)

In designing a combustion project, it is essential to first estimate the maximum possible air rate, V_{max} and F_c, since the air rate and fuel concentration will have a strong influence on the distance to which the front can be advanced. The following design steps are suggested:

1. Estimate $\bar{F}_c$ and R_{Ao} using Figure 13.11 and Eqs. (13.12), (13.13), and (13.6).

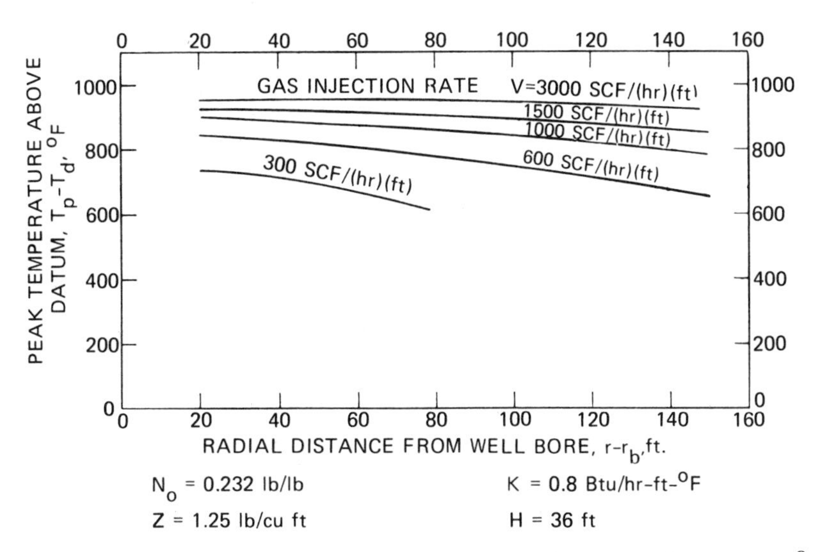

Figure 13.14. Effect of gas injection rate on center-plane peak temperature.[8] © 1963 SPE-AIME.

†The nomenclature used by Chu in the figures differs from our nomenclature; his nomenclature is indicated in square brackets.

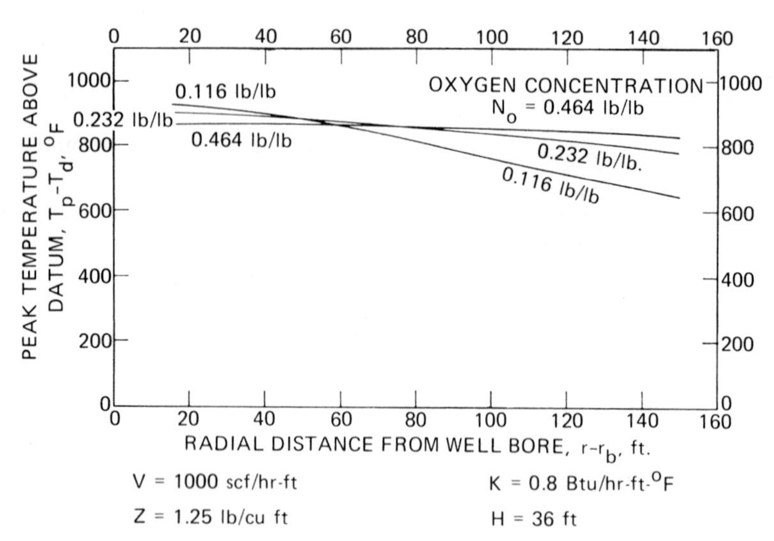

Figure 13.15. Effect of oxygen concentration on center-plane peak temperature.[8] © 1963 SPE-AIME.

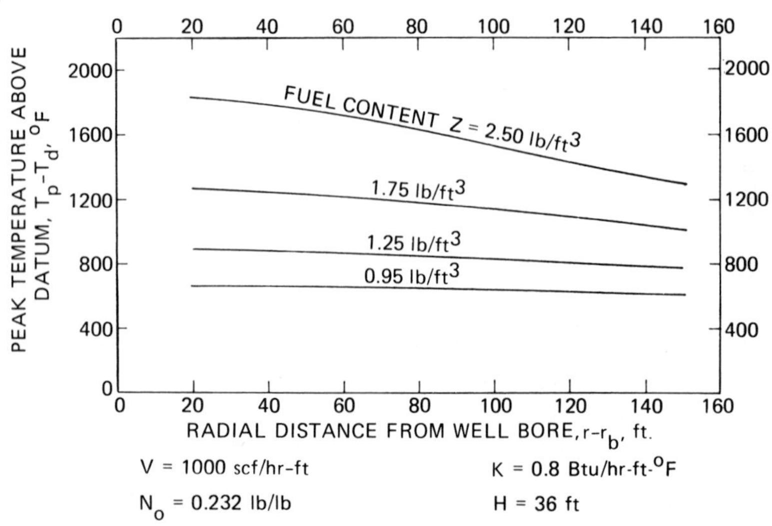

Figure 13.16. Effect of fuel concentration on center-plane peak temperature.[8] © 1963 SPE-AIME.

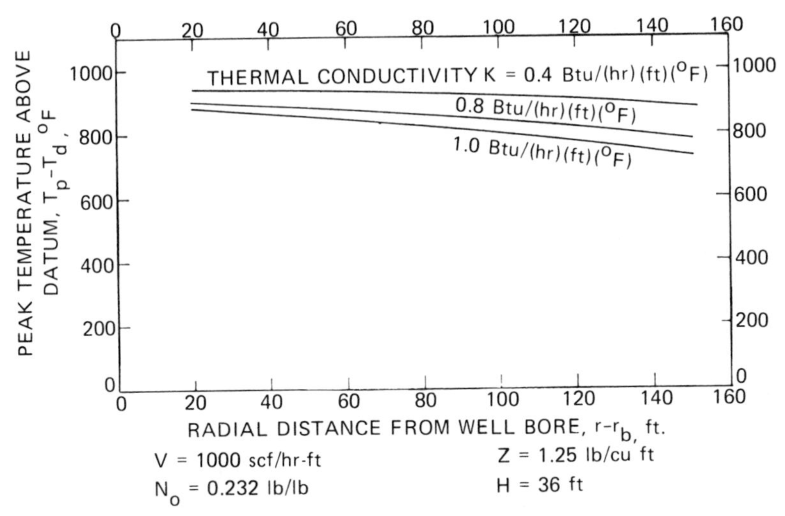

Figure 13.17. Effect of thermal conductivity on center-plane peak temperature.[8] © 1963 SPE-AIME.

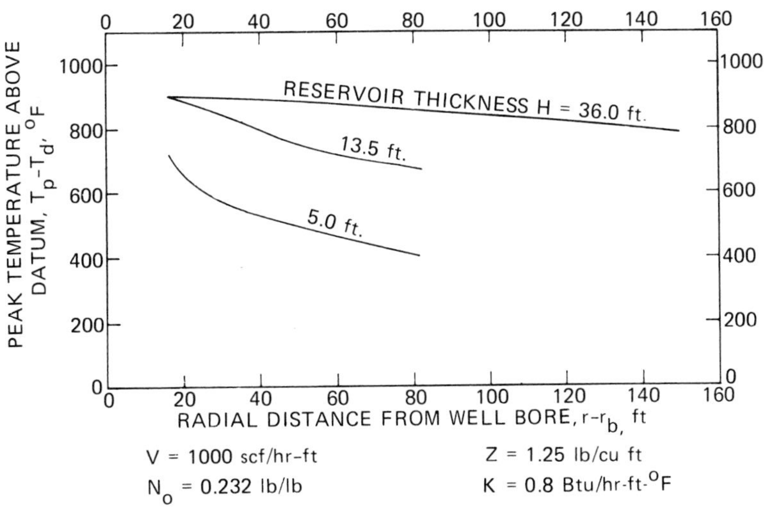

Figure 13.18. Effect of reservoir thickness on center-plane peak temperature.[8] © 1963 SPE-AIME.

2. Estimate V_{max} using the radial steady-state flow equation when the front is midway between injector and producer. Pressure drop in the burned zone may be neglected, and the oil zone pressure drop will be controlling for moderate to highly viscous oils. An estimated well spacing d is used if the wells are not already drilled. The injection pressure P_I is generally taken to be 100–200 psi below breakdown pressure. The bottom-hole producing well pressure P_w may be assumed to be about 100 psi for pumping wells.† For flowing wells, P_w may be estimated using a two-phase pressure drop calculation method such as that of Orkiszewski[24]:

$$V_{max} = \frac{2.946 \times 10^{-4} k_o (P_I - P_w)}{B_o \mu_o \ln(d/2r_w)} R_{Ao} \quad \text{(scf/hr-ft)} \tag{13.15}$$

3. Determine that for $r - r_b = \frac{1}{2}d$, V estimated by Figures 13.14–13.18 is less than V_{max} calculated by Eq. (13.15). The design V can be intermediate between these values. Unfortunately, for thin sands on spacings wider than 320 ft and fluxes higher than 3000 scf/hr-ft, we are out of the range of these figures and some rather crude extrapolations would have to be made.

Ramey's Solution

Ramey[9] obtained a solution for Eq. (13.14) under more limited assumptions than Chu. All assumptions are the same as in Chu's model, except that the effect of convection is taken into account in only an approximate fashion. In addition, conduction in only the vertical direction is taken into account. Results are plotted in Figures 13.19–13.22, where $(T_f - T_r)/(T_m - T_r)$ is plotted against r_f/a with parameters vr_f/α and va/α defined below:

T_f = temperature at combustion front (°F)
T_r = original reservoir temperature (°F)
T_m = maximum temperature attainable during passage of combustion front if behavior were adiabatic (°F)
r_f = radius of combustion front (ft)
a = half-thickness of sand $= \frac{1}{2}h$ (ft)
v = velocity of combustion front (ft/d)
α = thermal diffusivity, $K_h/(\rho c)_{R+F}$ (ft^2/d)

The parameter vr_f/α is related to the fuel concentration F_c and the air rate per foot of sand, V, by

$$\frac{vr_f}{\alpha} = \frac{V}{2\pi F_c C_A \alpha} \tag{13.16}$$

†High gas–oil ratio production attendant with in situ combustion operations may not be efficiently lifted by bottom-hole pumps, and experience may dictate assumption of a somewhat higher P_w than 100 psi, especially in the case of viscous oils.

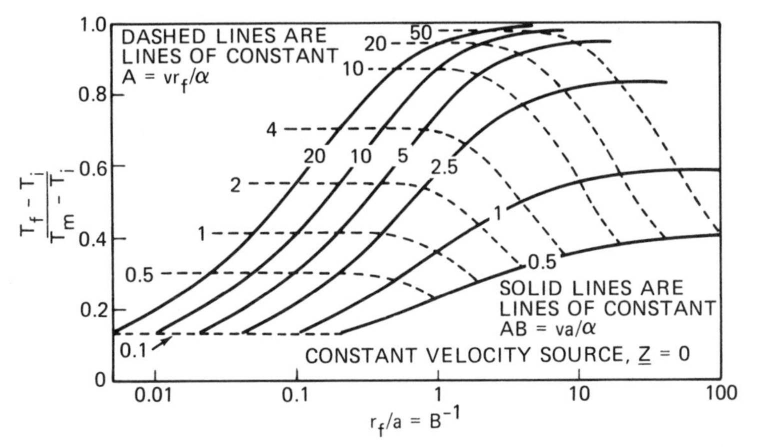

Figure 13.19. Temperature rise ratio at heat source of finite height moving at constant velocity; temperature at center-plane ($z = 0$).[9] © 1959 SPE-AIME.

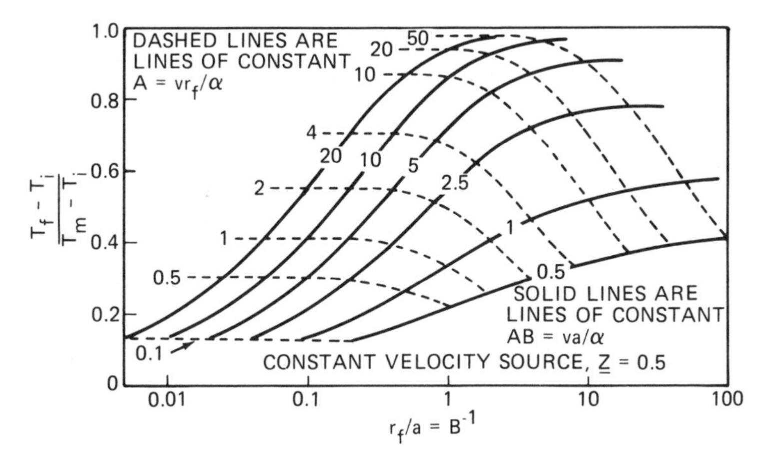

Figure 13.20. Temperature rise ratio at heat source of finite height moving at constant velocity; temperature half way between center-plane and edge ($z = 0.5$).[9] © 1959 SPE-AIME.

where C_A is the amount of air required to burn 1 lb of fuel in scf/lb.

Figures 13.19–13.22 give frontal temperatures at four different vertical positions ($z = y/a$): 0 (midplane), 0.5, 0.75, and 1.0 (edge of sand).

The suggested procedure for using these plots is:

1. Estimate $T_f - T_r$ for the minimum combustion temperature. Assume that $T_f = 600°F$ unless specific data for T_f are available.

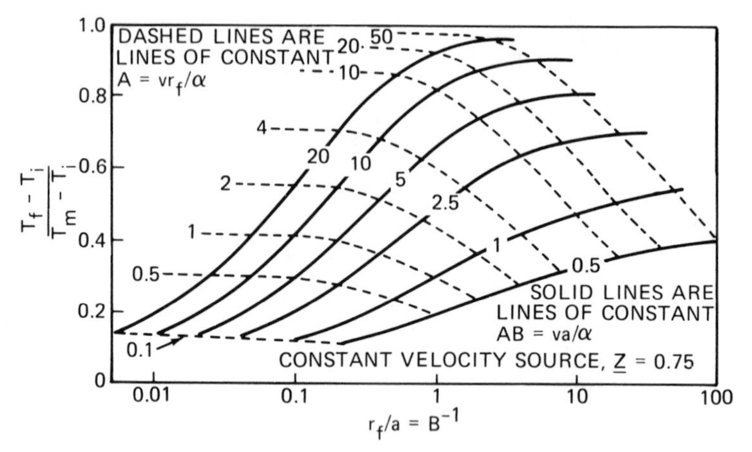

Figure 13.21. Temperature rise ratio at heat source of finite height moving at constant velocity; temperature three-quarters of distance from center-plane to edge ($z = 0.75$). © 1959 SPE-AIME.

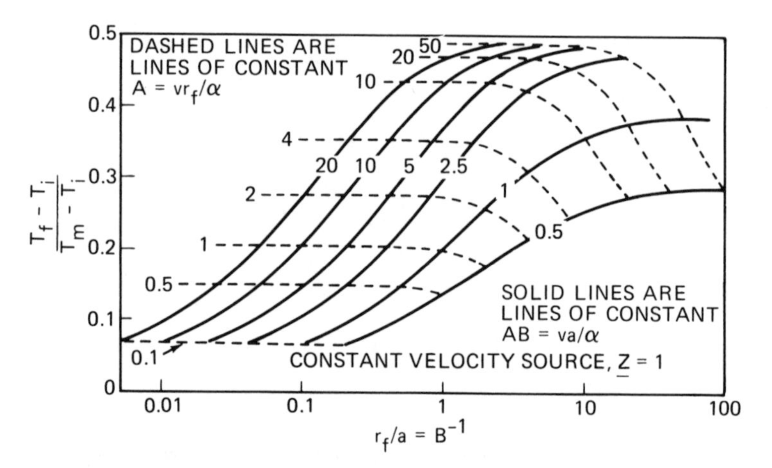

Figure 13.22. Temperature rise ratio at heat source of finite height moving at constant velocity; temperature at edge ($z = 1$).[9] © 1959 SPE-AIME.

2. Estimate the adiabatic temperature rise:

$$T_m - T_r = \frac{F_c[\Delta H + C_A \rho_g c_g (T_f - T_r)]}{(\rho c)_{R+F}} \tag{13.17}$$

where ΔH is the higher heating value of the fuel burned in Btu/lb.

3. Read vr_f/α or va/α corresponding to

$$\left(\frac{T_f - T_r}{T_m - T_r}\right)_{min}$$

and r_f/a from Figures 13.19–13.22. If no $[(T_f - T_r)/(T_m - T_r)]_{min}$ value can be read at $z = 1$, this probably means that some oxygen will always be produced or be consumed in low-temperature oxidation reactions. The vr_f/α (or va/α) values from these plots can be averaged. Alternatively, one might use the values of these parameters determined at $z = 0.5$ for an approximate estimate of $(vr_f/\alpha)_{avg}$.

4. Calculate V_{min}:

$$V_{min} = 2\pi F_c C_A \alpha \left(\frac{vr_f}{\alpha}\right)_{avg} \tag{13.18}$$

Comparison of Estimates Using Chu's and Ramey's Solutions

For an $F_c = 1.25$ lb/ft^3, $h = 36$ ft, and $K_h = 0.8$ Btu/hr-ft, Chu gives $T_f - T_r = 680$°F at the midplane ($z = 0$) at $r_f = 140$ ft for $V = 600$ scf/hr-ft.

Ramey's solution is employed using the above parameters to estimate V. Other parameters assumed by Chu needed for the Ramey solution are

$$\begin{aligned} \rho_g &= 0.07651 \text{ lb/scf} \\ c_g &= 0.24 \text{ Btu/lb-°F} \\ (\rho c)_{R+F} &= 30 \text{ Btu/ft}^3\text{-°F} \\ C_A &= 188 \text{ scf/lb fuel} \\ \Delta H &= 20000 \text{ Btu/lb} \end{aligned}$$

By Eq. (13.17),

$$T_m - T_r = 1.25\,\frac{20000 + 188(0.24)(0.0765)(680)}{30} = 931\text{°F}$$

$$\frac{T_f - T_r}{T_m - T_r} = \frac{680}{931} = 0.730$$

$$\frac{r_f}{a} = \frac{140}{18} = 7.8$$

From Figure 13.19, for $z = 0$,

$$vr_f/\alpha = 11$$

By Eq. (13.18),

$$V = 2\pi F_c C_A \alpha \frac{v r_f}{\alpha} = 6.28(1.25)(188)\frac{0.8}{30}(11)$$
$$= 433 \text{ scf/hr-ft}$$

The air rate required to produce the 680°F temperature rise is about 30% lower by Ramey's method than by the more complete analysis of Chu; probably this occurs because of the neglect of radial conduction in Ramey's model. In view of this, when using Ramey's solution, minimum air rates should probably be increased by 30% to provide a margin of safety.

COMBUSTION FRONT TEMPERATURES: THICK-SAND CASE

The estimation of overriding combustion front temperatures is complex and is beyond the scope of this book. Idealized treatments have been published by Gottfried[16] and Prats et al.[17] Prats' treatment predicts that frontal temperatures will drop to low values at the leading edge of the combustion front where it pinches out at the top of the sand owing to high heat losses. This implies that in an overriding front, some oxygen must be consumed by low-temperature reactions or be produced as free oxygen. Free oxygen is not often produced, however, because as high fluxes are reestablished near the producing well, spontaneous reignition will create a secondary combustion front. Should this occur, reignition could cause severe well completion damage.

IGNITION PHASE DESIGN

A comprehensive discussion on ignition for in situ burning operations is found in the literature.[25] When initiating a project, a decision must first be made concerning whether the use of an auxiliary ignition aid will be required or whether rapid spontaneous ignition will occur. Unless the formation temperature is sufficiently high that spontaneous ignition can occur within about 4 hr of the start of air injection, use of an auxiliary ignition aid is advisable. A long spontaneous ignition time is to be avoided because of auto-oxidation, the resulting buildup of tarry residuum and increased deposition of fuel, which must eventually be consumed. When combustion initiates, high bottom-hole temperatures, which can damage liners and downhole tubulars, are often experienced after this excessive coke buildup near the wellbore.

Spontaneous Ignition

Crude oils oxidize when exposed to air, and for many crudes, the reaction rate is reasonably rapid even at original reservoir temperatures. As reaction occurs, temperatures in the formation rise, and the reaction rate of oxidation increases exponentially until combustion occurs and carbon reacts to form CO_2. In general, the time to achieve an ignition temperature of 300–600°F is shorter for formations with higher initial temperatures.

An estimate of the time to achieve spontaneous ignition can be made if it is assumed that during the ignition phase, formation heat losses are negligible. An adiabatic energy balance gives

$$t_{ign} = \int_{\bar{T}_r}^{\bar{T}_{ign}} \frac{(\rho c)_{R+F}\, dT}{\Delta H_{ox} k_r e^{-E/RT}} \tag{13.19}$$

where t_{ign} = time to achieve ignition (hr)
T_{ign} = absolute ignition temperature (°R)
T_r = absolute original reservoir temperature (°R)
ΔH_{ox} = heat of reaction per mole O_2 reacted (Btu/lb-mol)
k_r = reaction rate constant (function of oxygen pressure, oil surface area, oil type, etc.) (lb-mol O_2 reacted/hr-ft^3)
E = activation energy (Btu/lb mol)
R = gas constant (1.987 Btu/lb mol-°R)

Laboratory tests must be made to determine k_r and E. Reported values, as an example, for a specific crude–sand system[25] are

$k_r e^{-E/RT} = e^{20-20000/T}$ lb-mol/hr-ft^3
$\Delta H_{ox} = 170{,}000$ Btu/lb-mol
$(\rho c)_{R+F} = 30$ Btu/ft^3-°F
$T_{ign} = 350°F = 810°R$

If $T_r = 65°F = 525°R$, then by Eq. (13.19),

$$t_{ign} = \int_{525}^{810} \frac{30 e^{20000/T}\, dT}{170{,}000 e^{20}} = 7740 \text{ days}$$

Obviously, this case would require artificial ignition. Table 13.2 gives observed ignition times for several projects.

A definite trend to shorter spontaneous ignition times as T_r increases can be seen. Ignition times measured in the laboratory for two crudes are shown in Figure 13.23. The effect of higher reservoir pressure is to shorten the ignition time.

Table 13.2 Spontaneous Ignition Times for Several Projects[a]

Project	Net Sand Thickness (ft)	Crude Gravity (°API)	T_r(°F)	Ignition Time t (days)	V_g (scf/hr-ft)
1	35	12.9	87	100	738
2	163	14.5	125	17	256
3	235	13.3	85	150	18[b]
4	220	14.0	113	62	379
5	40	12.1	140	13	190
6	51	11.3	125	9	735
7	110	16.0	95	24	121
8	90	16.1	75	47	602

[a]From Ref. 25. © 1964 *Petroleum Engineer*.
[b]Later increased to 213 scf/hr-ft.

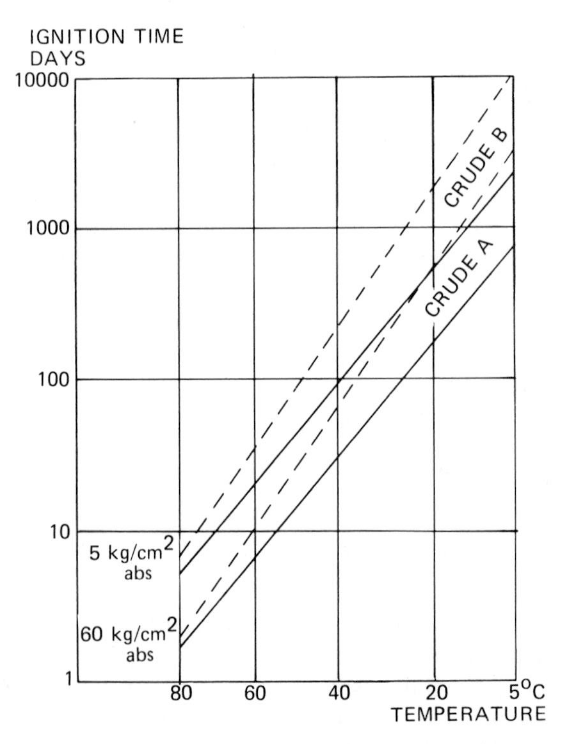

Figure 13.23. Ignition time as function of temperature at various air injection pressures.[2] © 1970 SPE-AIME.

Artificial Ignition

Ignition times can be greatly shortened by the following techniques:

(a) *Preheat Injected Air.* This is probably the most used method.

Downhole Gas Burner. A natural gas–air burner is lowered downhole and used to preheat injected air. A heat shield is normally required to prevent liner damage.

Downhole Electrical Heater. An electrical heater is lowered downhole and connected by a power cable to the surface. Excellent control of the amount of heat input is possible, but a lower rate of heat output than for the gas burner may require a longer time for the heater to be downhole.

(b) *Preheat the Formation.* A hot fluid is injected, such as steam, to heat the immediate region around the wellbore. Gas is generally unacceptable because of its low heat capacity, but hot water might be used if the formation is shallow. Mitigating wellbore heat losses and possible casing damage are the primary considerations in using this approach.

(c) *Improve Reactivity with Chemicals.* Numerous readily oxidizable materials and oxidizing agents have been proposed.[25] Among oxidizable materials, linseed oil has frequently been used. The use of oxidizing agents such as nitric acid is effective but dangerous because of the likelihood of downhole explosion and is not recommended.

Table 13.3 gives the heat supplied and other pertinent data for a number of projects in which burners, electrical heaters, and hot-fluid injection were employed.

Preheating the injected air is the generally recommended method of ignition, even if spontaneous ignition times are relatively short. Without preheating, ignition will occur some distance away from the injector, and reverse combustion will follow. At the wellbore, the direction of movement of the front will change, and forward combustion will begin. Where the direction of movement changes, the temperatures become extremely high, and damage to the injection well liner can result.

With moderate preheating, this effect is not nearly as great because the front ignites near the wellbore with little or no reverse burning. Interruption of air injection while the front is still near the injection well is also to be avoided, since oil can backflow into the wellbore and burn inside it when air injection is resumed.[2]

Table 13.3 Summary of Several Artificial Ignition Operations[a]

Project	Pay Thickness (ft)	Heat Supplied (10^6Btu/ft)	Ignition Type[b]	Rating of Heater	Crude Gravity (°API)	Pressure (psig)	Average Air Rate During Ignition (Mscf/d-ft)
Southern Oklahoma	20	1.6	E	45 kW	18.5	130	2.5
Delaware-Childers Field, OK	51	1.0	B	12 MBtu/d	35–37	—	9.4
Parker Pool, IL	35	—	H	43 MBtu/d	26.3	—	—
Southern Oklahoma	17	0.9	E	45 kW	18.4	150	10.7
Niitsu Field, Japan	33	2.2	B	16 MBtu/d[c]	18.2	—	10.1
Allegany Field, NY	34	2.3	E[d]	75 kw	43	—	4.2 (maximum)
Southeastern Kansas	9	—	B	—	23	—	—
Shannon Field, WY	33	0.5	B	10 MBtu/d[c]	25	—	12.2
Goodwill Hill Field, PA	30	3.3	B	6 MBtu/d[c]	43.9	185	7.7
"I"	60	0.4	E	45 kW	17.6	750	4.9
"J"	20	0.4	E	29 kW	21.4	360	10.0
"K"	18	0.4	E	24 kW	18.7	85	6.3
"L"	20	0.7	E	29 kW	14.5	215	2.2
"M"	[e]	4.0 (total)	E	24 kW	9.9	310	130 (total)
"N"	80	0.2	E	29 kW	17.6	830	7.1
"O"	60	0.3	B	10 MBtu/d[c]	17.6	920	7.7

[a]From Ref. 25. © 1964 *Petroleum Engineer*.
[b]E, electrical; B, downhole burner; H, hot-fluid injection.
[c]Calculated maximum rate, rating not given.
[d]Ignition not achieved.
[e]Single-entry through mechanically cut casing notch.

PRODUCTION CHARACTERISTICS OF IN SITU COMBUSTION WELLS

Depending on the air–oil ratio, gas–oil ratios produced in dry, forward combustion projects will range from high to extremely high by normal producing standards. The gas–oil ratio will be approximately related to the air–oil ratio by

$$R_g = 0.9 f_{cap} R_{Ao} \tag{13.20}$$

where f_{cap} is the fraction of flue gas captured. The 0.9 factor approximately accounts for the air held up behind the combustion front.

In addition to large gas volumes, a substantial amount of water will also be produced owing to the displacement of connate water from behind the combustion front and the water formed by the combustion reaction. An approximate relation of the producing water–oil ratio is given as

$$R_w = \frac{S_{wi} + 1.06 S_{of}(\rho_o/\rho_w)}{S_{oi} - S_{of}} \tag{13.21}$$

Equation (13.21) is only accurate when the hydrogen–carbon ratio n is 1.6.

Pumping problems owing to gas locking often require the use of gas anchors. In some cases, it is not feasible to pump and production is flowed from the well. High flowing bottom-hole pressures at producing wells are encountered in deeper wells, and this becomes an important design consideration as it affects injection–production rates as well as the degree of pattern confinement. The latter can be a serious problem when conducting a small project in a deep, low-pressure reservoir if flowing producing bottom-hole pressures are greater than starting formation pressure.

Sand cutting and corrosion of liners is a potential problem in unconsolidated sands. Gravel-packed slotted liners have been the most popular and successful completion. Stainless steel liners have been used in some projects where the production is quite corrosive. Most projects experience little corrosion, however, until heating occurs at the producing well, indicating proximity of the combustion front.

USE OF OXYGEN ENRICHMENT

In recent years, the use of oxygen enrichment to improve performance of dry, forward in situ combustion field projects has received increased attention. This has been partly in response to efforts by oxygen manufacturers to expand sales, but there are beneficial effects, such as reduced compression requirements, that deserve serious consideration. Some indication of the effect of oxygen enrichment on peak frontal temperatures can be seen from the work of Chu[8] (see Fig. 13.15).

Clearly, frontal temperatures can be kept higher at a greater distance from the injection well (for a given injection rate) the higher the oxygen concentration. Thus, oxygen enrichment should be useful in cold or in low-permeability formations and in deep formations where high well costs would necessitate use of wide well spacings. Special safety precautions must be taken when using oxygen in oil field operations. Great care in selection of piping and well tubulars is strongly recommended. Merington[26] and Howard[27] discuss solutions to the field-handling problems and plans for an experimental pilot involving nearly pure oxygen injection.

REFERENCES

1. Gates, C. F, and Sklar, I. "Combustion—A Primary Recovery Process, Moco Zone Reservoir—Midway-Sunset Field, Kern County, CA," *J. Pet. Tech.* (August 1971), pp. 981–986.
2. Dietz, D. N. "Wet Underground Combustion, State of the Art," *Trans. AIME*, **249,** 1–605 (1970).
3. Coats, K. H. "In Situ Combustion Model," *Soc. Pet. Eng. J.* (December 1980), p. 533.
4. Farouq Ali, S. M. "Multiphase, Multidimensional Simulation of In Situ Combustion," SPE 6896, presented at the Fifty-Second Annual Technical Conference and Exhibition of the Society of Petroleum Engineers, Denver, October 9–12, 1977.
5. Crookston, H. B., and Culham, W. E. "Numerical Simulation Model for Thermal Recovery Processes," *Soc. Pet. Eng. J.* (February 1979), pp. 37–58.
6. Youngren, G. K. "Development and Application of an In Situ Combustion Reservoir Simulator," *Soc. Pet. Eng. J.* (February 1980), pp. 39–51.
7. Binder, G. G., Elzinga, E. R., Tarmy, B. L., and Willman, B. T. "Scaled Model Tests of In Situ Combustion in Massive Unconsolidated Sands," *Proceedings of the Seventh World Petroleum Congress*, Mexico City, Vol. 3, Elsevier, 1968, p. 477.
8. Chu, C. "Two-Dimensional Analysis of a Radial Heat Wave," *J. Pet. Tech.* (October 1963), p. 1137.
9. Ramey, H. J., Jr. "Transient Heat Conduction During Radial Movement of a Cylindrical Heat Source—Application to the Thermal Recovery Process," *Trans. AIME*, **216,** 115 (1959).
10. Bailey, H. R., and Larkin, B. K. "Conduction–Convection in Underground Combustion," *Trans. AIME*, **219,** 320 (1960).

11. Thomas, G. W. "A Theoretical Analysis of Forward Combustion in Petroleum Reservoirs," Ph.D. Dissertation, Stanford University, February 1963.
12. Gottfried, B. S. "A Mathematical Model of Thermal Oil Recovery in Linear Systems," *Trans. AIME*, **234,** 2–196 (1965).
13. Chu, C. "The Vaporization–Condensation Phenomena in a Linear Heat Wave," *Trans. AIME*, **231,** 2–85 (1964).
14. Penberthy, W. L., Jr., and Ramey, H. J., Jr. "Design and Operation of Laboratory Combustion Tubes," *Trans. AIME*, **237,** 2–183 (1966).
15. Couch, E. J. "Effects of Porosity and Permeability in In Situ Combustion Fuel Consumption," SPE 2873, presented at the Forty-Fifth Annual Meeting of the Society of Petroleum Engineers, Houston, Texas, October 4–7, 1970.
16. Gottfried, B. S. "Some Theoretical Aspects of Underground Combustion in Segregated Oil Reservoirs," *Soc. Pet. Eng. J.* (September 1966), p. 281.
17. Prats, M., Jones, R. F., and Truitt, N. E. "In Situ Combustion Away from Thin, Horizontal Gas Channels," *Trans. AIME*, **243,** 2–18 (1968).
18. Wilson, L. S., Reed, R. L., Reed, D. W., Clay, R. R., and Harrison, N. H. "Some Effects of Pressure on Forward and Reverse Combustion," *Soc. Pet. Eng. J.* (September 1963), p. 127.
19. Parrish, D. R., and Craig, F. F., Jr. "Laboratory Study of a Combination of Forward Combustion and Waterflooding—The COFCAW Process," *Trans. AIME*, **246,** 1–753 (1969).
20. Showalter, W. E. "Combustion Drive Tests," *Soc. Pet. Eng. J.* (March 1963), p. 53.
21. Alexander, J. D., Martin, W. L., and Dew, J. N. "Factors Affecting Fuel Availability and Composition During In Situ Combustion," *J. Pet. Tech.* (October 1962), p. 1154.
22. Nelson, W. L. *Petroleum Refinery Engineering*, 3rd ed., McGraw-Hill, New York, 1949.
23. Orkiszewski, J. "Petroleum Recovery by In Situ Combustion," U.S. Patent 3,400,760, September 10, 1968.
24. Orkiszewski, J. "Predicting Two-Phase Pressure Drops in Vertical Pipe," *J. Pet. Tech.* (June 1967), p. 829.
25. Strange, L. K. "Ignition: Key Phase in Combustion Recovery," *Pet. Eng.* (November 1964), p. 105; (December 1964), p. 97.
26. Merington, L. "Designing for Safety in Enhanced Recovery Oxygen Fireflooding," *Oil Gas J.* (October 31, 1983), p. 124.
27. Howard, J. V. "In Situ Combustion," Presented at Enhanced Recovery Week's Moving Heavy Oil Conference, Los Angeles, California, June 7–8, 1982.
28. Neilson, W. T., and McNeil, J. S. "How to Engineer an In Situ Combustion Project," *Oil Gas J.* (June 5, 1961), pp. 58–65.

Chapter 14

DRY IN SITU COMBUSTION CASE HISTORIES

ABSTRACT

A large number of dry in situ combustion pilots have been conducted. Only a fraction of these has resulted in successful ventures as commercial projects. Nevertheless, future oil needs will undoubtedly see increased use of the method where other enhanced oil recovery (EOR) techniques are inapplicable. This chapter gives results for some of the better-documented pilots and commercial projects.

Details of the operating behavior for 13 projects having oil gravities ranging from 9 to 33°API are given where air–oil ratios ranged from 3.5 to 174 kscf/STB. Ignition problems and poor oxygen utilization caused excessive air requirements in one project (Pontotoc), and flue gas recycle was used in another (Humbolt–Chanute), greatly increasing compression requirements. The range of observed air–oil ratios for the more normal projects is 3.5–19 kscf/STB. Gas–oil ratios were generally lower than air–oil ratios, indicating that displaced fluid capture by the often small, unconfined pilots was incomplete. Assuming gas–oil ratios are indicative of air–oil ratios in a confined pattern, it can be concluded that the Orkiszewski correlation for the deposited fuel concentration presented in Chapter 13 is reasonably accurate for screening estimates of air–oil ratios. Best results in terms of low air–oil ratio occurred in the Midway-Sunset field in California where the thick section and steep dips permitted improved gravity control of the overriding combustion front.

Economic considerations, in particular, estimation of air compression requirements, are also covered. An alternative correlation of more recent combustion project data for estimating oil recovery and air requirements developed by Brigham is also discussed.

DISCUSSION OF EARLY PROJECTS

Although relatively little commercial oil production is being obtained by dry in situ combustion at present compared to steam stimulation and steam-

Table 14.1 Summary of Reservoir Data for Selected Early Dry, Forward Combustion Projects

Field	Depth (ft)	Sand Thickness Net (ft)	Sand Thickness Gross (ft)	Permeability (md)	Original Oil Viscosity (cp)	°API	Reservoir Temperature (°F)	Porosity (%)	Starting Oil Saturation (Fractional)	Starting Oil Saturation (bbl/acre-ft)
Humbolt-Chanute, KN	830	8.8	12	85	70	23	78	20.3	0.69	1089
South Belridge, CA	700	30	35	8000	2700	12.9	87	37.0	0.60	1720
Pontotoc, OK	195	17	—	7680	5000	18.4	66	27.2	0.64	1350
Shannon, WY	950	33	120	250	76	25	68	23.3	0.60	1084
Schoonebeek, Holland	2800	49	50	3000	175	25	100	29.5	0.84	1920
Fry, IL	888	50	56	320	40	29	65	19.7	0.68	1040
Delaware-Childers, OK	600	45	59	118	6	33	76	20.6	0.365	584
North Government Wells, TX	2320	18	60	800	10	22	120	32.0	0.32	800
Midway-Sunset, CA	2100–2700	129	500	1575	110	14	125	36.0	0.75	1980
Melones, Venezuela	4500	15	15	3500	400	9.5	145	35.0	0.94	2552
Delhi, LA	3400	8.3	—	1069	3	40	135	31.2	0.31	740
West Newport, CA (B zone)	868–1740	120	160	1070	700	15.2	105	37.0	0.67	1925

Table 14.2 Summary of Project Data for Early Dry, Forward Combustion Projects

Field	Operator (Ref.)	Pattern		Percent	Cumulative Incremental Recovery				Average Rates/ Injector or Producer	
		Type	Area (acres)	Oxygen Utilized	Oil/Pattern (kbbl/Pattern)[a]	Percentage of OOIP	Air/Oil (kscf/bbl)	Gas/Oil (kscf/bbl)	Air (kscf/d/well)	Oil (bbl/d/well)
Humbolt-Chanute, KS	Sinclair (1)	Line drive	11.4[a]	80	13	11	12.2	20[b]	300	2.4
S. Belridge, CA	G. Petr. (2)	Five-spot	2.5	95	74[c]	51	19.0[d]	6.4	3500	63
Pontotoc, OK	Mobil (3)	Five-spot	0.07	43	0.071	51	44.0	35.0	76	2
Shannon, WY	Pan Am (4)	Five-spot	1.32	50	35	74	12.0	9.3	500	8.4
Schoonebeek, Holland	NAM (5)	Seven-spot	29.0	96	55	2	16.8	10.5	3000	15.0
Fry, IL	Marathon (6, 7)	Five-spot	3.3	87	42.6[e]	25	11.5	9.0	1520	13.0
Delaware-Childers, OK	Sinclair (8)	Five-spot	2.22	80	12.4	21	174.0	84.9	1410	2.0
North Government Wells, TX	Mobil (9)	Line drive	122.0[a]	96	83.0[a]	4.7	16.8	—	2500	5.0
Midway-Sunset, CA	Mobil (10)	[f]	50.0[a]	95	1500	20	3.5	—	1380	60
Melones, Venezuela	Mene Grande (11)	Two-spot	—	95	—	—	7.7[g]	8.0	1100	160[h]
Delhi, LA	Sun (12)	Five-spot	40.0	92	163	66	14.3	11.3	1822	32
West Newport (B zone), CA	Kadane (13)	Five-spot	3.8	95	94[i]	11[j]	10.8	—	830	22

[a]Per injection well.
[b]790 Mscf of recycle gas were injected in addition to 970 Mscf air.
[c]Through November 15, 1957.
[d]Pattern leak occurred. Pattern also unconfined.
[e]Including additional production from outlying wells, total incremental recovery was 100.6 kbbl.
[f]Crestal injection on anticline for pressure maintenance.
[g]Forward drive phase only.
[h]Total oil production rate, bbl/d.
[i]Through October 1964.
[j]Incremental Recovery (six affected patterns).

flooding, the process has been widely field tested. Basic reservoir and performance data for a number of early projects are listed in Tables 14.1 and 14.2. A brief description of these projects follows.

Humbolt-Chanute, Kansas

This project was operated with selected wells as injectors in a random line drive.[1] The reservoir is a long, narrow, isolated sand. An air–oil ratio of 12.2 kscf/bbl was obtained, but individual producing well rates were quite low. Despite the low air injection rates (360 kscf/d per injector) and thin sand (8.8 ft), the oxygen utilization was fairly good (80%). The produced gas–oil ratio was higher than the air–oil ratio because the operator recycled flue gas to attempt to drive additional heat forward. Chu's calculations (see Fig. 13.15) show that beyond a certain distance, dilution of injected oxygen will lower combustion front temperatures. The additional compression required to recycle flue gas is generally not warranted.

South Belridge, California

This project was operated on close spacing in a 30-ft sand.[2] The effects of rapid channelling owing to combustion front override were demonstrated in this pilot. Severe corrosion required producing well replacement at one location. Well productivity, however, was enhanced because of the early heating at producing wells resulting from early combustion front breakthrough.

Extended hot–well operation was demonstrated in this project. Loss of air from the unconfined, inverted 5-spot pattern caused the injected air–oil ratio to be much higher than the produced gas–oil ratio. There may also have been some leakage to a higher horizon. Probably the air injection rate was too high for the small pattern. This had the effect of driving the overriding combustion front well beyond the pattern boundaries.

Pontotoc, Oklahoma

This was a research pilot drilled on very close spacing.[3] Oxygen utilization was poor (43%) owing to the low air injection rate and low initial reservoir temperature. The high in-place oil viscosity (5000 cp) was also a detrimental factor.

Shannon, Wyoming

Oxygen utilization was poor in this project, which was operated on close spacing.[4] Some evidence of air channelling was noted in the shaly sand in which the project was operated. Despite this, the project had a fairly good air–oil ratio; possibly some benefit was gained by gas drive in addition to combustion drive, owing to the relatively low oil viscosity (76 cp). The oil recovery appears anomalously high, suggesting that some oil was captured from outside the pattern.

Schoonebeek, Holland

Three commercial-scale inverted patterns were operated in this project.[5] The combustion front rapidly channelled through a high-permeability streak at the top of the sand. Well completions were severely damaged by corrosion, high temperature, and sand erosion owing to the high air injection rates used. This project later employed water injection with air injection at reduced rates, and more satisfactory performance was achieved. The project was later converted entirely to a waterflood.

Fry, Illinois

Originally operated as a small, inverted five-spot, this project has been expanded, and nearly all wells completed in the sand have been affected by the injected air.[6,7] During the pilot stage, an air–oil ratio of 11.5 kscf/bbl was achieved, including additional oil recovery from wells outside the pattern. This relatively low air–oil ratio may in part have resulted from effective gas drive of the medium-viscosity (40 cp) oil. The operator also measured a much lower fuel deposition than is predicted by the Orkiszewski correlation (see Table 14.3).

The combustion front did not override the sand but moved preferentially through the center zone, giving the burning front roughly a parabolic shape. Nearly all wells had to be recompleted to accommodate the higher reservoir pressures associated with this project. Because of this and unusual research expenses and compressor downtime, the economics were unfavorable. However, in later stages, the project showed an operating profit on a daily basis.[7]

Delaware-Childers, Oklahoma

This was a pioneering project to examine the potential of in situ combustion as a tertiary light-oil recovery process.[8] Because of the low initial oil saturation (584 bbl/acre-ft) and a fairly high fuel concentration (275 bbl/acre-ft), the process was uneconomic, as compression requirements were much too high.

The injected air–oil ratio for the unconfined project was 174 kscf/bbl; had the pattern been confined, the injection requirement would have been lower, probably on the order of 60–90 kscf/bbl (based on observations of

Table 14.3 Comparison of Calculated and Actual Fuel Concentrations and Air–Oil Ratios

Field	Fuel (bbl/acre-ft)		Estimated Confined Air–Oil Ratio (kscf/bbl)	
	Measured	Calculated[a]	Corrected Observed[b]	Calculated[c]
South Belridge	253	205	7.11	9.3
Pontotoc	236	237	39	32
Fry	139	229	10	21
Delaware-Childers	275	192	94	40
North Government Wells	212	168	16.8[d]	18.0
Midway-Sunset	286	184	3.5[d]	7.0
Melones	274	214	7.7[d]	6.2
Delhi	118	115	12.6	13.0
West Newport (B zone)	265	266	10.8	11.0

[a] Using Orkiszewski correlation.
[b] Corrected for unconfinement. Estimated as $R_g/0.9$.
[c] Using calculated fuel saturation from Orkiszewski correlation and Eq. (13.6a).
[d] Actual air-oil ratio (project confined by reservoir boundaries).

the gas–oil ratio). One detrimental factor was the low original reservoir temperature, which probably caused increased fuel deposition.

North Government Wells, Texas

This was a commercial project conducted in a nearly-depleted, strong water-drive, medium oil gravity field.[9] Injection wells were located low on the structure (Fig. 14.1) to utilize the invading water and achieve some of the benefits of wet combustion. The project was primarily a dry-combustion project, however. The combustion front moved preferentially up-structure. As it passed producing wells, they were converted to injection wells. Faults made it necessary to locate additional injectors in several areas of the field.

Despite the low initial oil saturation (800 bbl/acre-ft), the project operated at an economically favorable air-oil ratio of 17 kscf/bbl. A fairly large steam zone ahead of the combustion front indicates that at least some favorable utilization of encroaching water was occurring. The air–oil ratio was slightly lower than indicated by Eq. (13.6a) for terminal conditions using the fuel concentration determined by the Orkiszewski correlation (see Table 14.3). In addition to the large steam zone, a fairly effective gas drive appeared to occur in the unheated part of the reservoir. The vertical sweep of the thin sand by the combustion front was reported to be good.

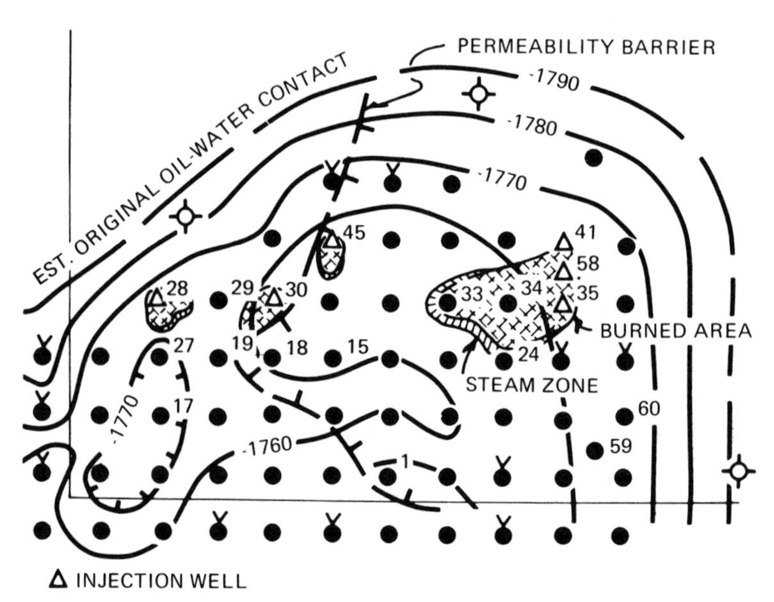

Figure 14.1. Structural map drawn on top of Government Wells sand showing estimated burned and steam zones.[9] © 1971 SPE-AIME.

Midway-Sunset, California

Mobil's Midway-Sunset project was a combination pressure maintenance–in situ combustion project in which air was injected high on the structure in a steeply dipping anticline (Fig. 14.2).[10] The flanks have dips of 20°–45° (see Fig. 14.3). The project was highly successful, recovering 9 Mbbl, or 24%, of the oil in place since its inception in 1959. Heating by combustion improved the effectiveness of the pressure maintenance by promoting improved gravity drainage of the oil (original viscosity 110 cp).

The operator estimated that 7.5 Mbbl were incrementally produced because of in situ combustion, giving this project a low incremental air–oil ratio of 3.5 kscf/bbl. This compares to the 7 kscf/bbl calculated using the Orkiszewski correlation [Table 14.3 and Eq. (13.6a)]. It should be pointed out, however, that the assumptions inherent in deriving Eq. (13.6a) are not valid in this case because they do not account for the substantial gravity drainage of oil beneath the combustion front. The combustion front appeared to be overriding in several of the individual sand members.

Well completion problems occurred as wells became heated but were minimized by using 316 stainless liners and downhole circulation of cooling water. Damage of injection well liners occurred frequently immediately after ignition because of high temperatures generated near the wellbore.

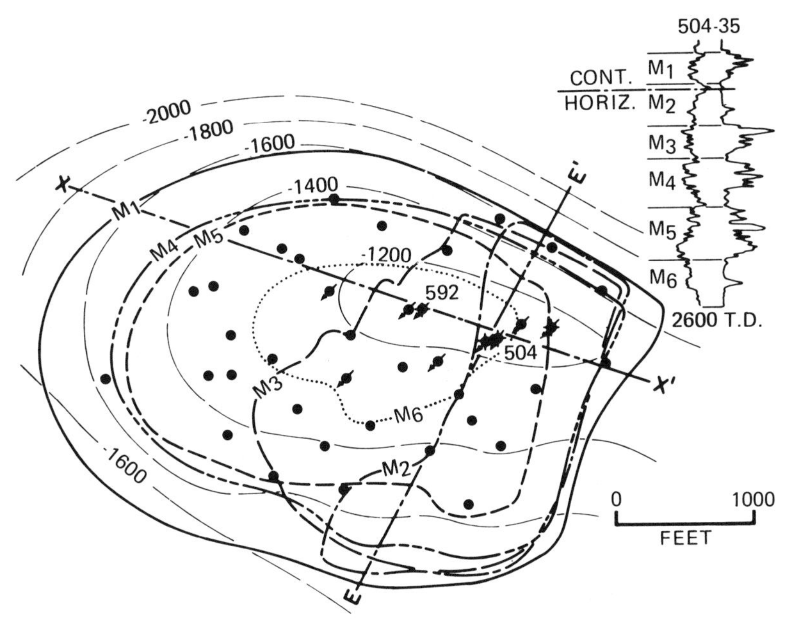

Figure 14.2. Moco zone sand distribution and structure (Midway-Sunset).[10] © 1971 SPE-AIME.

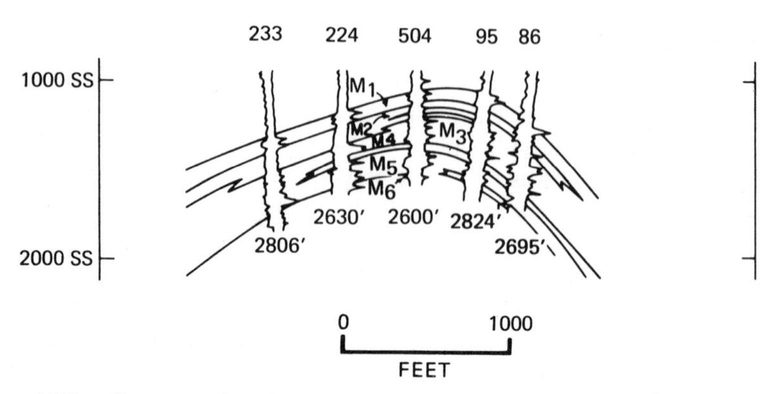

Figure 14.3. Cross section, E–E′, Moco zone (Midway-Sunset).[10] © 1971 SPE-AIME.

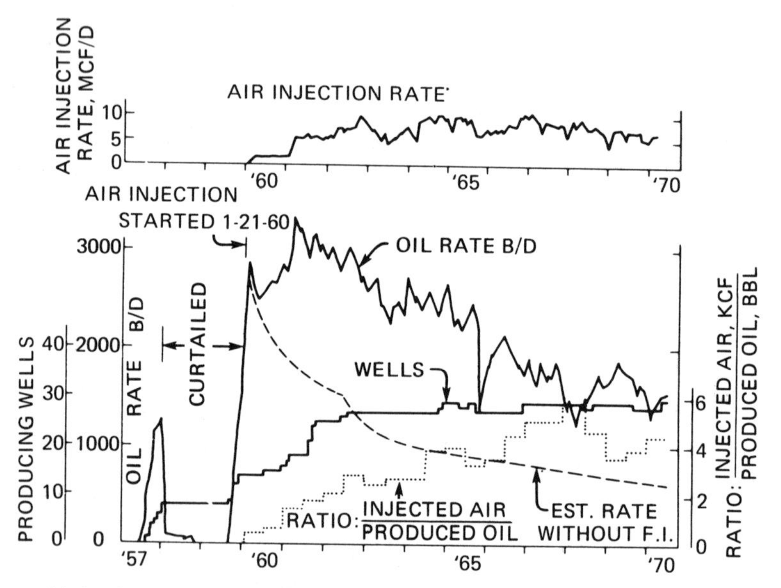

Figure 14.4. Moco zone performance (Midway-Sunset).[10] © 1971 SPE-AIME.

Little oil was produced prior to inception of the project, so it can be considered essentially a primary project. The injection–production history is given in Figure 14.4.

Melones, Venezuela

This was a two-well project operated in a thin, deep, isolated sand lens.[11] It also was essentially a primary recovery project. The data reported in Table

14.2 are for the forward-combustion phase. The observed air–oil ratio is in reasonable agreement with the calculated value given in Table 14.3. The low air–oil ratio resulted mainly from the high initial oil saturation (2552 bbl/acre-ft) and the high reservoir temperature, which reduced the consumption of injected oxygen in low-temperature oxidation reactions. Following arrival of the combustion front, air was injected intermittently.

Improved drawdown was possible when air was not being injected. Bottom-hole pressures were high in production wells during air injection periods because of inefficient pumping and the necessity of flowing the production. The bulk of the oil was produced during this later period when air was not being injected. The project's overall incremental air–oil ratio was approximately 4.9 kscf/bbl.

Delhi, Louisiana

This project was one of the more successful attempts to use in situ combustion as a tertiary light-oil recovery process.[12] The initial oil content was substantially higher (740 bbl/acre-ft) than in the Delaware-Childers project (584 bbl/acre-ft). In addition, the fuel concentration was much less (118 bbl/acre-ft compared to 275 bbl/acre-ft). The higher initial reservoir

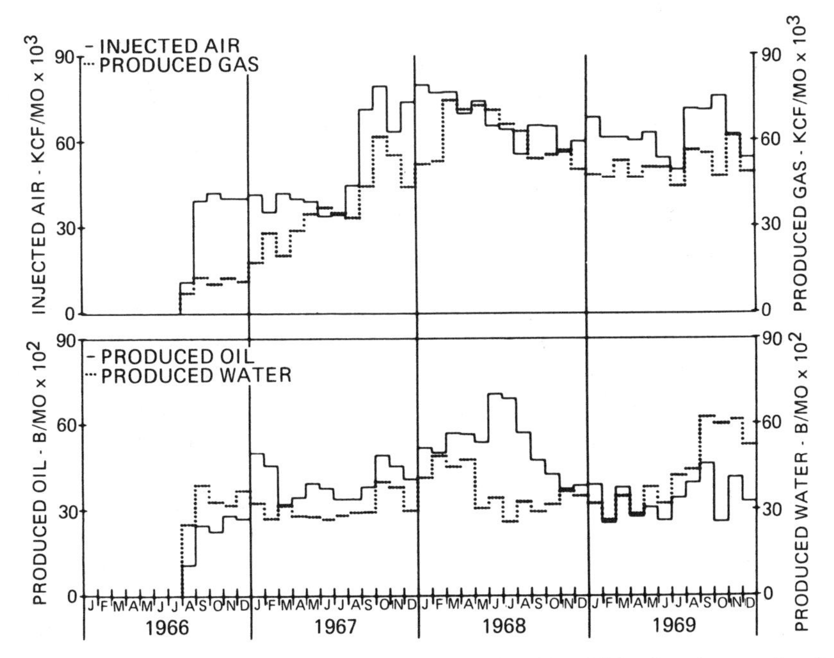

Figure 14.5. Injection and production history for May-Libby in situ combustion project (Delhi, LA)[12] © 1972 SPE-AIME.

temperature and higher injection rate permitted by better sand permeability aided the Delhi project performance. The observed fuel deposition agrees well with the value obtained from the Orkiszewski correlation, and the predicted and observed air–oil ratios are in good agreement (Table 14.3).

As of January 1970, oil recovery from the 40-acre pattern was excellent, about 66% of the original oil in place. The vertical sweep in the thin (8.3-ft) sand was essentially 100%. Injected air and produced oil, water, and gas rates for the pattern are shown in Figure 14.5. Following about 1 year of air injection, water injection was started. Because of this, the project has some aspects of wet combustion. However, the cumulative air–water ratio is quite high (8.5 kscf/bbl) compared to ratios normally considered for wet-combustion operations (1–3 kscf/bbl). Therefore, this project appears to be essentially a dry-combustion project. Reduced air injectivity was observed during the wet-combustion phase.

The operator reported a correlation of the product of volumetric average boiling point and oil viscosity at 350°F with burning characteristics of various crudes. According to this correlation, when this product (termed the *combustion index*) exceeds 350°F-cp, the crude should be capable of sustaining combustion. Delhi crude has a combustion index of 411°F-cp.

West Newport, California

A major commercial application of in situ burning was initiated in early 1958 in the B zone of the Banning lease, West Newport field, near Newport Beach, California.[13] As of late 1964, the project had been expanded to include 6 air injection wells affecting about 38 producing wells out of a total of 89 wells completed in the B zone. At that time, there were 25 wells immediately surrounding the 6 injection wells. The predominant pattern is a 5-spot, although one of the patterns is an 8-spot (one injector, seven producers).

The earliest area developed was a $1\frac{1}{4}$-acre inverted 5-spot that served as the pilot for later expansions of the project. Air was injected into the two lowest sands of the B zone while producing wells were completed over the entire four-sand B unit. Shale between the lower two and upper two sands aided in preventing rapid override of the combustion front early in the project. In addition, moderate injection rates were employed (1–2 Mscf/d). These rates on a per-injection-well basis were reduced to below 1 Mscf/d as additional injection wells were added in later expansions.

After 80 months of operation on the 200-ft spacing of the original test area, the combustion front had not yet reached the producing wells despite the substantial thickness of the B zone. This behavior was largely attributed to the use of low air injection rates, just slightly above theoretical values, to maintain efficient oxygen utilization. In the early stages of injection, the highest injection pressures were encountered, requiring a compressor horsepower of 240 BHP/Mscf-d of air injected. As the burned-out zone expanded and air injectivity improved, compressor horsepower requirements 3 years later had decreased to 180 BHP/Mscf-d of air injected.

After about 4 months of operation, API gravity of produced oil increased significantly in the pilot pattern, rising from the original crude gravity of 15.2°API (sulfur 2.2 wt %) to 17°API by the end of 1959. Subsequently, some wells produced 20°API oil (sulfur 1.8 wt %) as a result of thermal cracking and utilization of the tarry, or cokelike, residuum as the fuel. This produced crude quality improvement, which has occurred generally in all forward-combustion projects, provides a small economic plus because of the higher crude value.

To minimize air pollution, objectionable components in the produced flue gas (CO and H_2S are present in 1–2% concentration) were completely oxidized before being vented to the atmosphere. The combustion gases were gathered and flowed to specially designed gas treaters where natural gas and flue gas (heating value 60–250 Btu/scf) were blended and burned before venting. Noise-muffled equipment was employed extensively since the project was located near a residential area.

Later, the operator increased the use of steam stimulation to improve well productivities and reduce costs of air compression. As of 1965, approximately 35% of the total oil production of the B zone under the Banning lease was attributed to the combustion operation. The increased production was estimated at 562 kbbl over an 80-month period at an incremental cumulative air–oil ratio of 10.8 kscf/bbl.

Miga Field, Eastern Venezuela

In 1964, Mene Grande Oil Company started a dry, forward-combustion project in the $P_{2\text{-}3}$ sand reservoir in the Miga field of Eastern Venezuela.[14]

The original oil in place in the 25-ft sand was estimated at 23.2 Mbbl. The depth of the reservoir ranges from 4000 to 4350 ft KB with a dip of about 2°. The sand is loosely consolidated, with a porosity of 22.6% and an estimated average permeability of 5 darcies. Connate water saturation was about 22%. Oil gravity ranges from 13 to 14°API, with a viscosity range of 280–430 cp at 146°F reservoir temperature.

To mid-1973, an additional 2.6 Mbbl, or more than twice the primary production, had been recovered by dry in situ combustion. The air injection rate averaged about 10 Mcf/d over the first 9 years of operation. The average air–oil ratio was 11 kscf/bbl.

No serious operating problems were encountered. The loosely consoli-

Table 14.4 In Situ Combstion Test General Characteristics[a]

Number	Field	Location	Company	Formation	Start Date	Pilot (P), Fieldwide (F)	Dry (D), Wet (W)	Test Pattern Type	Area (acres)	Thickness (ft)	Depth (ft)	ϕ (%)	k (md)
1	Glen Hummel	TX	Sun	Semiconsolidated sand	1/68	F	D	Line	544	8.11	2432	36	1000
2	Moco	CA	Mobil	Unconsolidated sand	1/60	F	D	Irregular	150	129	2400	36	1575
3	Heidelberg	MS	Gulf	Consolidated sand	12/71	F	D	Irregular	400	40	11,200	16.4	39
4	Trix Liz	TX	Sun	Consolidated sand	9/68	F	D	Line	250	9+	3650	28	500
5	Gloriana	TX	Sun	Semiconsolidated sand	5/69	F	D	Line	534	4.4	1600	35	1000
6	W. Newport	CA	GEKS	Unconsolidated sand	3/56	P,F	D	—	23	120	1000	37	1070
7	B. Olinda	CA	Union	Sand	3/72	F	D	Irregular	31	150	3550	29	300
8	N. Tisdale	WY	Conoco	Sand	5/59	P,F	D	Line	12.3	50	929	24.5	1034
9	W. Casablanca	TX	Mobil	Consolidated sand	2/67	F	D	Irregular	250	12.3	1030	34	600
10	Fry	IL	Marathon	Sand	10/61	P,F	D	Irregular	160	50	875	19.8	354
11	Miga	Venezuela	Mene Grande	Consolidated sand	4/64	F	D	Irregular	—	20	4050	22.6	5000
12	N. Government	TX	Mobil	Sand	9/62	F	D	Line	728	18	2320	32	800
13	S. Belridge	CA	Mobil	Unconsolidated sand	9/63	P	D	Five-spot	2.75	30	700	36	8000
14	Bellevue	LA	Getty	Unconsolidated sand	9/63	P,F	D,W	Line	2.8	74	360	38	500

Number	Field	Initial Reservoir Conditions T (°F)	P (psi)	Oil μ at Reservoir T (cp)	Oil Density (°API)	At Start S_o (%)	S_w (%)	Fuel Consumed (bbl/acre-ft)	Air Requirement (Mscf/acre-ft)	Oxygen Utilization (%)	Average AOR (kscf/bbl)	Previous Recovery (%OOIP)	Combustion Recovery (%OOIP)
1	Glen Hummel	115	800	74	21.9	65	30	137	11.0	98.5	2.9	10.4	13.5
2	Moco	110	850	110	14.5	75	25	258	18	100	3.1	0.4	11.3
3	Heidelberg	221	1500	4	20	79	15	—	—	—	2.64	6.7	2.0
4	Trix Liz	138	200	26	24	56	35	136	10.9	98	8.14	14	7.7
5	Gloriana	112	300	174	20.8	52.8	41.5	153	11.1	86.9	5	11.4	10.0
6	W. Newport	105	85	700	15.2	69	—	268	—	—	13	<15	22
7	B. Olinda	135	—	20	22	50	—	169	11.1	100	—	27.8	7.6
8	N. Tisdale	73	290	175	21	61.4	35.4	230	15.9	72	14	5	1.36
9	W. Casablanca	100	30	<150	20.4	45	30	218	15.1	91	12.1	28	7.3
10	Fry	65	20	40	28.5	68	20	127	—	90	17.5	—	—
11	Miga	146	—	355	13	75	22	180	—	100	11	3.3	12
12	N. Government	120	330	10	22	36	41	212	—	95.5	15.5	39.5	<5
13	S. Belridge	87	220	2700	12.9	60	37	256	27	100	20	2.61	52.1
14	Bellevue	75	—	500	19.5	51	—	90.2	—	—	16	5	45

[a]From Ref. 16. © 1980 SPE-AIME.

dated sand was controlled through the use of gravel-packed slotted liners. Corrosion was not a problem. No water injection was needed for producing well cooling, although a light oil was used for downhole blending to increase the producing rates and facilitate surface handling of the viscous crude.

In 1970, one air injector was converted to water injection to utilize stored heat and prevent resaturation of the region with oil. Water channelling to an offset producer occurred after a few months, and water injection was stopped after injecting 556 kbbl.

Past performance and model studies indicated that in situ combustion could result in the production of 50% of the original oil in place, whereas the ultimate primary recovery would have been only 5%. Experience in this and similar reservoirs had shown that gas drive and waterflooding would be completely ineffective owing to the high oil viscosity. Additional data for this project are found in Table 14.4.

COMPARISON OF OBSERVED AND CALCULATED AIR–OIL RATIOS

Table 14.3 compares observed air–oil ratios, corrected for unconfinement, for several of these early projects and the air–oil ratio calculated using the Orkiszewski correlation and Eq. (13.6a). Details of the calculations are given in Tables 14A.1 and 14A.2 in the Appendix. For most projects, the agreement was excellent. The projects where agreement is poor can usually be explained as resulting from additional benefits of gas drive and accelerated gravity drainage (see the earlier discussion of the Fry and Midway-Sunset projects).

The reason for the poor agreement in the case of the Delaware-Childers project is not apparent. The operator estimated that the producing gas–oil ratio would have been 57 kscf/bbl had the project been confined. This would give an air–oil ratio of 63 kscf/bbl, which is still substantially higher than the calculated 40 kscf/bbl. It may be that the 584 bbl/acre-ft initial oil saturation is overstated for this project.

ECONOMIC CONSIDERATIONS

The major economic consideration in an in situ combustion project is the air compression requirement.

Compressor prices depend on whether the units are high-speed packaged units or are integral units with accessories. Additional items such as automation, gas coolers, and noise abatement will add approximately an additional 10% of the installed cost.

The major operating cost item is fuel gas. Gas requirements will vary

Table 14.5 Approximate Fuel Requirements of Compressors

Compressor Type	Gas Requirement (scf/hp-hr)
Integral reciprocating	7.0–8.5
High-speed direct-coupled reciprocating	9.5–10.0
Gas-turbine-driven centrifugal	11.0–14.0

depending on the type of installation. Table 14.5 gives approximate fuel requirements for various compressor types.

The theoretical adiabatic horsepower to compress 1 Mscf/d of air for a per stage compression ratio R is

$$\frac{\text{Adiabatic BHP}}{\text{Mscf/d}} = 152.5(R^{0.285} - 1) \tag{14.1}$$

The actual horsepower for each stage of compression is estimated by[15]

$$\text{BHP} = \left(\frac{\text{Adiabatic BHP}}{\text{Mscf/d}}\right)\left(\frac{P_b}{14.4}\right)\left(\frac{T_s}{520}\right) Q \left(\frac{1}{\text{OE}}\right)\left(\frac{Z_s + Z_d}{2}\right) \tag{14.2}$$

where P_b = base reference pressure for standard conditions (usually 14.65 psia) (psia)
T_s = suction temperature (°R)
Q = volume to be compressed per day (Mscf/d)
OE = overall mechanical efficiency (from Fig. 14.6)
Z_s, Z_d = air compressibilities at suction and discharge conditions (close to 1.0 for air)

The compression ratio R per stage is obtained by considering the maximum discharge temperature from each stage:

$$R = (T_d/T_s)^{3.5} \tag{14.3}$$

where T_d = temperature at discharge (°R)
T_s = temperature at suction (°R)

It is inadvisable to exceed $T_d = 300–350°F$ because of the danger of overheating lubricants. Special high-boiling or silicone lubricants *must* be used with air compression. Disastrous explosions have resulted where these precautions were not taken. Interstage cooling and piping will cause about a 5–15 psi reduction in pressure between stages.

Example Problem. Estimate the compression horsepower to a total discharge pressure (P_{td}) of 1000 psia at 60°F for 2 Mscf/d. Slow-speed reciprocating compression is assumed.

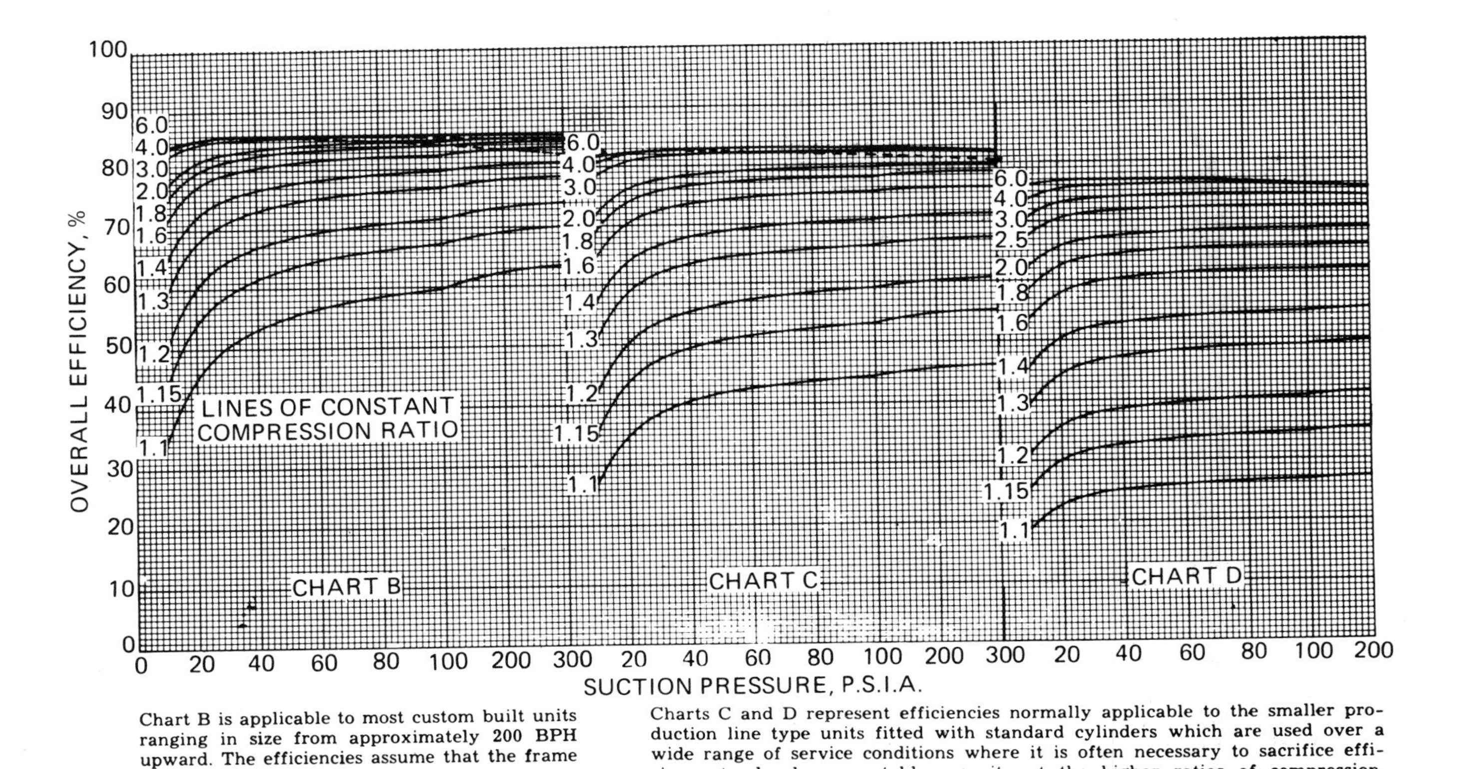

Chart B is applicable to most custom built units ranging in size from approximately 200 BPH upward. The efficiencies assume that the frame and running parts are loaded to or near the point of maximum mechanical efficiency.

Charts C and D represent efficiencies normally applicable to the smaller production line type units fitted with standard cylinders which are used over a wide range of service conditions where it is often necessary to sacrifice efficiency to develop acceptable capacity at the higher ratios of compression. Field gas gathering units and close coupled high speed (900 to 1000 RPM) units are usually in this category. Interpolation may be made between Charts C and D for intermediate efficiencies.

Figure 14.6. Overall efficiency versus suction pressure (all charts based on 93% mechanical efficiency).[15] © 1966 Natural Gas Processors Association.

1. Estimate the number of stages of compression; design for $T_d = 300°F$:

$$R = \left(\frac{300 + 460}{60 + 460}\right)^{3.5} = 3.77$$

Note that

$$P_{td}/14.7 = R^n$$

Then

$$n = \frac{\log(P_{td}/14.7)}{\log R} \tag{14.4}$$

where n = number of stages
$= \log(1000/14.7)/\log(3.77)$
$= 1.832/0.58 = 3.16$

Hence, three stages of compression will probably be adequate.

2. Estimate the brake horsepower:

$$R = \left(\frac{P_{td}}{14.7}\right)^{1/n} = \left(\frac{1010}{14.7}\right)^{1/3} = 4.09$$

(*Note*: 10 psia is added to account for the pressure drop in lines and interstage coolers.) For each stage,

$$\frac{\text{Adiabatic BHP}}{\text{Mscf/d}} = 152.5(4.09^{0.285} - 1) = 75.3 \text{ hp}$$

$$\text{Total Adiabatic BHP/Mscfd} = 3(75.3) = 226 \text{ hp}$$

From Fig. 14.6 (chart C), the overall efficiency is obtained: OE = 0.82. Assume that

$$T_s = 60°\text{F} + 460 = 520°\text{R}$$
$$P_s = 14.65 \text{ psia}$$
$$Z_s = Z_d \cong 1.0$$

By Eq. (14.2),

$$\text{BHP} = (226)(14.65/14.4)(520/520)(2)(1/0.82) = 561 \text{ hp}$$

Added to the compression cost would be cost of new wells, additional separation equipment to handle high gas–oil-ratio production, and added lease and well costs (particularly if hot well production is encountered). Ignition costs must also be included, but usually these are a small part of the overall cost.

SUMMARY OF MORE RECENT FULL-FIELD PROJECTS

A number of more recent projects (and continuation of some of the earlier projects already discussed) have been reported in the literature.[16–18] Twelve of these projects, which are summarized in Table 14.4, were full-field dry-combustion projects and were included by Brigham et al.[16] in an empirical correlation relating porosity, volume of air injected, oxygen utilization, amount of fuel burned, starting oil saturation, formation thickness, and oil viscosity to oil recovery by burning.

The first 12 of the projects listed in Table 14.4 were the full-field dry-combustion projects included in Brigham's correlation. In this table, the air–oil ratios reported are not corrected for possible losses from the oil sand or from the project area and thus differ from the estimated confined project air–oil ratios based on gas–oil ratio behavior that have been reported for the other projects discussed in this chapter.

The correlation proposed by Brigham, which should only be used for oils above 10 cp viscosity, is shown in Figure 14.7. This correlates the data for

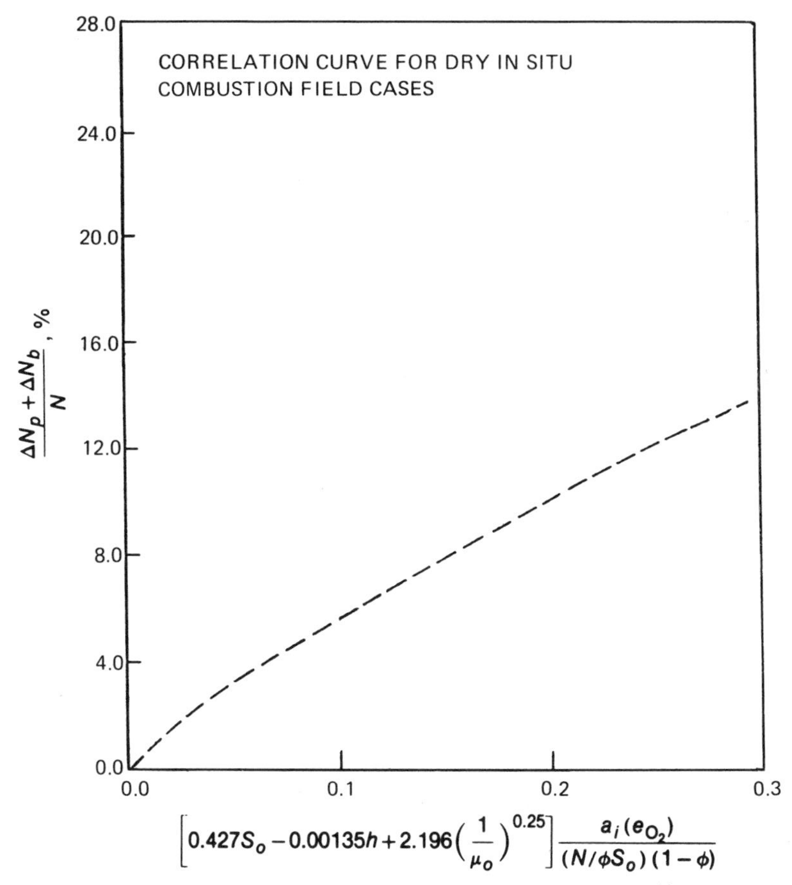

Figure 14.7. Correlation of Brigham et al. (second correlation).[16] © 1980 SPE-AIME.

the 12 projects to a standard deviation of 9% with a maximum error of 14%. His first correlation (not shown) is for oil below 10 cp viscosity.

These correlations plot fuel burned plus oil produced as a percentage of oil in place at project start, $[(\Delta N_p + \Delta N_b)/N] \times 100\%$, as a function of the group:

$$[0.427S_o - 0.00135h + 2.196(1/\mu_o)^{0.25}]\,\frac{a_i(e_{O_2})}{(N/\phi S_o)(1-\phi)}$$

In the correlation, the following terms are defined:

N = oil in place at start of combustion test (bbl)
ΔN_b = fuel burned (bbl)
ΔN_p = cumulative incremental oil production (bbl)
S_o = oil saturation (fraction)
μ_o = oil viscosity (cp)
a_i = cumulative air injection (scf)
e_{O_2} = oxygen utilization (fraction)
h = net formation thickness (ft)

More detailed data for specific dry-combustion projects can be found in the literature.[1–13,14,19–22]

APPENDIX

Table 14A.1 Estimate of Fuel Concentration from Orkiszewski Correlation

			lb Fuel/100lb Sand				lb Fuel/ft^3	bbl/acre-ft
	T_r(°F)	°API	F_1	F_2	$\bar{F}_c$	ϕ	F_c[a]	$\hat{S}_{of}$[b]
South Belridge	87	12.9	0.89	0.71	1.6	0.37	1.65	205
Pontotoc	66	18.4	0.71	0.89	1.6	0.272	1.90	237
Fry	65	29	0.50	0.90	1.4	0.197	1.84	229
Delaware-Childers	76	33	0.41	0.78	1.19	0.206	1.54	192
North Government Wells	120	22	0.62	0.59	1.21	0.32	1.35	168
Midway-Sunset	125	14	0.84	0.57	1.41	0.36	1.48	184
Melones	145	9.5	1.06	0.56	1.62	0.35	1.72	214
Delhi	135	40	0.26	0.56	0.82	0.312	0.92	115
West Newport (B zone)	105	15.2	0.80	0.64	1.44	0.37	2.14	266

[a] $F_c = 1.635(1-\phi)\bar{F}_c$.
[b] $\hat{S}_{of} = 124.5F_c$; circumflex (^) denotes saturation expressed as bbl/acre-ft.

Table 14.A.2 Estimate of Confined Air–Oil Ratio

	bbl/acre ft				
	$\hat{S}_{of}$	$\hat{S}_{oi}$	$\hat{S}_{oi} - \hat{S}_{of}$	ϵ	R_{Ao}[a](scf/bbl)
South Belridge	205	1720	1515	0.95	9300
Pontotoc	237	1350	1113	0.43	32,000
Fry	229	1040	811	0.87	21,000
Delaware-Childers	192	584	392	0.80	40,000
North Government Wells	168	800	632	0.96	18,000
Midway-Sunset	184	1980	1796	0.95	7000
Melones	214	2552	2338	0.95	6000
Delhi	115	740	625	0.92	13,000
West Newport (B zone)	266	1925	1659	0.95	11,000

[a] $R_{Ao} = 65{,}036\hat{S}_{of}/\epsilon(\hat{S}_{oi} - \hat{S}_{of})$.

REFERENCES

1. Emery, L. W. "Results from a Multi-Well Thermal Recovery Test in Southwestern Kansas," *J. Pet. Tech.* (June 1962), p. 671.
2. Gates, C. F., and Ramey, H. J. "Field Results of South Belridge Thermal Recovery Experiment," *Trans. AIME*, **213,** 236 (1958).
3. Moss. J. T., White, P. D., McNeil, J. S., Jr. "In Situ Combustion Process—Results of a Five-Well Field Experiment in Southern Oklahoma," *Trans. AIME*, **216,** 55 (1959).
4. Parrish, D. R., Rausch, R. W., Beaver, K. W., and Wood, H. W. "Underground Combustion in the Shannon Pool, Wyoming," *J. Pet. Tech.* (February 1962), p. 197.
5. Dietz, D. N. "Wet Underground Combustion State of the Art," *J. Pet. Tech.* (May 1970), p. 605.
6. Clark, G. S., Jones, R. G., Kinney, W. L., Schilson, R. E., Surkalo, H., and Wilson, R. S. "The Fry In Situ Combustion Test Performance," *J. Pet. Tech.* (March 1965), p. 348.
7. Earlougher, R. C., Jr., Galloway, J. R., and Parsons, R. W. "Performance of the Fry In Situ Combustion Project," *J. Pet. Tech.* (May 1970), p. 551.
8. Barnes, A. L. "Results from a Thermal Recovery Test in a Watered-Out Reservoir," *J. Pet. Tech.* (November 1965), p. 1343.
9. Casey, T. J. "A Field Test of the In Situ Combustion Process in a Near-Depleted Water Drive Reservoir," *J. Pet. Tech.* (February 1971), pp. 153–160.
10. Gates, C. F., and Sklar, I. "Combustion—A Primary Recovery Process, Moco Zone Reservoir—Midway-Sunset Field, Kern County, California," *J. Pet. Tech.* (August 1971), pp. 981-986.

11. Bowman, C. H. "A Two-Spot Combustion Recovery Project," *J. Pet. Tech.* (September 1965), p. 994.
12. Hardy, W. C., Fletcher, P. B., Shepard, J. C., Dittman, E. W., and Zadow, D. W. "In Situ Combustion Performance in a Thin High Gravity Oil Reservoir, May-Libby Reservoir, Delhi Field," *J. Pet. Tech.* (February 1972), pp. 199–208.
13. Koch, R. L. "Practical Use of Combustion Drive at West Newport Field," *Pet. Eng.* (January 1965), p. 72.
14. Terwilliger, P. L., Clay, R. R., Wilson, L. A., and Gonzales-Gerth, E. "Fireflood of the P_{2-3} Sand Reservoir in the Miga Field of Eastern Venezuela," *J. Pet. Tech.* (January 1975), pp. 9–14.
15. *Engineering Data Book*, 8th ed., Natural Gas Processors Suppliers Association, Tulsa, OK, 1966, pp. 37-48.
16. Brigham, W. E., Satman, A., and Soliman, M. Y. "Recovery Correlations for In Situ Combustion Field Projects and Application to Combustion Pilots," *J. Pet. Tech.* (December 1980), p. 2132.
17. Farouq Ali, S. M. "A Current Appraisal of In Situ Combustion Field Tests," *J. Pet. Tech.* (April 1972), p. 477.
18. Chu, C. "A Study of Fireflood Field Projects," *J. Pet. Tech.* (February 1977), pp. 111–120.
19. Gates, C. F., and Ramey, H. J. "A Method for Engineering In Situ Combustion Oil Recovery Projects," *J. Pet. Tech.* (February 1980), pp. 285–294.
20. Widmyer, R. H. "The Charco Redondo Thermal Recovery Pilot," *J. Pet. Tech.* (December 1977), pp. 1522–1532.
21. Gates, C. F., Jung, K. D., and Surface, R. A. "In Situ Combustion in the Tulare Formation, South Belridge Field, Kern County, California," *J. Pet. Tech.* (May 1978), pp. 798–806.
22. Gadelle, C. P., Burger, J. G., Bardon, C. P., Machedon, V., Carcoana, A., and Petcovici, V. "Heavy-Oil Recovery by In Situ Combustion—Two Field Cases in Rumania," *J. Pet. Tech.* (November 1981), pp. 2057–2066.

Chapter 15

INTRODUCTION TO WET COMBUSTION PRINCIPLES

ABSTRACT

Injecting water simultaneously or alternately with air has the effect of reducing air compression requirements associated with dry combustion by advancing the heat normally left behind the combustion front so that it can aid oil displacement by steam bank formation. The simplified theory behind wet combustion is discussed, including that for the extreme case of partially quenched combustion where not all of the fuel deposited on the sand grains is burned off.

Results of laboratory tests are described that are helpful in estimating combustion front velocities, air–water ratios and unrecovered fuel concentration for the partially quenched combustion case. Results of field applications of wet combustion are discussed for pilots at Schoonebeek and Tia Juana, where some benefit was observed but where gravity segregation of injected air and water reduced the effectiveness of the approach. Gravity segregation was less detrimental in the case of a deep, thin-sand project conducted in the Sloss field, but poor sweep efficiency reduced the effectiveness of this process in a similar fashion to that observed for many dry-combustion projects.

Texaco is conducting a nearly 3000-bbl/d wet combustion project in the Bellevue, Louisiana, field.[1] This is the largest wet combustion project undertaken in the United States. In addition to air–water gravity segregation and low sweep efficiency, corrosion and low injectivity with attendant formation plugging and emulsion formation have been observed as operating problems in wet combustion projects.

INTRODUCTION

Theoretically, the simultaneous injection of air and water should improve utilization of energy left behind the combustion front. Moving this heat forward should permit a large steam zone to be created ahead of the combustion front, reducing fuel requirements and causing more oil to be displaced per unit volume of air injected. The desirability of reducing air compression requirements is evident from the previous chapter, which discussed costs of compression and the high air–oil ratios associated with most dry-combustion projects.

SIMPLIFIED ONE-DIMENSIONAL DESCRIPTION OF WET COMBUSTION

Wet combustion is a complex process, and a complete analysis of its behavior is not yet available. We can, however, discuss the various types of wet combustion for the idealized case of one-dimensional flow in a perfectly insulated (adiabatic) tube. The terminology of Dietz[2] is used.

Figure 15.1 shows idealized plots of temperature and oil saturation (shaded) as a function of distance from the inlet end of the tube for the case of *dry combustion*. The cooling effects of the injected air and the small steam zone have been neglected. Most of the displacement occurs close to the combustion front.

Figure 15.2 shows the situation for *normal wet combustion*. Injected water has moved some heat forward to form a steam condensation front that moves well ahead of the combustion front. Most of the displacement occurs at the condensation front.

Figure 15.3 indicates how normal wet combustion can be *optimized* under ideal circumstances to utilize essentially all the energy left behind the combustion front. In this case, the evaporation front would be essentially coincident with the combustion front, and the condensation front would be moved to its maximum extent ahead of the combustion front.

Figure 15.4 indicates the temperature–saturation distribution for *partly* quenched combustion. This condition occurs when the water–air ratio is

Figure 15.1. Dry combustion.[2] © 1968 SPE-AIME.

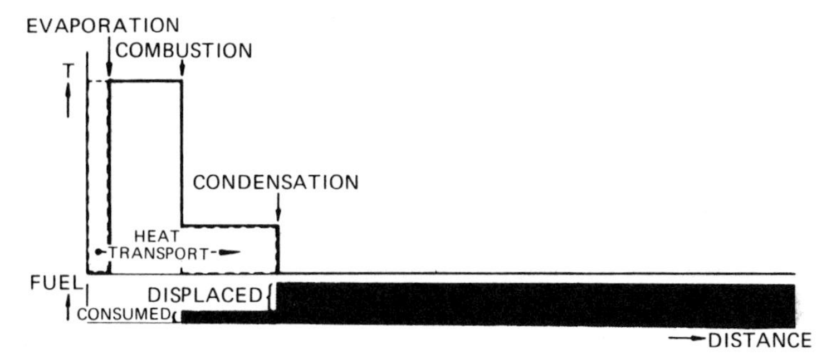

Figure 15.2. Normal wet combustion.[2] © 1968 SPE-AIME.

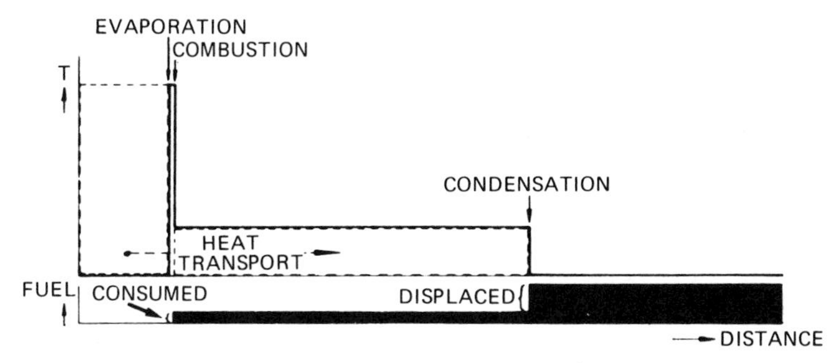

Figure 15.3. Optimal normal wet combustion.[2] © 1968 SPE-AIME.

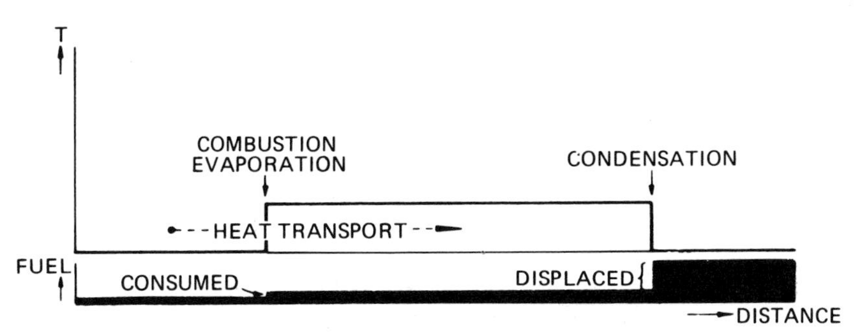

Figure 15.4. Partially quenched combustion.[2] © 1968 SPE-AIME.

increased to a point (usually between 500 and 1800 bbl/10^6 scf) at which the evaporation front merges with the combustion front; this reduces temperature enough that combustion is partially quenched. Once this condition is reached, no more heat need be generated than is required to make up for heat losses from the heated zone.

Note that a water saturation pressure of 2000 psia implies a maximum temperature of 636°F, which is considerably less than that considered necessary for vigorous combustion. Oxidation rates become slower as more water is added, and therefore, energy release rates are reduced. The combustion front becomes ill-defined as low-temperature reactions become more important. Energy release occurs over an increasingly wide expanse of the reservoir.

Low air requirements associated with partially quenched combustion are very desirable. However, no successful field tests have been run under these conditions. Uneven water and air distribution leads to pockets of essentially dry combustion, regions of normal wet combustion, and regions of quenched combustion in the reservoir. This reduces oil recovery. In addition, in a thick sand, gravity segregation may be so complete that only a waterflood occurs at the base of the reservoir. A lower water–air ratio than for partially quenched conditions is recommended for field operations to avoid completely quenching the front in some places.

RESULTS OF LABORATORY TESTS

The feasibility of partially quenched combustion has been confirmed in laboratory tube experiments.[2–5] A heat wave at steam temperature is observed (see Fig. 15.5). Preferential combustion of hydrogen occurs (hydrogen stripping), and a carbonaceous residue is left in the formation (see Fig. 15.4).

Results of tube experiments with partially quenched wet combustion are reported by Parrish and Craig[3] and Dietz.[2] Figure 15.6 gives the ratio of the combustion front to the air injection velocity (v_b/u_a) in the tube as a function of the injected water–air ratio (u_w/u_a) for Schoonebeek crude. The dry-combustion value for v_b/u_a occurs where the u_w/u_a ratio is zero. Run 3 occurs approximately at the point of optimal normal wet combustion where partially quenched conditions are incipient. Points to the right of run 3 are in the partially quenched range. This occurs at about 0.003 m^3/sm^3 or an air–water ratio of 1870 scf/bbl under tube conditions. The effect of heat losses is to lower v_b/u_a. Hence, partially quenched combustion will start at a somewhat higher air–water ratio in the field than is indicated in Figure 15.6.

The data of Parrish and Craig[3] (for various crude-oil systems) fall generally in line with the data of Dietz and Weijdema.[2] Figure 15.7 shows

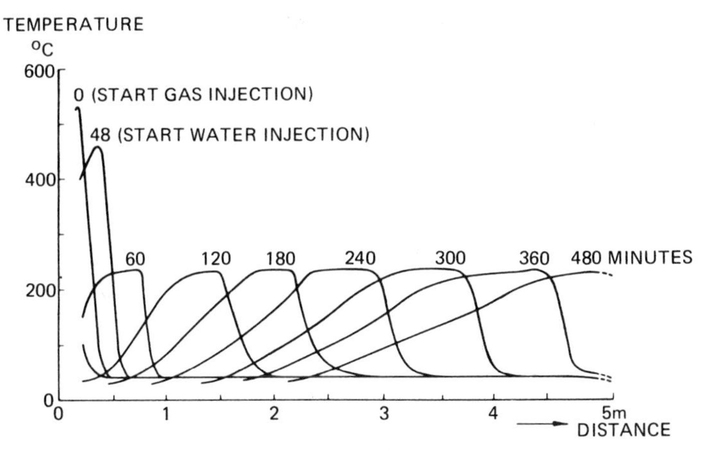

Figure 15.5. Experimental temperature profiles in partially quenched combustion.[2] © 1968 SPE-AIME.

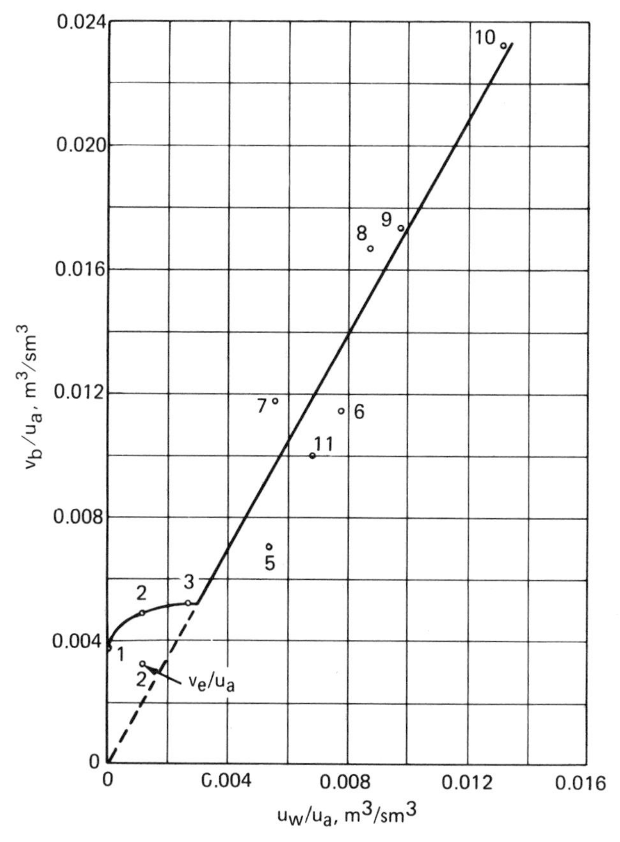

Figure 15.6. Combustion front velocity as function of specific water throughput.[2] © 1968 SPE-AIME.

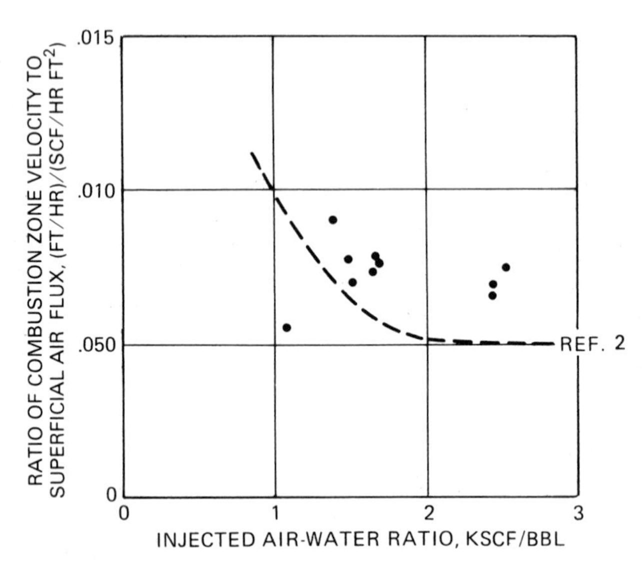

Figure 15.7. Relation between velocity of trailing edge of combustion zone and injected air–water ratio.[3] © 1969 SPE-AIME.

the ratio of combustion front velocity to the superficial air flux as a function of the air–water ratio. The dashed curve corresponds to the data of Figure 15.6.

Later findings by Garon and Wygal[4] agree well with those of Dietz. At 2000 psi, Garon obtained more nearly constant burning velocities at all water–air ratios. He observed decreasing fuel availability and oxygen consumption with increasing water–air ratio. Oxygen consumption decreased less rapidly than fuel burned, a phenomenon Garon attributed to low-temperature oxidation reactions. Figure 15.8 compares dry- and wet-combustion recovery curves for a core initially at connate water conditions. At a water–air ratio of 9 ft^3/kscf (1600 bbl/10^6 scf), combustion is partially quenched. The difference in the time to reach ultimate recovery is considerable.

The results of Garon and Wygal[4] and Burger and Sahuquet[5] are summarized in Table 15.1, which expresses the ratio

$$R = \frac{\text{fuel burned wet}}{\text{fuel burned dry}} \cong \frac{\text{air consumed wet}}{\text{air consumed dry}}$$

for three experimental water–air ratios. This ratio will determine the difference in burning front velocity for dry combustion relative to wet combustion at a given water–air ratio. It *does not* indicate that ultimate recovery for wet combustion will be greater.

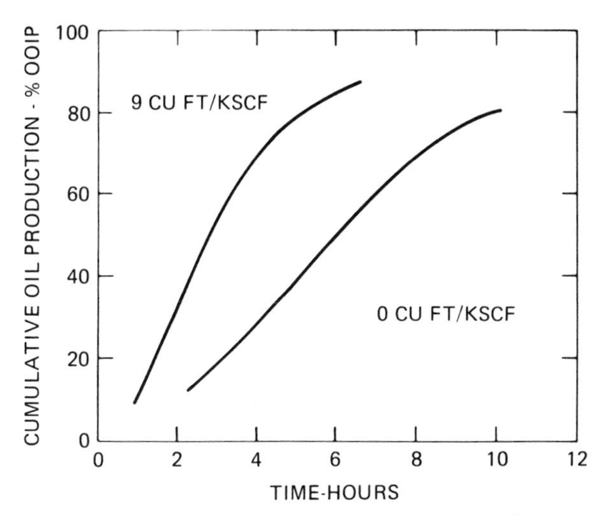

Figure 15.8. Effect of water–air ratio on time of recovery.[4] © 1974 SPE-AIME.

Table 15.1 Effect of Water–Air Ratio on Ratio of Wet–Dry Fuel Requirement

Water-Air Ratio	Air-Water Ratio	$R = \frac{\text{wet consumption of fuel or air}}{\text{dry consumption of fuel or air}}$	
(ft^3/kscf)	(scf/STB)	Garon[a]	Burger[b]
3.74	1500	0.67	~0.6
5.61	1000	0.58	~0.3
11.2	500	0.47	

[a]Reference 4.
[b]Reference 5.

A residual unburned fuel is left in the case of wet combustion that is a function of the water–air ratio. The Table 15.1 data are for representative reservoir and fluid properties. The question of accurately quantifying fuel and air reduction for a particular set of conditions cannot be answered until more data are available.

RESULTS OF MATHEMATICAL ANALYSIS OF PARTIALLY QUENCHED (SUPERWET) COMBUSTION

Selected results of a mathematical analysis of the partially quenched (superwet case) have been presented.[6] The analysis assumes

1. one-dimensional flow in a linear system,
2. a reservoir pressure of 710 psia,
3. a vertical combustion front, and
4. heat losses by conduction to cap and base rock.

These assumptions are more realistic for the case of a thin sand than for a thick one, as we have previously noted. Results for the combustion front velocity v_b as a function of air flux u_a and water flux u_w are given in Figure 15.9 for a sand 2 m thick (6.56 ft). As the heated zone grows in size, it will approach an equilibrium length L at which v_b, the combustion front velocity, equals v_c, the velocity of the condensation front.

Figure 15.10 gives equilibrium length L as a function of u_a and u_w. Dietz presented no mathematical details or basic data and gives no explanation of the meaning of a finite burning front velocity at a zero air input. Presum-

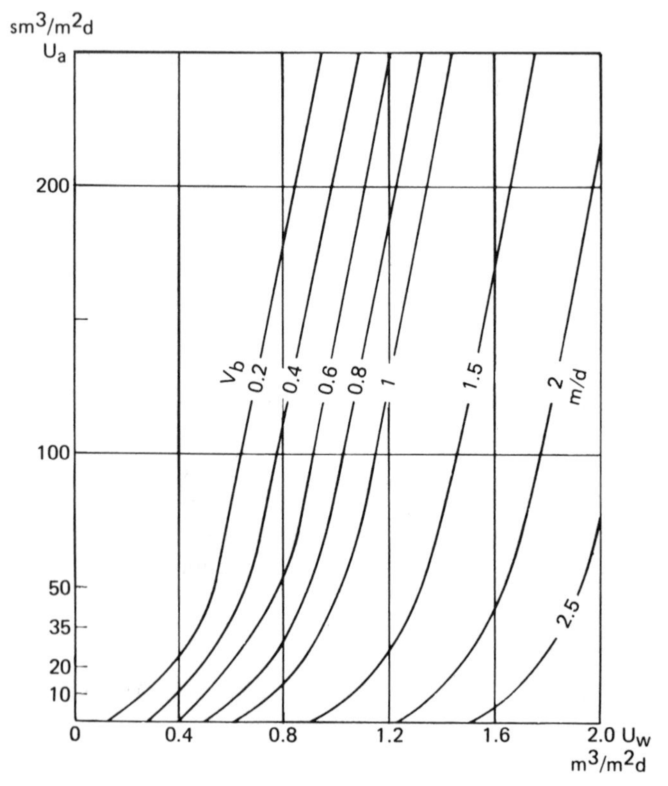

Figure 15.9. Combustion front velocities according to Baskir et al. Formation thickness 2 m (6.56 ft).[6] © 1970 SPE-AIME.

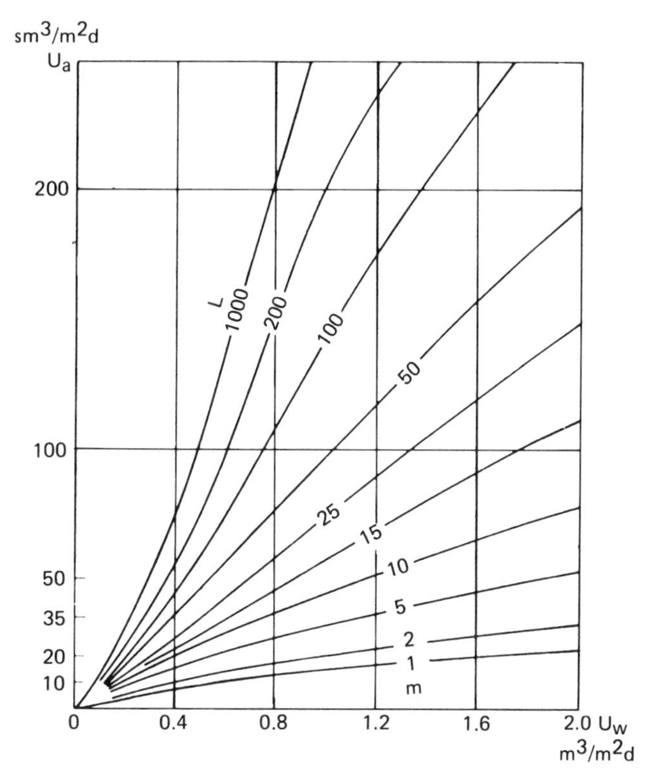

Figure 15.10. Lengths of heat wave according to Baskir et al. Formation thickness 2 m (6.56 ft).[6] © 1970 SPE-AIME.

ably, it is meant that although combustion has ceased, the heat front (velocity v_c equal to v_b) moves at increased velocity as water injection rates increase.

RESIDUAL HYDROCARBON NOT DISPLACED BY PARTIALLY QUENCHED COMBUSTION

Limited information is available concerning the unrecovered oil saturation (burned plus that left as residuum). Figure 15.11 gives published data for a number of crudes.[3] The unrecovered oil is seen to be a function of the API crude gravity and air–water ratio; it is higher at higher air–water ratios and for lower gravity crudes. There is a trend of lower unrecovered oil saturation at lower air–water ratios. Straight-line extrapolation of the lines drawn would be incorrect. Upper and lower limits would be determined by

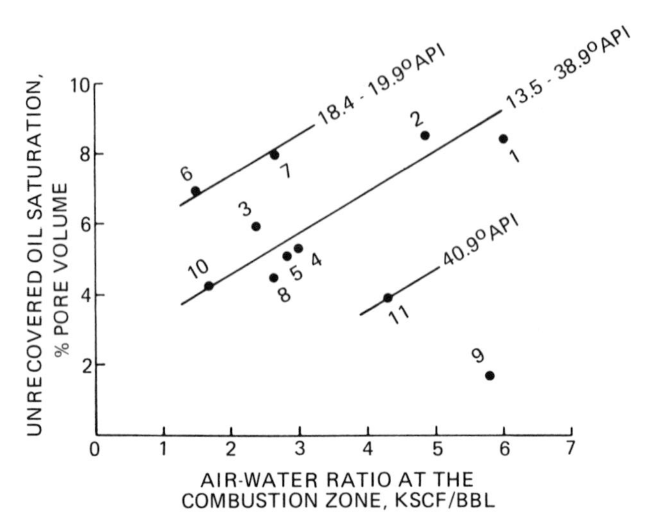

Figure 15.11. Relation between unrecovered oil saturation and air–water ratio at combustion zone (numbers shown at points are run numbers).[3] © 1969 SPE-AIME.

the waterflood residual oil saturation (for the case of a zero air–water ratio) and by the fuel saturations for dry combustion (for the case of infinite air–water ratio). Garon[4] observed no residual fuel behind the combustion front at high pressures.

EARLY FIELD TEST RESULTS

Field results for early tests of wet combustion demonstrated the feasibility of the process but also pointed out some problems. Chapter 13 discussed the North Government Wells and Delhi projects, which had some aspects of wet combustion, but essentially they were dry-combustion projects since the air–water ratios were not low enough for these projects to be in the partially quenched range.

Two other projects, one at Schoonebeek and one at East Tia Juana, Venezuela,[6] were conducted. However, these were in thick sands, and it would appear that a large amount of injected water underran the oil column, so that the full effects of wet combustion were not realized. The wet-combustion project at Schoonebeek was an extension of the dry-combustion project discussed in Chapter 13. Data for the East Tia Juana Project are given in Table 15.2.

A comparison of the total air–oil ratios for the Schoonebeek and the East Tia Juana projects together with the corresponding injected air–water ratios used is given in Table 15.3.

Table 15.2 Reservoir Data, East Tia Juana Wet Combustion Project

Pattern	Inverted 7-spot
Area, acres	11.4
Net sand (upper sand), ft	39
Net sand (lower sand), ft	89
Dip	4°
Initial oil in place, Mbbl	3.248
Permeability, darcies	5
Distance between injector and producer, ft	436
Initial oil saturation, %	73
Initial oil saturation, bbl/acre-ft	2226

Table 15.3 Field Test Air–Water and Air–Oil Ratios for Two Wet-Combustion Projects.

	Air–Water Ratio (scf/bbl)	Air–Oil Ratio[a] (scf/bbl)
Schoonebeek, Holland		
Dry phase	—	16,830
Wet phase	1603	5610
East Tia Juana, Venezuela		
Wet phase	801	937

[a]Based on total oil, not incremental over primary.

The exceptionally low air–oil ratio obtained at East Tia Juana undoubtedly included some primary production. Nevertheless, it is clear that in these two reservoirs, which had moderately viscous oils and high initial oil saturations, increasing the water input relative to air had a decidedly beneficial effect.

A potential use of wet combustion lies in the area of tertiary light-oil recovery in thin sands where no other thermal drive technique will have sufficiently high thermal effeciency to permit economic operation.

One of the largest such applications of wet combustion occurred in the Sloss, Nebraska Field.[7] Reservoir data for this project are given in Table 15.4.

This project was initiated as a 40-acre 5-spot pilot and later expanded to the 960-acre multi-5-spot project shown in Figure 15.12. Each 5-spot had an 80-acre area. The thin (14.5-ft) sand had been previously waterflooded. The initial oil saturation was approximately 30%, which is exceptionally low for a combustion project. Had this been a dry-combustion project, the air–oil ratio would have been about 32 kscf/bbl, as predicted by the Orkiszewski correlation.

The actual air–oil ratio was 19.6 kscf/bbl, with 701 kSTB being produced

Table 15.4 Average Reservoir Data, Sloss Field Wet Combustion Project[a]

Depth, ft	6200
Porosity, %	19.3
Air permeability, md	191
Oil saturation at project start, %	30 ± 10
Bottom-hole pressure, psig	
1968	2274
1970	3158
Bottom-hole temperature, °F	200
Oil gravity[b], °API	38.8
Oil viscosity,[b] cp	0.8
Reservoir volume factor, RB/STB	1.05
Full-scale project:	
Area, acres	960
Net pay thickness, ft	14.3
Oil-in-place	
bbl/acre-ft	427
million barrels	5.9

[a]Reference 7.
[b]Prior to project start.

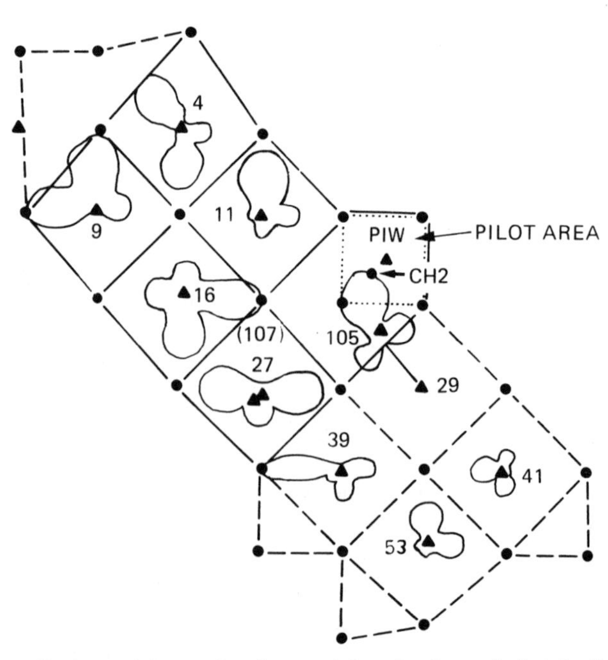

Figure 15.12. Estimated burned volume at termination of air injection Sloss wet combustion project, Sloss field, Nebraska.[8] © 1974 SPE-AIME.

since project start. It was estimated that an additional 210 kSTB could have been recovered from the vented flue gas. The overall injected air–water ratio was 1271 scf/bbl. This was higher than the optimum ratio of 750 scf/bbl indicated by design calculations. However, water injectivity was too low to permit achieving this ratio at the design air injection rate.

Areal sweep was poor in the Sloss project. After heat breakthrough occurred at a producing well, it had to be shut in because erosion and corrosion became too severe. This resulted in the other 5-spot quadrants drained by that well being incompletely swept. Figure 15.12 shows the estimated burned area for each 5-spot at termination of air injection. This problem may be an inherent defect of pattern burning. The use of a line drive with injection on the crest appears to have avoided this problem in several dry-combustion projects.[4]

Operating data for the Sloss project is given in Figure 15.13. Additional follow-up waterflood data are presented elsewhere.[8] Postproject coring was conducted and showed that burning took place in the upper, more permeable part of the pay. The average vertical sweep of the greater-than-700°F

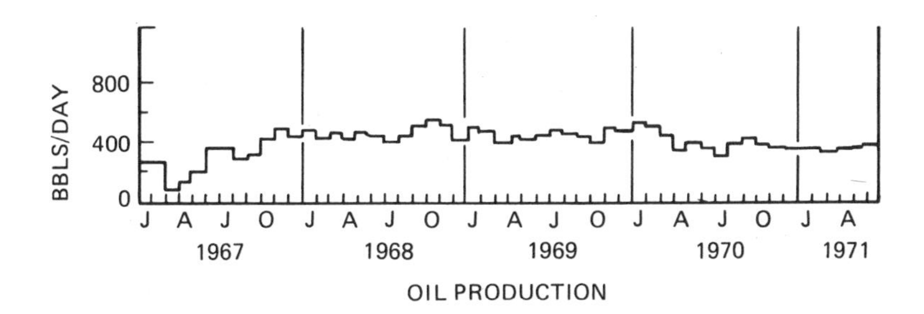

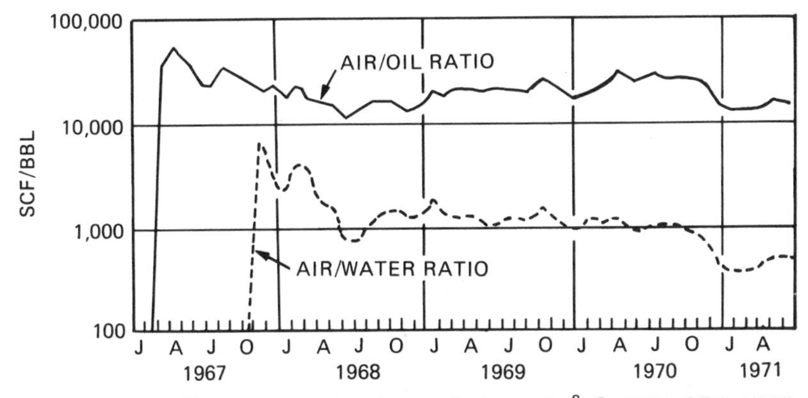

Figure 15.13. Sloss wet combustion project results.[8] © 1974 SPE-AIME.

zone was only 12%. Vertical sweep of the greater-than-350°F zone was 28%. Areal sweep was also poor but could have approached 50%, giving a volumetric sweep of only 14%. Significant oil production occurred after air injection was stopped.

Texaco's (formerly Getty's) Bellevue project[1] is a more recent application of wet combustion in a thick, shallow, high-permeability sand. Data for this project are given in Table 15.5, and the field oil production rate is plotted in Figure 15.14. At 2700 bbl/d, oil production from this project amounts to over 25% of the roughly 10,000 bbl/d production from all U.S. in situ combustion projects combined. The shallow depth reduces air injection pressures and horsepower requirements for compression. Reservoir pressure initially was only 40 psi, and the compressors discharge at about 100 psi. Initially, operations were piloted in the thicker part of the reservoir (70–90 ft), but later, expansions were made into the thinner sections (20–25 ft). Piloting was conducted using inverted 5-spots with a very close, 0.695-acre/well density. Later expansions have used a 9-spot pattern (three producers per injector) on a 1.25–2-acre/well density.

A field comparison of wet and dry combustion conducted by Cities Service in the Bellevue field was published in 1980.[9] Comparison of the two companion pilots was difficult, but wet combustion results were sufficiently encouraging that this method was selected for project expansion.

Texaco has also experimented with wet versus dry combustion. For many years, their design practice was to inject enough air to burn 50% of the 185-acre-ft in a pattern and then switch to water injection. Later, field experiments in multiple patterns compared their past practice of 2 years of dry combustion followed by water to 2 years of dry combustion followed by 2 years of wet combustion followed by water. The wet combustion patterns declined more slowly than those in which only dry combustion was used. Thus, dry followed by wet combustion followed by water became the

Table 15.5 Bellevue Project Data[a]

Reservoir	Crude
Depth, 350 ft	API gravity, 19.5
Net sand, 90 or less feet	μ_o, 675 cp at 75°F
Net/gross, 0.75	
Permeability, 50–1000 md	
Porosity, 38%	
Oil saturation, 1530 bbl/acre-ft ($\phi S_o = 0.20$)	
Temperature, 75°F	
Pressure, 40 psig	
Dip, negligible	

[a]Reference 1.

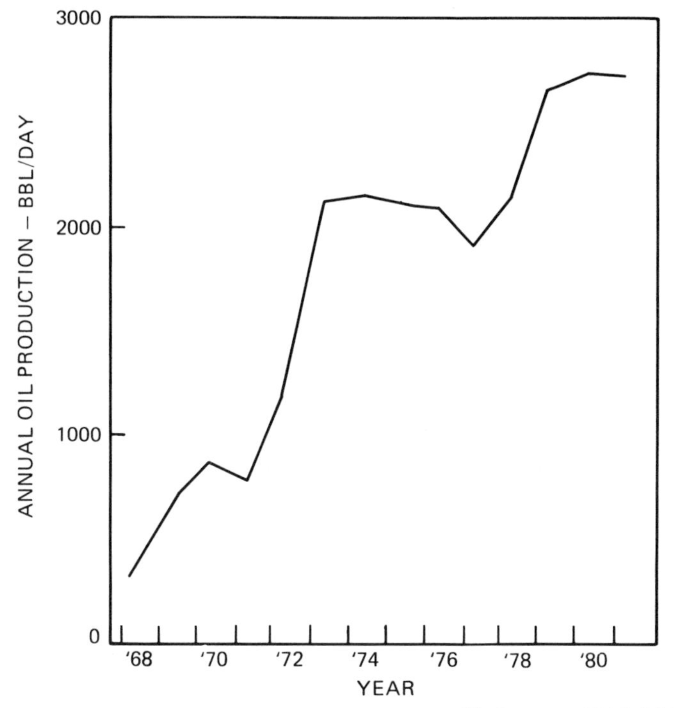

Figure 15.14. Texaco's Bellevue production.[10] © 1982 SPE-AIME.

preferred operating plan for thinner parts of the reservoir. In thick-sand areas with higher air injection rates, wet combustion is initiated later than in thin-sand areas and is continued past a calculated 50% volumetric burn. Texaco's experimental results on wet combustion somewhat parallel those obtained by Cities Service, who showed that water improved heat transfer and performance in the lower part of the reservoir.

Long and Nuan[1] describe Texaco's experience with severe sanding and downhole corrosion/erosion of producing well tubulars. It was found that they cannot protect downhole equipment against hot-gas blasting and enter a well only when it will not produce oil satisfactorily. Multiple workovers on producing wells are usually required. During the combustion phase, workovers to shut off gas are necessary in the top of the section. During follow-up water injection, water shutoffs are performed at the bottom of the section. The 300–400-ft depth is the key to economically successful workover operations.

A recovery of about 60% of the original oil in place is expected with consumption of another 6–15% of the oil in place as fuel. This would correspond to a field fuel consumption of about 1.5–1.8 lb/ft^3 for 20°API crude. Thus, it appears that fuel consumption is greater than suggested by literature correlations.

Through perseverance over many years, a highly successful and profitable operation has been developed at Bellevue. However, considering the shallow depth, it is possible that steamflooding would have even superior economics, at least in the thicker sand.

OPERATING PROBLEMS

In addition to the problems of poor sweep efficiency also noted for forward dry combustion, four additional problems occur for wet combustion.[10]

1. *Corrosion*. Truly simultaneous injection of air and water would be highly corrosive. Great care is usually taken to remove all traces of oxygen in the case of water injection for conventional waterflooding. At Schoonebeek and East Tia Juana, air and water injection were alternated in 1-week cycles. Injected volumes during these cycle times are small relative to the storage volume of the reservoir.

Operators may also use plastic-lined tubing and stainless–steel liners to further minimize the chance of products of corrosion being carried into the formation. Corrosion products are highly amorphous and can have a severe plugging effect. An alternative and probably better solution would be to use separate wells for injection of air and water. Economics will dictate the best approach for a given situation.

Corrosion at producing wells has also been observed in wet-combustion projects because of the excessive amounts of acids produced by low-temperature oxidation reactions when burning is partially quenched. This effect will be pronounced in the case of projects where overburning occurs and the combustion front is close to producing wells throughout most of the project. The severity of corrosion at producing wells can be reduced by the following:

(a) Operating in reservoirs that have a high neutralizing capacity per unit volume of oil sand. Tests in the laboratory should be made to determine the amount of acid generated per standard cubic foot of air by low-temperature oxidation of the crude in question and the neutralizing capacity of the oil sand per unit volume.

(b) Operating in thinner sands and on wider spacings. This will minimize the danger of override and the chance of having to operate producing wells at elevated temperatures at which corrosion rates would be quite rapid.

(c) Injecting at a higher water–air ratio. Increased water volumes will dilute the produced acids and reduce producing well temperatures.

Water containing buffering salts such as sodium carbonate can be injected to raise the pH of the acidic produced water. However, as buffered water is vaporized and salt is deposited in the reservoir, the salt can reduce permeability. In addition, less soluble salts such as magnesium, calcium, and barium carbonate can precipitate. Soluble salts can be adsorbed by the clays in the reservoir as well. Hence, use of concentrated buffering solutions is usually inadvisable.

2. *Formation Plugging*. Poor injectivity has been observed in most wet-combustion projects. This may in part be a result of asphaltene and paraffin wax precipitation plus the deposition of solid salts and corrosion products. Additional pressure drop (compared to dry combustion) will be caused by the simultaneous flow of air and water in the burned-out zone, which will contain some unburned hydrocarbon not present in the dry-combustion case.

3. *Emulsion Formation*. Combustion projects, especially those involving wet combustion, have encountered difficult-to-break emulsions. These form as a result of low-temperature oxidation products that are produced in increased quantity at the lower temperatures characteristic of wet combustion.

4. *Gravity Segregation of Water and Air*. In reservoirs thicker than about 15 ft, this can be a significant problem, especially if vertical permeability is high. In such situations, the process can revert to essentially a dry-combustion override in the top of the sand and a conventional waterflood in the lower part of the sand. In thinner, lower permeability sands, capillary effects can help to prevent gravity segregation.

CONCLUSIONS

1. In a thin sand where gravity segregation of air and water is not extremely important, wet combustion can have substantially lower air requirements than dry combustion. Just how much lower is a strong function of the oil type and the nature of the reservoir.

2. The use of a high water–air ratio and a high air injection rate can improve performance. How high the water–air ratio can become before combustion is extinguished is not yet well defined. In combustion tubes, 2500 bbl/10^6 scf appears to be the absolute upper limit. In the field, 500 bbl/10^6 scf appears to be a practical upper limit.

3. Wet combustion appears to be better suited for thin sands because, unlike the case for steam drive, the heated zone extends through only a portion of the reservoir. For this reason, heat losses have a lesser effect on the rate of advance of the heated zone for wet combustion than they do for steam drive.

4. The main operating problems are corrosion of tubular goods, low injectivity, difficult-to-break produced water–oil emulsions, and gravity segregation of air and water. None of these appears to be a sufficient problem to preclude at least consideration of this method as a recovery process, especially in medium- to light-oil reservoirs where steamflooding would be uneconomic. However, the difficulties of well operation after thermal breakthrough often result in premature shut in, and consequently, poor volumetric sweep efficiencies can be expected, especially in pattern projects.

REFERENCES

1. Long, R. E., and Nuan, M. F. "A Study of Getty Oil Company's Successful In Situ Combustion Project in the Bellevue Field," SPE/DOE 10708, Third Joint EOR Symposium, Tulsa, Oklahoma, April 4–7, 1982.
2. Dietz, D. N., and Weijdema, J. "Wet and Partially Quenched Combustion," *J. Pet. Tech.* (April 1968), p. 411.
3. Parrish, D. R., and Craig, F. F. "Laboratory Study of a Combination of Forward Combustion and Waterflooding—The COFCAW Process," *J. Pet. Tech.* (June 1969), p. 753.
4. Garon, A. M., and Wygal, R. J., Jr. "A Laboratory Investigation of Wet Combustion Parameters," *Soc. Pet. Eng. J.* (December 1974), pp. 537–544.
5. Burger, J. G., and Sahuquet, B. D. "Laboratory Research on Wet Combustion," *J. Pet. Tech.* (October 1973), p. 1137.
6. Dietz, D. N. "Wet Underground Combustion, State of the Art," *J. Pet. Tech.* (May 1970), p. 605.
7. Parrish, D. R., Pollock, C. B., Ness, N. L., and Craig, F. F. "A Tertiary COFCAW Pilot Test in the Sloss Field, Nebraska," *J. Pet. Tech.* (June 1974), pp. 667–675.
8. Buxton, T. S., and Pollock, C. B. "The Sloss COFCAW Project—Further Evaluation of Performance During and After Air Injection," *J. Pet. Tech.* (December 1974), pp. 1439–1448.
9. Joseph, C., and Pusch, W. H. "A Field Comparison of Wet and Dry Combustion," *J. Pet. Tech.* (September 1980), pp. 1523–1528.
10. Buchwald, R. W., Jr., Hardy, W. C., and Neinast, G. S. "Case Histories of Improved Recovery Projects," SPE 3781, presented at SPE Symposium on Improved Oil Recovery, Tulsa, Oklahoma, April 16–19, 1972.

Chapter 16

THERMAL PROCESS SELECTION

ABSTRACT

Selection of the best depletion scheme is the most important question faced when making a decision about a given reservoir. This chapter reviews the simplified calculations presented throughout this book as it leads one through a suggested set of screening calculations for steam stimulation, steamflooding, and dry forward in situ combustion for a hypothetical heavy-oil reservoir. In addition, screening by inspection of pertinent data is demonstrated for a list of reservoirs that differ markedly from one another. The reader can test his ability to make an intelligent choice without making the detailed calculations. Finally, the published screening criteria for in situ combustion and steamflooding are reviewed. In some cases, such as the NOIL reservoir example, process choice can be a difficult one, requiring consideration of nonthermal methods as well.

INTRODUCTION

This chapter reviews the simplified calculations described in earlier chapters that are most helpful in making the selection of the appropriate thermal process for a specific situation. Then we will examine the data for several potential reservoir candidates to see if we can make the proper selection by inspection.

First, we will assume that the selection must be made between steam stimulation, steamflooding, and forward in situ combustion. We will not attempt in this level of screening to distinguish between dry and wet combustion. As we saw in the last chapter, there are pros and cons to wet combustion. Wet combustion would be more energy efficient but more troublesome to operate and may in many situations have little overall economic advantage over dry combustion. The choice between wet or dry would require a much more exhaustive study than covered here.

Table 16.1 Basic Data: The NOIL Sand

Depth, 3000 ft	Permeability, 1000 md
Net thickness, 30 ft	Oil viscosity at T_r, 100 cp
Gross thickness, 40 ft	API gravity, 20°
ϕ, 0.3	Thermal conductivity (K_h), 1.4 Btu/hr-ft-°F
S_{oi}, 0.6	$(\rho c)_{R+F}$, 35 Btu/ft^3-°F
S_{wi}, 0.4	X_{surf}, 0.8 lb vapor/lb total water
B_o, 1.0 RB/STB	T_{surf}, 70°F
Casing o.d., 7 in.	T_r, 120°F
Grade J-55	P_r, 200 psia
Tubing o.d., $2\frac{7}{8}$ in.	
r_w, 0.165 ft	
Breakdown gradient, 0.6 psi/ft	

SCREENING CALCULATIONS FOR NOIL RESERVOIR

To begin our review assume we have the following hypothetical reservoir, which we will call the NOIL sand. Table 16.1 contains the basic data for this old, depleted heavy-oil field that produced by solution gas drive during primary production. Currently, the field is drilled on a 20-acre spacing that would be too wide for steamflooding. We will assume that for steamflooding, infill drilling would be required. For our steamflooding calculations, we will assume that a 5-acre 5-spot can be developed by infill drilling.

At the outset and by inspection, the NOIL sand does not appear to be a good steam stimulation candidate mainly because it is pressure depleted, and initial well productivity is too low. In addition, the sand is rather thin (30 ft) and deep (3000 ft), which will further detract from any kind of steam injection operation. Nevertheless, let us consider steam stimulation first.

STEAM STIMULATION

1. What kind of a first-cycle oil–steam ratio would be obtained using steam stimulation?

$$\frac{khP}{\mu_o} = \frac{(1000)(30)(200)}{100} = 60{,}000 \text{ md-ft-psi/cp}$$

From Figure 7.1, the oil–steam ratio is 1.2 STB/bbl. This is an excellent oil–steam ratio if it could be sustained. It would have to be further discounted if the initial production were at an elevated water cut, however. The first-cycle oil–steam ratio could be substantially lower if the initial water cut were above 50%. If the water cut were low initially, steam stimulation would be a possibility for this sand.

However, since the static pressure is only 200 psi, and the oil saturation is high, we can assume this is a solution gas drive reservoir. The sand thickness is too low for gravity drainage to be a significant mechanism, and unless the dip is great (this is not reported), gravity drainage must be assumed negligible. Assuming a minimum producing bottom-hole pressure of 50 psi, an undamaged, unstimulated well can produce with this drawdown only about 4 STB/d (assuming pseudo-steady-state flow). Table 4.1 shows that a steam stimulated, undamaged well would only sustain a rate about three times the unstimulated, or 12 STB/d, which would be too little benefit to be economic.

It is noteworthy that the reservoir properties of NOIL sand are similar to those of the Schoonebeek (The Netherlands) reservoir in the thinner portions of the pressure-depleted solution gas drive area of that reservoir as it existed in about 1960. Steam stimulation results observed in that field were, as predicted here, generally disappointing.

STEAMFLOODING

2. For steamflooding, let us assume injection of steam is carried out at 80% of breakdown pressure; what is the steam injection temperature at bottom hole? Then obtain enthalpies H_{wv} and H_w at T_s and H_w at T_r:

$$P_{break} = (0.6)(3000) = 1800 \text{ psi}$$

$$P_{inj} = (0.8)(1800) = 1440 \text{ psia}$$

From the steam tables (Table 1.2) at 1440 psia,

$$T_s = 590°\text{F} \qquad H_{wv} = 567 \text{ Btu/lb} \qquad H_{ws} = 604 \text{ Btu/lb}$$

At $T_r = 120°\text{F}$,

$$H_{wr} = 120 - 32 = 88 \text{ Btu/lb}$$

3. Estimate the injectivity for steam if the entire flow resistance occurs around the producing wells in a 5-acre 5-spot pattern. Assume a producing-well bottom-hole pressure of 140 psia and an average produced oil–injected steam ratio of 0.2 STB/bbl as an initial guess. If 1 bbl of steam is 350 lb, calculate the steam injection rate in pounds per hour:

$$\text{Distance (injection} - \text{production)} = 0.707\sqrt{(43560)5} = 330\text{ ft}$$

$$q_o = \frac{0.00708 k_o h \, \Delta P}{B_o \mu_o \ln(d/2r_w)} \tag{9.11}$$

$$\Delta P = 1440 - 140 = 1300\text{ psi}$$

$$q_o = \frac{(0.00708)(1000)(30)(1300)}{(1.0)(100)(2.3)\log_{10}[330/2(1.165)]}$$

$$= 400\text{ bbl/d}$$

Since $R_{os} = 0.2$ is estimated, we calculate

$$q_s = \frac{400}{0.2} = 2000\text{ bbl/d}$$

$$m_s = \frac{(2000)(350)}{24} = 29{,}200\text{ lb/hr}$$

4. What is the steam quality at bottom hole? Use the Huygen–Huitt charts at 1 year. Using Figure 2.8 for injection through tubing,

$$Q = 780\text{ Btu/hr-ft}$$

From Eq. (2.5),

$$X_i = 0.8 - \frac{(780)(3000)}{(29{,}200)(567)} = 0.66$$

5. What is the casing temperature? From Figure 2.8,

$$T_{casing} = 480°\text{F}$$

6. Rework steps 4 and 5 using the Ramey–Willhite method. Assume a packer is used and that the annulus is vented to the atmosphere†:

(a) Estimate U_{to} (use Fig. 2.4 in Chapter 2 for $P = 0$ psig in annulus, mill scale):

$$U_{to} = 4.0\text{ Btu/hr-ft}^2\text{-°F}$$

†It is probably not practical to maintain an atmospheric pressure annulus because packer materials cannot withstand this pressure differential at elevated temperatures for an extended injection period. This calculation would serve to define an optimistic upper limit of the insulating effect to be obtained by this approach.

(b) Calculate T_h using Eqs. (2.2) and (2.3):

$$\alpha = \frac{1.4}{35} = 0.04 \text{ ft}^2/\text{hr}$$

Assuming a hole (cement sheath) diameter of $9\frac{5}{8}$ in:

$$r_h = 9.625/(2)(12) = 0.401 \text{ ft}$$

$$t = 1 \text{ yr} = (365)(24) = 8760 \text{ hr}$$

$$f(t) = \ln\left(2\frac{\sqrt{\alpha t}}{r_h}\right) - 0.29$$

$$= \ln\left[2\frac{\sqrt{(0.04)(8760)}}{0.401}\right] - 0.29 = 4.246$$

$$\gamma = (r_{to}U_{to}/K_h)^{-1} = \left(\frac{(2.875)(4)}{(2)(12)(1.4)}\right)^{-1} = 2.922$$

$$T_e = \tfrac{1}{2}(70 + 120) = 95°\text{F}$$

$$T_h = \frac{(590)(4.246) + (2.922)(95)}{4.246 + 2.922} = 388°\text{F}$$

(c) Calculate Q using Eq. (2.1):

$$Q = 2\pi K_h(T_h - T_e)/f(t) = (6.28)(1.4)(388 - 95)/4.246$$
$$= 607 \text{ Btu/hr-ft}$$

(d) Calculate the bottom-hole steam quality:

$$X_i = 0.8 - \frac{(607)(3000)}{(29{,}200)(567)} = 0.68$$

(e) Calculate the casing temperature T_{ci}. Assume K_{cem} = 0.51 Btu/hr-ft-°F:

$$T_{ci} = T_h + \frac{r_{to}U_{to}\ln r_h/r_{co}}{K_{cem}}(T_s - T_e)$$

where

$$r_{co} = 7/(2)(12) = 0.292 \text{ ft}$$

$$T_{ci} = 388 + \frac{(0.120)(4)\ln(0.401/0.292)}{0.51}(590 - 95)$$
$$= 536°\text{F}$$

(f) Will the casing fail? From Eq. (2.8),

$$S_c \sim 200\ \Delta T$$

where

$$\Delta T = T_{ci} - T_e$$
$$= 536 - 95 = 441°\text{F}$$
$$S_c = 88{,}200 \text{ psi}$$

Note that Figures 2.22 and 2.23 give the J-55 yield point at 55,000 psi and ultimate strength at 120,000 psi at 536°F. These figures are for the pipe body. Joint strengths are only ~70% of these values (i.e., 38,500 and 84,000 psi). Hence, at this temperature J-55 joints would probably *fail*. Use of tubing insulation or redrilling and using N-80 or stronger casing would be required for wells receiving injected steam.

7. At the above injection rate, what will be the radius of the steam zone after 1 year of injection? Assume a value of 0.67 for the average bottom-hole steam quality. Use the simplified Marx–Langenheim method of Chapter 9:

$$\tau = \frac{4K_h t}{h_t^2(\rho c)_{R+F}} = \frac{(4)(1.4)(24)(365)(1)}{(40)^2(35)} = 0.876$$

$$\sqrt{\tau} = 0.936$$

From the table in Appendix A in Chapter 3 or Figure 3.6,

$$\xi = 0.505$$

From Eq (3.6a) and assuming $(\rho c)_{R+F} = (\rho c)_{ob}$,

$$A_h = \frac{m_s H_t h \xi_s / \theta}{4K_h\ \Delta T}$$

where

$$H_t = \bar{X}_i H_{wv} + H_{ws} - H_{wr} = (0.67)(567) + 604 - 88$$
$$= 896 \text{ Btu/lb}$$
$$\Delta T = T_s - T_r = 590 - 120 = 470°\text{F}$$

then

$$A_h = \frac{(29{,}200)(896)(40)(0.505)}{(4)(1.4)(470)}$$
$$= 2.008 \times 10^5 \text{ ft}^2$$

and

$$r_h = \sqrt{\frac{2.008 \times 10^5}{3.14}}$$
$$= 253 \text{ ft}$$

8. What fraction of the injected heat has been lost to unproductive rock? Use Eq. (9.6):

$$E_h = \xi / \tau = 0.505/0.876 = 0.576$$
$$\text{Fraction lost} = 1 - 0.576 = 0.424$$

9. If steam breakthrough occurred at one year, what would be the oil–steam ratio? Assume the steam zone residual oil saturation (S_{ors}) is 0.15. Use Eq. (9.9). Assume that the steam quality at 1 year approximates the average steam quality for the flood up to this point. (It would be higher but only slightly; see Fig. 2.18.) Then the oil–steam ratio R_{os} is calculated:

$$R_{os} = \frac{62.4\phi h E_h[\bar{X}_i H_{wv}(S_{oi} - S_{ors})]}{h_t(T_s - T_r)(\rho c)_{R+F}}$$
$$= 62.4 \frac{(0.3)(30)(0.576)[0.67(567)(0.6 - 0.15)]}{(40)(470)(35)}$$
$$= 0.084 \text{ bbl/bbl}$$

This is such a low value for R_{os} that it is not worthwhile to proceed to iterate to obtain a final estimate (using the method outlined in Chapter 9). Steamflooding will be uneconomic, in general, if R_{os} is below about 0.15 bbl/bbl. If horizontal fracturing were feasible in the NOIL sand, an approach similar to that applied by Conoco in South Texas (see Chapter 10) should be considered to see if this more rapid and efficient steamflooding approach could raise R_{os} into the economic (>0.15 bbl/bbl) range.

10. What would be the cost of steam per barrel of oil produced using lease crude valued at $20/bbl? Assume the lease and well cost is $1/bbl steam. Then the cost per barrel of steam (C_s) as a function of the oil cost per barrel of oil would be (see p. 97)

$$C_s = 1.00 + 0.071 C_o$$

Then,

$$C_s = 1.00 + (0.071)(20) = \$2.42/\text{bbl steam}$$

If the oil–steam ratio is 0.084 STB/bbl, the steam cost per barrel of oil can be calculated:

$$\text{Steam cost/bbl of oil displaced} = 2.42/0.084 = \$28.80/\text{STB}$$

Obviously, conventional steamflooding would not be economic for this situation if lease crude at $20/STB were used as fuel. Use of a cheaper energy source (gas, coal) may reduce costs somewhat. Possibly coupled with cogeneration of electric power, a cheaper energy source might put steamflooding in the economic range, but it is highly unlikely in this case.

DRY IN SITU COMBUSTION

11. Next let us consider dry in situ combustion. What would be the fuel concentration for forward dry combustion? Use Figure 13.11, Eq. (13.13). Assume the reservoir rock density is 163.5 lb/ft^3:

$$\text{For 20°API:} \quad F_1 = 0.67$$
$$\text{For 120°F:} \quad F_2 = \underline{0.59}$$
$$\bar{F}_c = 1.26\ \text{lb}/100\ \text{lb rock}$$

$$\rho_f = \frac{141.5\rho_w}{131.5 + °\text{API}}$$

$$\rho_f = \frac{(141.5)(62.4)}{151.5} = 58.3\ \text{lb/ft}^3 \quad \text{for 20°API oil}$$

$$S_{of} = \frac{\bar{F}_c(1-\phi)\rho_r}{100\phi\rho_f}$$

$$= \frac{(1.26)(1-0.3)(163.5)}{(100)(0.3)(58.3)} = 0.0825$$

Converting to lb fuel/ft^3 oil sand,

$$F_c = \rho_f \phi S_{of} = (58.3)(0.3)(0.0825) = 1.44\ \text{lb/ft}^3$$

12. What would be the air–oil ratio if combustion were 90% efficient? Using Eq. (13.6a),

$$R_{Ao} = \frac{(65036)(0.0825)}{(0.9)(0.6 - 0.0825)}$$
$$= 11520\ \text{scf/bbl}$$

13. Assuming an air injection expense of \$1/kscf, what is the cost per barrel of oil production?

$$\text{Cost/bbl} = (11.52)(\$1.00) = \$11.52/\text{STB}$$

14. Assuming injection at 80% of formation breakdown pressure, what will the maximum permissible air injection rate be for a 20-acre line drive? Assume $P_w = 140$ psi. Use Eq. (13.15):

$$d = \sqrt{(43{,}560)(20)} = 933 \text{ ft}$$

$$P_I = 1440 \text{ psia}$$

$$P_w = 140 \text{ psia}$$

$$V_{max} = \frac{(2.946 \times 10^{-4})(1000)(1300)(11{,}520)}{(1.0)(100)(2.3)\log_{10}[933/2(0.165)]}$$

$$= 5.56 \times 10^3 \text{ scf/ft-hr}$$

$$q_{max} = 5.56 \times 10^3 \times 24 \times 30 = 4.0 \text{ Mscf/d}$$

15. What is the minimum air rate per foot required to maintain combustion (maintain at least 600°F at the combustion front) for this sand? Assume that the combustion front is halfway between injector and producer.

Assume that

$$c_g = 0.24 \text{ Btu/lb-°F}$$

$$\rho_g = 0.0765 \text{ lb/scf}$$

$$(\rho c)_{R+F} = 35 \text{ Btu/ft}^3\text{-°F}$$

$$\Delta H = 20{,}000 \text{ Btu/lb}$$

$$C_A = 188 \text{ scf/lb}$$

(a) $T_f - T_r = 600 - 120 = 480°\text{F}$

(b)

$$T_m - T_r = F_c \frac{\Delta H + C_A \rho_g c_g (T_f - T_r)}{(\rho c)_{R+F}} \qquad (13.17)$$

$$T_m - T_r = 1.44 \frac{20{,}000 + (188)(0.0765)(0.24)(480)}{35}$$

$$= 891°\text{F}$$

$$\frac{T_f - T_r}{T_m - T_r} = \frac{480}{891} = 0.54$$

$$\frac{r_f}{a} = \frac{(933)(2)}{(2)(30)} = 31$$

From Figure 13.20 (vertical position $z = 0.5$),

$$vr_f/\alpha = 30$$

$$V_{min} = 2\pi F_c C_A \alpha (vr_f/\alpha)$$

$$= (6.28)(1.44)(188)(1.4/35)(30) = 2.04 \times 10^3 \text{ scf/ft-hr}$$

Since the rate with the combustion front at the midway point is less than half that attainable, there should be no difficulty sustaining combustion in this sand.

FINAL PROCESS SELECTION

16. Would steam stimulation, steam drive, or forward combustion be selected for the NOIL sand?

Forward combustion appears to be the only method of the above applicable to this reservoir. Steamflooding appears clearly uneconomic unless a basal horizontal fracturing approach were feasible to speed the heating of the oil sand. Steam stimulation probably would not be attractive considering the state of pressure depletion, the depth, and the low strength casing in existing wells. Nevertheless, considering the problems encountered with in situ combustion in actual operation, it is suggested that it be considered a last resort and other approaches be considered first. Conventional waterflooding may have merit since the oil viscosity is not too high (100 cp). Thickened waterflooding with polymers might be considered. Hot waterflooding is likely to fare even worse than steamflooding because of the heat loss problem. The basal horizontal fracture approach as well as cyclic steam stimulation employing a tubing insulation to protect the wells probably should still be piloted so as not to write off steam-based processes prematurely. Compression requirements would likely favor the economics for wet combustion or a mixture of dry and wet combustion as Texaco is using at Bellevue, Louisiana. The final decision should be made after piloting the alternatives, making an exhaustive geologic and reservoir engineering study, and a complete economic analysis.

FAST SCREENING OF THERMAL PROCESS CANDIDATES

Six sets of reservoir data are listed below. By inspection, let us decide which thermal recovery process (steam stimulation, steamflood, or in situ combustion) to apply in each reservoir.

	A	B	C	D	E	F
Depth, ft	2300	4200	2200	1500	800	1400
Sand thickness, ft	140	17	120	20	56	140
k, md	1500	700	1500	300	2200	1500
ϕ, per cent	30	26	30	20	32	33
Oil Gravity, °API	9	22	26	36	16	10
μ_o at T_r, cp	2000	22	160	6	1600	100,000
S_o	75	50	70	26	55	65
Reservoir pressure, psi	1400	500	500	500	200	450
Cold producing rates						
Oil, STB/d	300	8	50	7	2	0
Water, bbl/d	0	200	400	100	50	0
Well spacing, acre	20	10	10	20	2.5	4

Reservoir A

Steam stimulation would be chosen. Note that

$$khP/\mu_o = (1500)(140)(1400)/2000 = 1.47 \times 10^5 \text{ md-ft-psi/cp}$$

$$R_{os} = 2.3 \text{ bbl/bbl} \quad \text{(from Fig. 7.1)}$$

Obviously, such a high oil-steam ratio would have superior economics to that attainable with steamflooding or in situ combustion.

Reservoir B

In situ combustion would be chosen for this reservoir because it is deep (4200 ft) and the sand is thin (17 ft).

Air requirements and fuel consumption would be lower for wet combustion. Assume that an air–water injection ratio of 1000 scf/bbl is used. From Table 15.1, using Garon's data,

$$\frac{\text{fuel burned in wet combustion}}{\text{fuel burned in dry combustion}} = 0.6$$

Reservoir C

Because of its thickness (120 ft), good oil saturation and porosity ($\phi S_o = 0.21$), and reasonably shallow depth, this reservoir should be a good steamflood candidate. The water–oil ratio is too high for it to be a good steam stimulation candidate.

Reservoir D

This reservoir is too thin to be a candidate for steamflooding. It has an outside chance of being a candidate for in situ combustion, but its oil saturation and porosity are likely too low ($\phi S_o = 0.052$) for an economic thermal process of any kind.

Reservoir E

The shallow depth (800 ft), good thickness (56 ft), high permeability (2200 md), and high ϕS_o product (0.176) make this a good steamflood candidate. The high initial water cut would mitigate against this reservoir being a good steam stimulation candidate.

Reservoir E is in fact a typical California heavy-oil sand as might be found in the Kern River area. In situ combustion may also work here but would likely not be as economically attractive as steamflooding because of the high costs associated with air compression and well completions in in situ combustion operations.

Reservoir F

If it were not for the extremely high oil viscosity in this reservoir, it would be a good steamflooding candidate. As it is, no thermal flood (steam or combustion) would be practical on commercial well spacing because of the low formation transmissibility. Steam stimulation will be effective here, however, provided the steam is injected at above formation breakdown pressure. If sufficient interwell heating can occur after many cycles of steam stimulation, eventually a steamflood may become possible. This reservoir is in fact the Mannville Clearwater sand at Cold Lake, Alberta, Canada, where cyclic steam stimulation is being applied commercially at the present time.

PUBLISHED SCREENING GUIDES

Tables 16.2 and 16.3 give published screening guidelines for both in situ combustion and steamflooding.[1–9] It should be emphasized that these guidelines are very approximate and should be used with caution. The preferred approach is to make a complete calculation of the oil–steam or air–oil ratio as outlined earlier and compute rough economics for these processes based on prevailing costs and crude prices.

Table 16.2 Publishing Screening Criteria for In Situ Combustion

Source	Gravity (°API)	μ_o (cp)	S_o (%)	Depth (ft)	Thickness (ft)	k (md)	ϕ (%)	ϕS_o (fraction)	kh/μ_o (md-ft/cp)	Spacing (acres)	Other
Poettman[a]	—	—	—	—	—	>100	>20	>0.10	—	—	—
Geffen[b]	<45	—	—	>500	>10	—	—	>0.05	>100	<40	$P>250$
Lewin[c]	10–45	Not critical	>50	>500	>10	Not critical	—	>0.05	>20	<40	—
Iyoho[d]	10–40	<1000	>50	>200	5–50	>300	>20	>0.08	>20	<40	Sandstone
Chu[e]	<40	—	>35	—	—	>100	>16	>0.10	—	—	—
NPC[f]	10–35	$\leqslant 5000$	—	$\leqslant 11500$	$\leqslant 20$	>35	$\geqslant 20$	$\geqslant 0.08$	$\geqslant 5$	—	$P\leqslant 2000$

[a]Reference 7.
[b]References 4 and 5.
[c]References 2 and 3.
[d]Reference 6.
[e]References 8 and 9.
[f]Reference 10.

Table 16.3 Steam Drive Screening Criteria[a]

Parameter	NPC[b]	NPC[c]	Lewin[d]
Oil gravity, °API	≤25	10–34	>10
Oil viscosity, cp	≥20	≤15,000	—
Depth, ft	200–5000	≤3000	<5000
Zone thickness, ft	>20	≥20	>20
Transmissibility, md-ft/cp	>100	≥5	>100
Minimum ϕS_o, fraction	>0.065	≥0.10	>0.065
Reservoir pressure, psia	—	≤2000	—

[a]First NPC guidelines are from 1976 report (Ref. 1). The second are from the 1984 NPC report (Ref. 10) for the implemented technology case.
[b]From Ref. 1.
[c]From Ref. 10.
[d]From Refs. 2 and 3.

REFERENCES

1. Haynes, H. J., Thrasher, L. W., Katz, M. L., and Eck, T. R. *Enhanced Oil Recovery*, National Petroleum Council, Industry Advisory Council to the U.S. Department of the Interior, Washington, DC, 1976.
2. *Research and Development in Enhanced Oil Recovery, Final Report, The Methodology*, prepared by Lewin and Associates, Inc. for ERDA, Washington, DC, ERDA Report 77-20/3, December 1976, pp. 5–42.
3. Brashear, J. P., and Kuuskraa, V. A. "The Potential and Economics of Enhanced Oil Recovery," *J. Pet. Tech.* (September 1978), pp. 1231–1239.
4. Geffen, T. M. "Oil Production to Expect from Known Technology," *Oil Gas J.* (May 1973), pp. 66–76.
5. Geffen. T. M. "Improved Oil Recovery Could Help Ease Energy Shortage," *World Oil* (October 1983), pp. 84–88.
6. Iyoho, A. W. "Selecting Enhanced Oil Recovery Processes," *World Oil* (November 1978), pp. 61–64.
7. Poettman, F. H. "In Situ Combustion: A Current Appraisal," *World Oil* (May 1964), pp. 95–98.
8. Chu, C. "A Study of Fireflood Field Projects," *J. Pet. Tech.* (February 1977), pp. 111–119.
9. Chu, C. "State-of-the-Art Review of Fireflood Field Projects,"*J. Pet. Tech.* (January 1982), pp. 19–36.
10. National Petroleum Council. *Enhanced Oil Recovery*, National Petroleum Council, Washington, DC, June 1984.

Chapter 17

PROPERTIES OF FLUID AND ROCK

ABSTRACT

This chapter summarizes fluid and rock properties as a reference for making thermal reservoir engineering calculations. The order of this material is as follows:

1. fluid density change with temperature and pressure;
2. enthalpy and specific heat data for water and petroleum fluids;
3. specific heat and thermal conductivity of fluid-saturated rock;
4. thermal properties of other materials such as steel, insulation, and various fluids;
5. viscosity data for steam, water, and petroleum fluids; and
6. heats of combustion and thermal cracking data for petroleum hydrocarbons.

VARIATION OF DENSITY OF STEAM, WATER, AND OIL WITH TEMPERATURE AND PRESSURE

Steam and Water

Specific volume (reciprocal density) for water and steam as a function of saturation pressure is given in Figure 17.1. Table 17.1 gives specific volume data for steam and hot water over the primary range of interest.

Petroleum Hydrocarbons

Figure 17.2 gives the specific gravity of petroleum liquids as a function of temperature and the specific gravity at 60°F. The density (in lb/ft^3) is

Table 17.1 Thermodynamic Properties of Saturated Steam.[a]

Abs. Press. Lb./Sq. In.	Temp. Fahr.	Specific Volume Sat. Liquid.	Specific Volume Sat. Vapor	Enthalpy Sat. Liquid	Enthalpy Evap.	Enthalpy Sat. Vapor
P	T	v_w	v_s	H_w	H_{wv}	H_s
170	368.41	0.01822	2.675	341.09	854.9	1196.0
172	369.35	0.01823	2.645	342.10	854.1	1196.2
174	370.29	0.01824	2.616	343.10	853.3	1196.4
176	371.22	0.01825	2.587	344.09	852.4	1196.5
178	372.14	0.01826	2.559	345.06	851.6	1196.7
180	373.06	0.01827	2.532	346.03	850.8	1196.9
182	373.96	0.01829	2.505	347.00	850.0	1197.0
184	374.86	0.01830	2.479	347.96	849.2	1197.2
186	375.75	0.01831	2.454	348.92	848.4	1197.3
188	376.64	0.01832	2.429	349.86	847.6	1197.5
190	377.51	0.01833	2.404	350.79	846.8	1197.6
192	378.38	0.01834	2.380	351.72	846.1	1197.8
194	379.24	0.01835	2.356	352.64	845.3	1197.9
196	380.10	0.01836	2.333	353.55	844.5	1198.1
198	380.95	0.01838	2.310	354.46	843.7	1198.2
200	381.79	0.01839	2.288	355.36	843.0	1198.4
205	383.86	0.01842	2.234	357.58	841.1	1198.7
210	385.90	0.01844	2.183	359.77	839.2	1199.0
215	387.89	0.01847	2.134	361.91	837.4	1199.3
220	389.86	0.01850	2.087	364.02	835.6	1199.6
225	391.79	0.01852	2.0422	366.09	833.8	1199.9
230	393.68	0.01854	1.9992	368.13	832.0	1200.1
235	395.54	0.01857	1.9579	370.14	830.3	1200.4
240	397.37	0.01860	1.9183	372.12	828.5	1200.6
245	399.18	0.01863	1.8803	374.08	826.8	1200.9
250	400.95	0.01865	1.8438	376.00	825.1	1201.1
255	402.70	0.01868	1.8086	377.89	823.4	1201.3
260	404.42	0.01870	1.7748	379.76	821.8	1201.5
265	406.11	0.01873	1.7422	381.60	820.1	1201.7
270	407.78	0.01875	1.7107	383.42	818.5	1201.9
275	409.43	0.01878	1.6804	385.21	816.9	1202.1
280	411.05	0.01880	1.6511	386.98	815.3	1202.3
285	412.65	0.01883	1.6228	388.73	813.7	1202.4
290	414.23	0.01885	1.5954	390.46	812.1	1202.6
295	415.79	0.01887	1.5689	392.16	810.5	1202.7
300	417.33	0.01890	1.5433	393.84	809.0	1202.8
310	420.35	0.01894	1.4944	397.15	806.0	1203.1
320	423.29	0.01899	1.4485	400.39	803.0	1203.4
330	426.16	0.01904	1.4053	403.56	800.0	1203.6
340	428.97	0.01908	1.3645	406.66	797.1	1203.7
350	431.72	0.01913	1.3260	409.69	794.2	1203.9
360	434.40	0.01917	1.2895	412.67	791.4	1204.1
370	437.03	0.01921	1.2550	415.59	788.6	1204.2
380	439.60	0.01925	1.2222	418.45	785.8	1204.3
390	442.12	0.01930	1.1910	421.27	783.1	1204.4
400	444.59	0.0193	1.1613	424.0	780.5	1204.5
410	447.01	0.0194	1.1330	426.8	777.7	1204.5
420	449.39	0.0194	1.1061	429.4	775.2	1204.6
430	451.73	0.0194	1.0803	432.1	772.5	1204.6
440	454.02	0.0195	1.0556	434.6	770.0	1204.6
450	456.28	0.0195	1.0320	437.2	767.4	1204.6
460	458.50	0.0196	1.0094	439.7	764.9	1204.6
470	460.68	0.0196	0.9878	442.2	762.4	1204.6
480	462.82	0.0197	0.9670	444.6	759.9	1204.5
490	464.93	0.0197	0.9470	447.0	757.5	1204.5

Abs. Press. Lb./Sq. In.	Temp. Fahr.	Specific Volume Sat. Liquid.	Specific Volume Sat. Vapor	Enthalpy Sat. Liquid	Enthalpy Evap.	Enthalpy Sat. Vapor
P	T	v_w	v_s	H_w	H_{wv}	H_s
500	467.01	0.0197	0.9278	449.4	755.0	1204.4
520	471.07	0.0198	0.8915	454.1	750.1	1204.2
540	475.01	0.0199	0.8578	458.6	745.4	1204.0
560	478.85	0.0200	0.8265	463.0	740.8	1203.8
580	482.58	0.0201	0.7973	467.4	736.1	1203.5
600	486.21	0.0201	0.7698	471.6	731.6	1203.2
620	489.75	0.0202	0.7440	475.7	727.2	1202.9
640	493.21	0.0203	0.7198	479.8	722.7	1202.5
660	496.58	0.0204	0.6971	483.8	718.3	1202.1
680	499.88	0.0204	0.6757	487.7	714.0	1201.7
700	503.10	0.0205	0.6554	491.5	709.7	1201.2
720	506.25	0.0206	0.6362	495.3	705.4	1200.7
740	509.34	0.0207	0.6180	499.0	701.2	1200.2
760	512.36	0.0207	0.6007	502.6	697.1	1199.7
780	515.33	0.0208	0.5843	506.2	692.9	1199.1
800	518.23	0.0209	0.5687	509.7	688.9	1198.6
820	521.08	0.0209	0.5538	513.2	684.8	1198.0
840	523.88	0.0210	0.5396	516.6	680.8	1197.4
860	526.63	0.0211	0.5260	520.0	676.8	1196.8
880	529.33	0.0212	0.5130	523.3	672.8	1196.1
900	531.98	0.0212	0.5006	526.6	668.8	1195.4
920	534.59	0.0213	0.4886	529.8	664.9	1194.7
940	537.16	0.0214	0.4772	533.0	661.0	1194.0
960	539.68	0.0214	0.4663	536.2	657.1	1193.3
980	542.17	0.0215	0.4557	539.3	653.3	1192.6
1000	544.61	0.0216	0.4456	542.4	649.4	1191.8
1050	550.57	0.0218	0.4218	550.0	639.9	1189.9
1100	556.31	0.0220	0.4001	557.4	630.4	1187.8
1150	561.86	0.0221	0.3802	564.6	621.0	1185.6
1200	567.22	0.0223	0.3619	571.7	611.7	1183.4
1250	572.42	0.0225	0.3450	578.6	602.4	1181.0
1300	577.46	0.0227	0.3293	585.4	593.2	1178.6
1350	582.35	0.0229	0.3148	592.1	584.0	1176.1
1400	587.10	0.0231	0.3012	598.7	574.7	1173.4
1450	591.73	0.0233	0.2884	605.2	565.5	1170.7
1500	596.23	0.0235	0.2765	611.6	556.3	1167.9
1600	604.90	0.0239	0.2548	624.1	538.0	1162.1
1700	613.15	0.0243	0.2354	636.3	519.6	1155.9
1800	621.03	0.0247	0.2179	648.3	501.1	1149.4
1900	628.58	0.0252	0.2021	660.1	482.4	1142.4
2000	635.82	0.0257	0.1878	671.7	463.4	1135.1
2100	642.77	0.0262	0.1746	683.3	444.1	1127.4
2200	649.46	0.0268	0.1625	694.8	424.4	1119.2
2300	655.91	0.0274	0.1513	706.5	403.9	1110.4
2400	662.12	0.0280	0.1407	718.4	382.7	1101.1
2500	668.13	0.0287	0.1307	730.6	360.5	1091.1
2600	673.94	0.0295	0.1213	743.0	337.2	1080.2
2700	679.55	0.0305	0.1123	756.2	312.1	1068.3
2800	684.99	0.0315	0.1035	770.1	284.7	1054.8
2900	690.26	0.0329	0.0947	785.4	253.6	1039.0
3000	695.36	0.0346	0.0858	802.5	217.8	1020.3
3100	700.31	0.0371	0.0753	825.0	168.1	993.1
3200	705.11	0.0444	0.0580	872.4	62.0	934.4
3206.2	705.40	0.0503	0.0503	902.7	0	902.7

[a]From Ref. 2. After J. H. Keenan, and F. G. Keyes, *Thermodynamic Properties of Steam* 1st ed., © 1936 John Wiley & Sons.

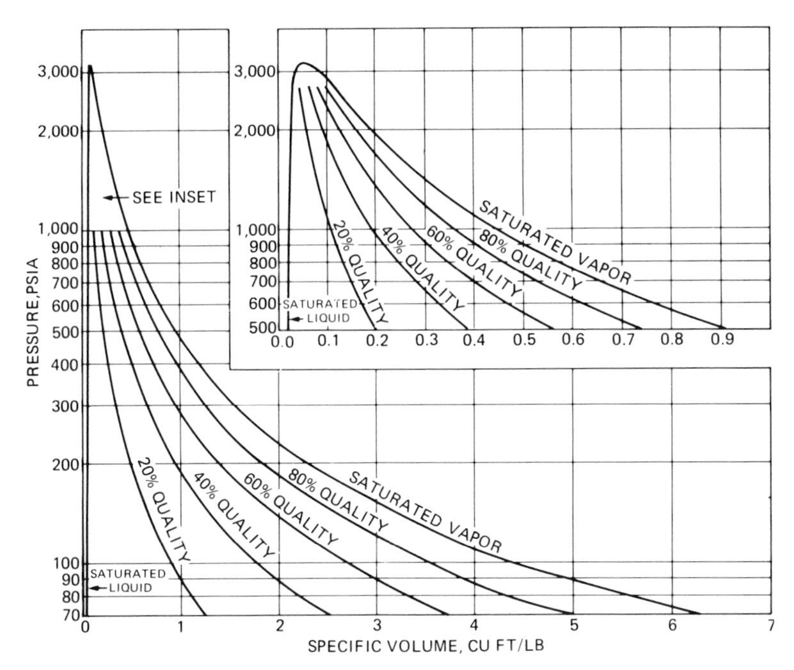

Figure 17.1. Pressure-specific volume chart for saturated steam showing steam quality lines.[1] © 1965 *Oil & Gas Journal.*

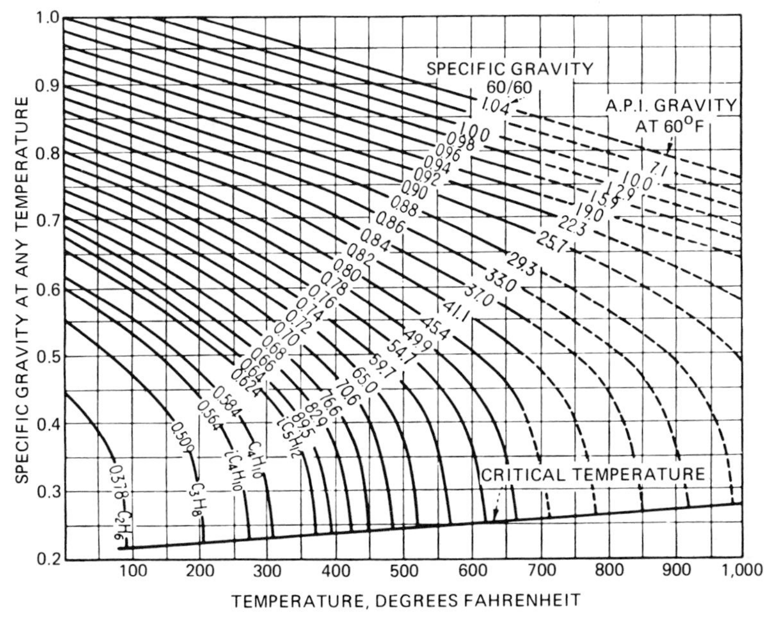

Figure 17.2. Approximate change of specific gravity of petroleum liquids with temperature.[3] © 1938 *Oil & Gas Journal.*

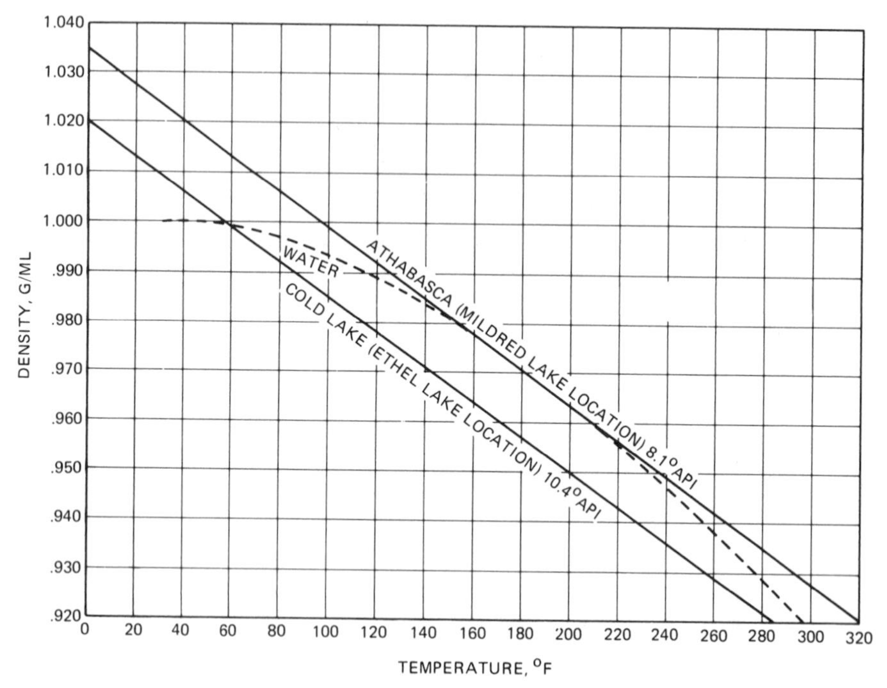

Figure 17.3. Density versus temperature for Cold Lake and Athabasca oil.[4]

obtained by multiplying the specific gravity by 62.4. The density-versus-temperature curves for two major potential thermal prospects (Cold Lake and Athabasca oil) are given in Figure 17.3.

ENTHALPY OF STEAM–WATER SYSTEMS AND SPECIFIC HEATS OF WATER, PETROLEUM OILS, AND GASES

Steam and Water

Table 17.1 gives enthalpy–pressure–temperature data for saturated steam over the primary range of interest in reservoir applications. Figure 17.4 gives pressure–enthalpy data over a wide range. In Table 17.1 and Figure 17.4, zero enthalpy corresponds to liquid water at 32°F and saturation vapor pressure. Figure 17.4 also gives temperature and specific volume data.

In order to calculate the specific enthalpy (Btu/lb) of wet steam injected into a formation at initial temperature T_r,

$$H_t = X_i H_{wv} + H_{ws} - H_{wr}$$

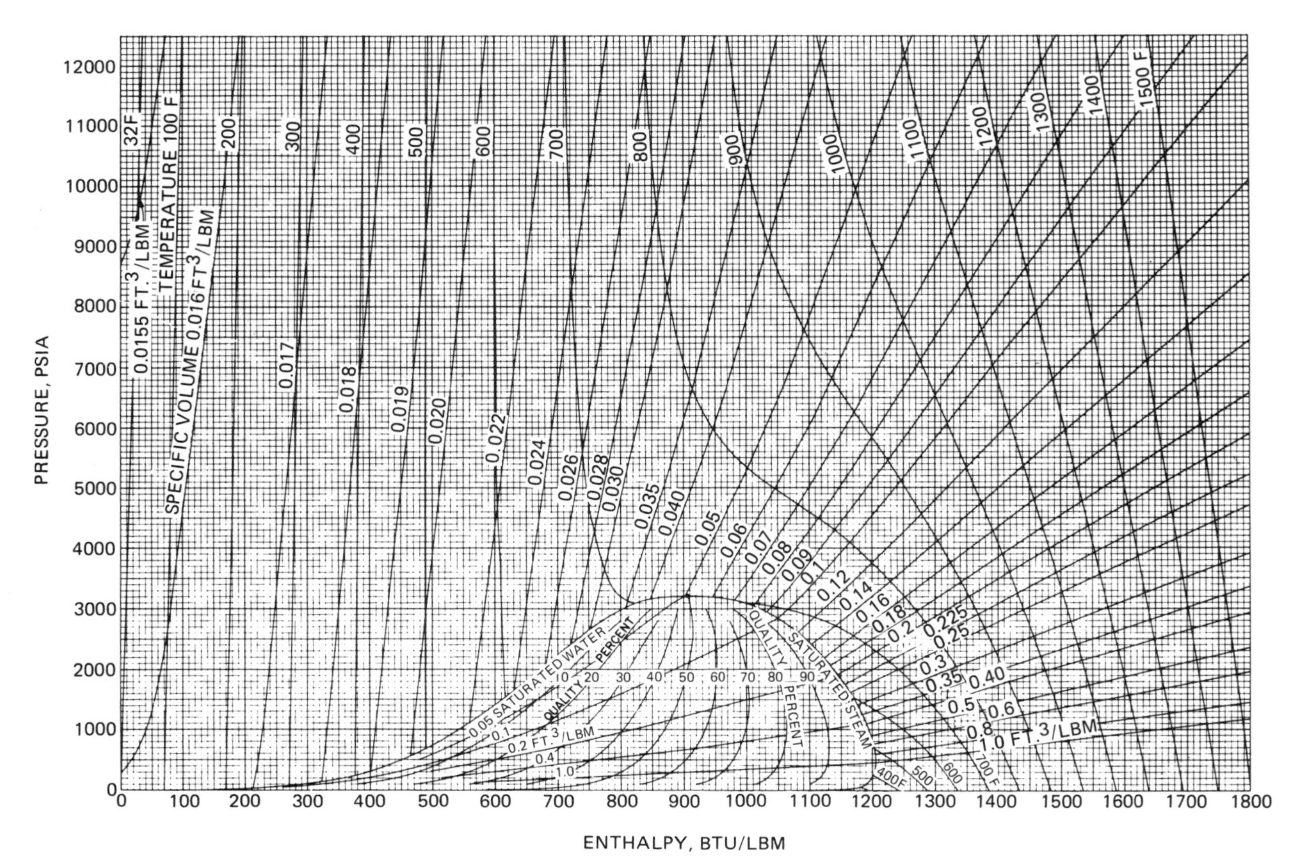

Figure 17.4. Pressure–enthalpy chart for steam and water.[5] © 1967 ASME.

where H_t = total specific enthalpy injected above T_r (Btu/lb)
X_i = injected steam quality (lb steam vapor/lb vapor plus liquid water at steam temperature)
H_{wv} = latent enthalpy of vaporization at T_s (Btu/lb)
H_{ws} = enthalpy of liquid water at T_s above 32°F (Btu/lb)
H_{wr} = enthalpy of liquid water at T_r above 32°F (Btu/lb)
T_s = saturated steam temperature at steam injection pressure (°F)
T_r = original reservoir temperature (°F)

The specific heat of liquid water as a function of pressure and temperature is given in Figures 17.5 and 17.6.

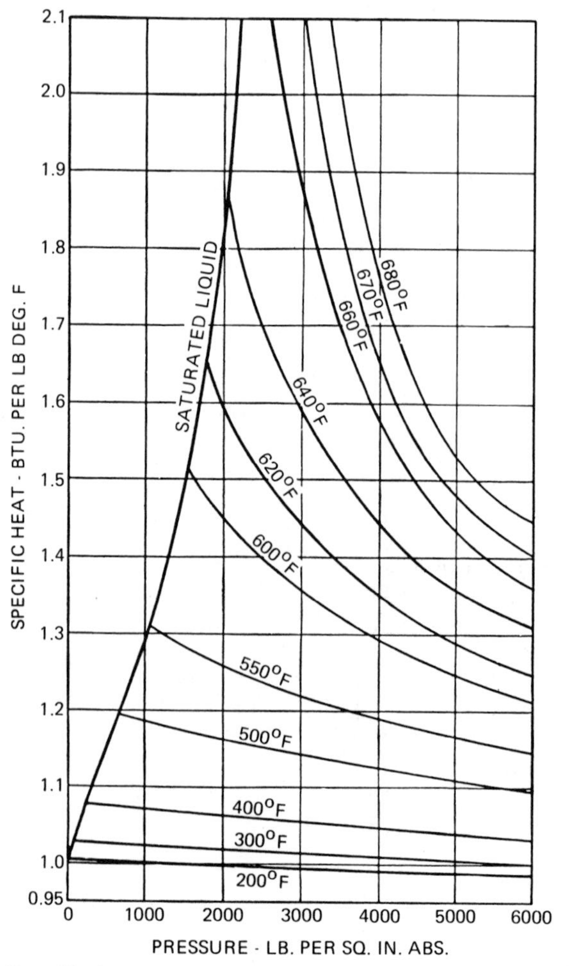

Figure 17.5. Specific heats of liquid water for constant pressure.[2] © 1936 John Wiley & Sons.

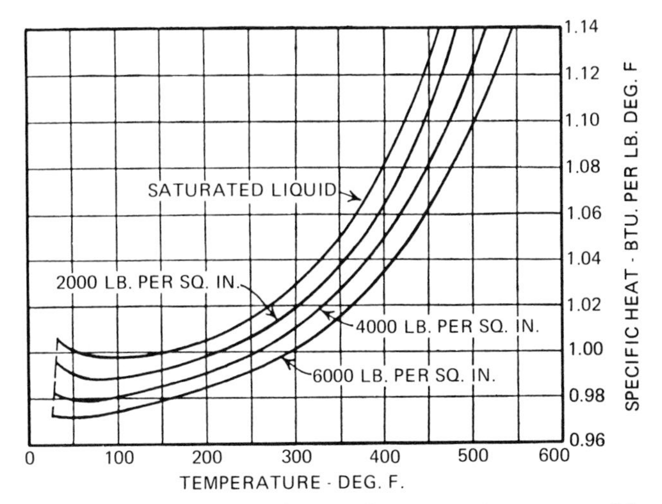

Figure 17.6. Specific heats of liquid water for constant pressure at low temperatures.[2] © 1936 John Wiley & Sons.

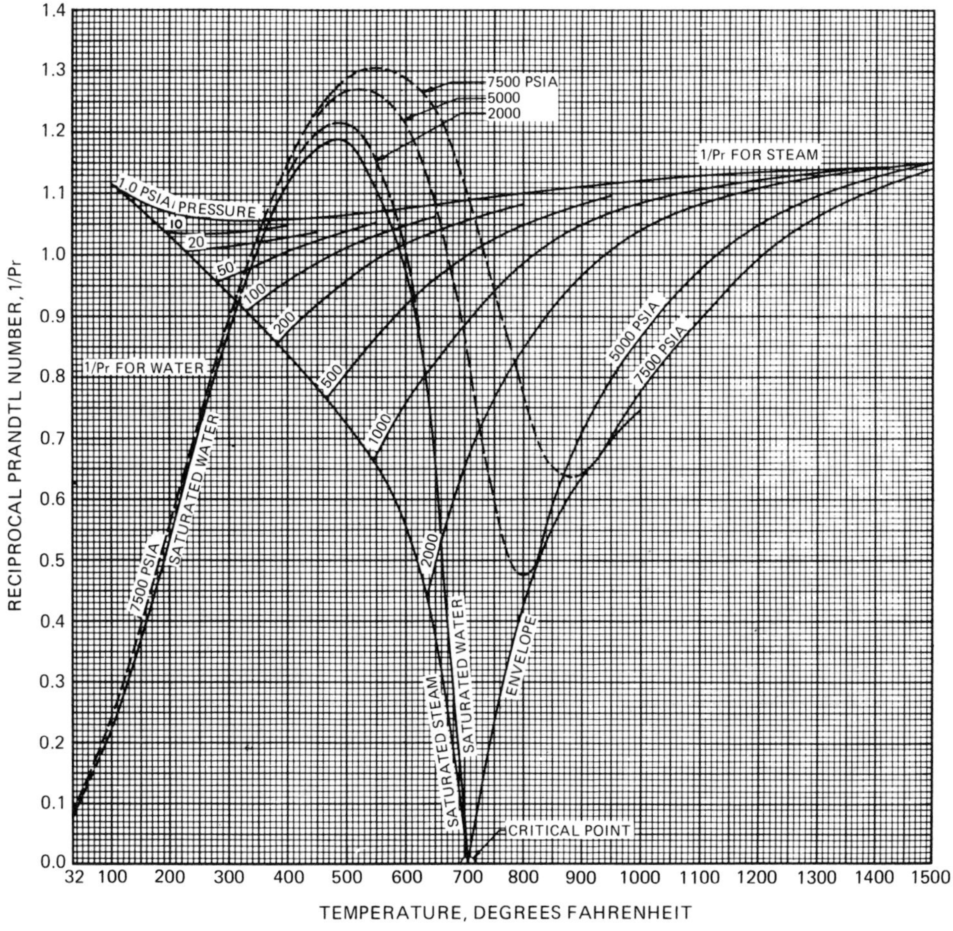

Figure 17.7. Reciprocal Prandtl number 1/Pr for steam and water.[5] © 1967 ASME.

Table 17.2 Prandtl Numbers $C_p\mu/K$ for Gases and Vapor at 1 atm and 212°F[a]

	$C_p\mu/K$	$(C_p\mu/K)^{0.3}$	$(C_p\mu/K)^{1/3}$	$(C_p\mu/K)^{0.4}$	$(C_p\mu/K)^{2/3}$
Air, hydrogen	0.69	0.894	0.884	0.866	0.781
Ammonia	0.86	0.956	0.951	0.941	0.904
Argon	0.66	0.883	0.871	0.847	0.759
Carbon dioxide, methane	0.75	0.917	0.909	0.891	0.826
Carbon monoxide	0.72	0.906	0.896	0.877	0.803
Helium	0.71	0.902	0.892	0.872	0.796
Nitric oxide, nitrous oxide	0.72	0.906	0.896	0.877	0.803
Nitrogen, oxygen	0.70	0.899	0.888	0.867	0.789
Steam (low pressure)	1.06	1.018	1.020	1.024	1.040

[a]From Ref. 6.

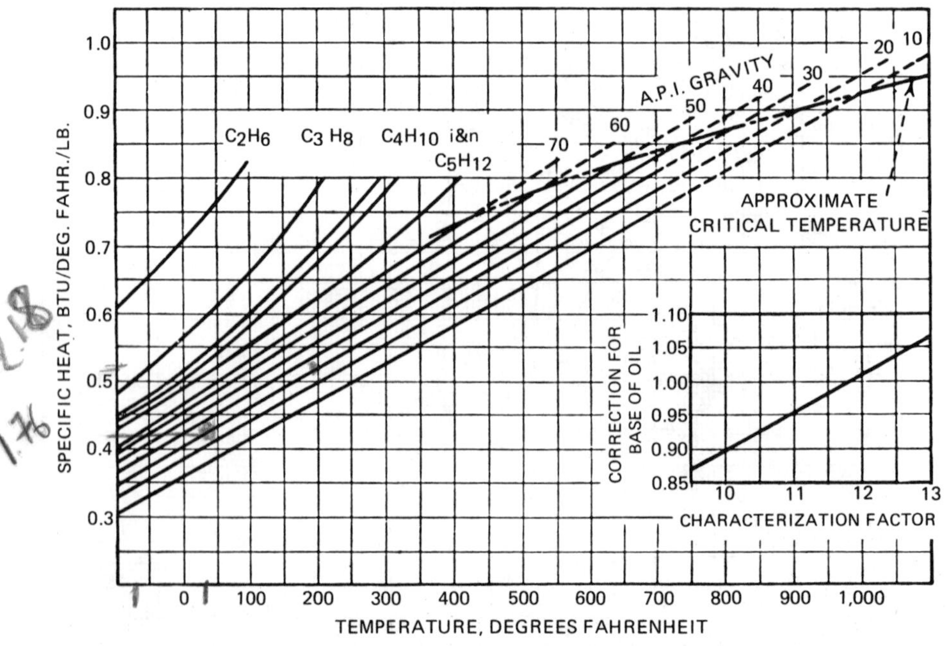

Figure 17.8. Specific heat of midcontinent liquid oils with correction factor for other bases of oils.[7] © 1949 McGraw-Hill.

Values of the Prandtl number $C_p \mu / K$ for steam, air, and other gases at 1 atm are given in Table 17.2. This group is used in evaluating the contribution of convection in the tubing–casing annulus to the overall wellbore heat transfer coefficient. For the case of a high-pressure annulus containing steam, Figure 17.7 may be used.

Petroleum Liquid Specific Heats

Specific heats at constant pressure for petroleum liquids are given in Figure 17.8 as a function of temperature and API gravity.[3] A correction factor can

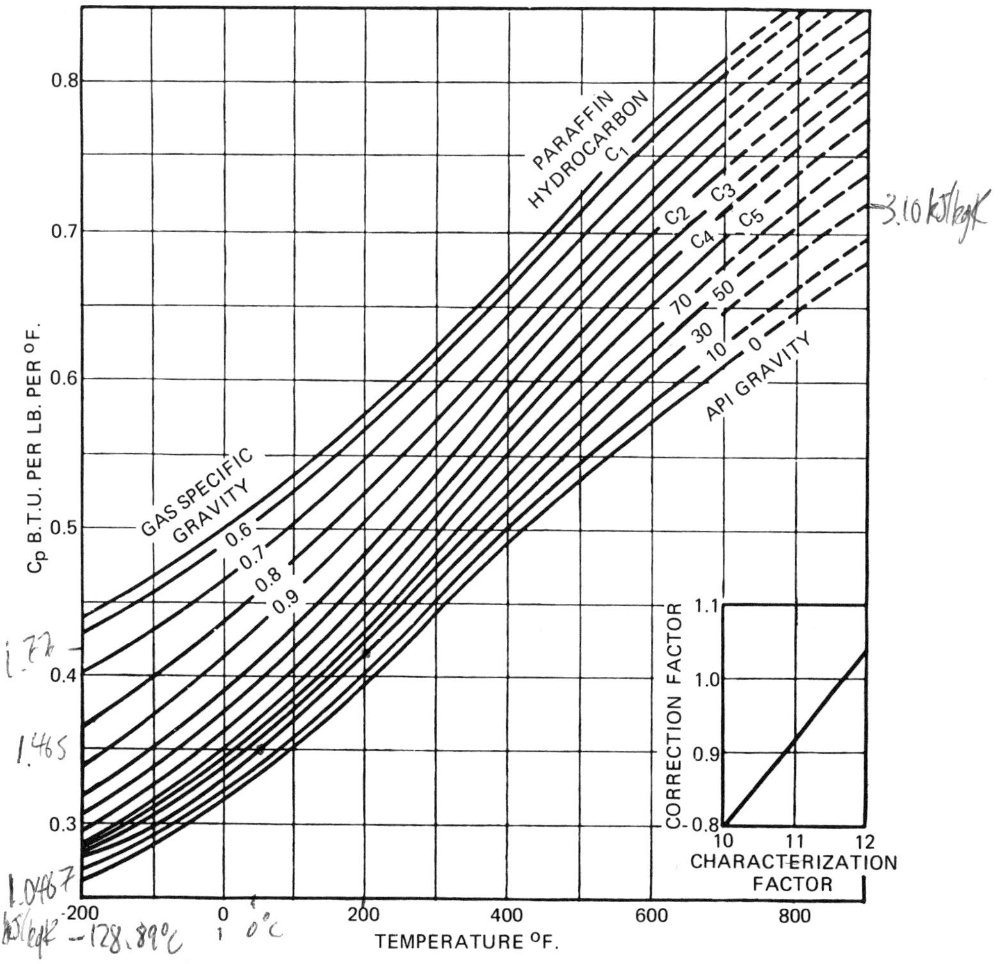

Figure 17.9. Specific heats of midcontinent oil vapors with correction factor for other bases of oils.[8] Reprinted with permission from D. E. Holcomb and G. G. Brown, *Ind. Eng. Chem.*, **34,** 590. Copyright 1942 American Chemical Society.

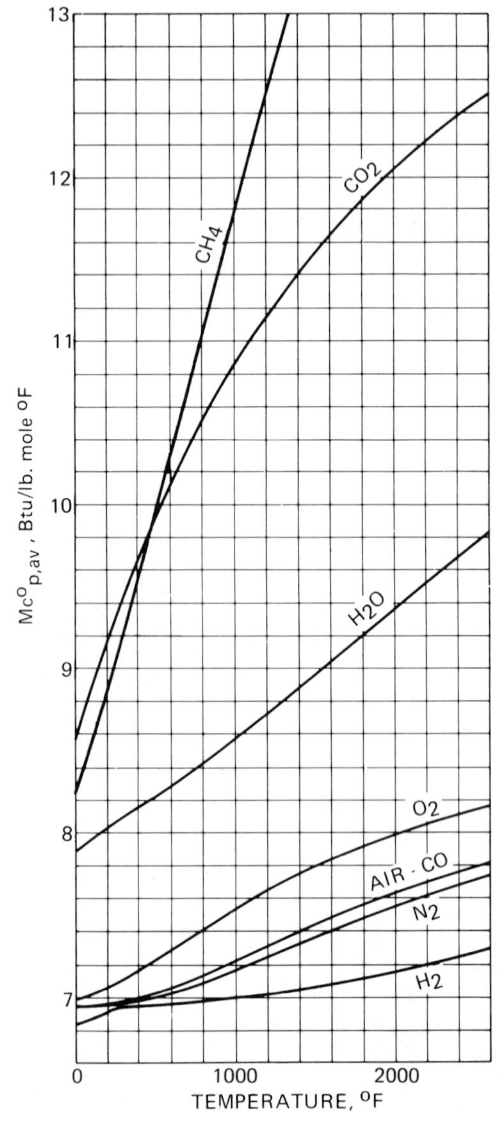

Figure 17.10. Average molal heat capacity of gases at zero pressure $Mc^{\circ}_{p,\mathrm{av}}$, between 60°F and abscissa temperature.[9] © 1954 McGraw-Hill.

be applied to correct values read for various degrees of paraffinicity. This factor is a function of the characterization factor, defined by

$$\text{characterization factor} = \sqrt[3]{T_B/S}$$

where T_B = average boiling point (°R)
S = specific gravity at 60°F

Specific Heats of Petroleum Vapors and Other Gases

Figure 17.9 gives specific heats of midcontinent (U.S.) oil vapors and a correction factor plot as a function of the characterization factor.

Figure 17.10 gives the *average* molal heat capacity of various gases at low pressure between 60°F and the abscissa temperature. Ordinate values must be divided by M, the molecular weight of the gas, to obtain the specific heat.

SPECIFIC HEAT AND THERMAL CONDUCTIVITY OF FLUID–SATURATED RESERVOIR ROCK

Variation of Rock Specific Heat (c_r) with Temperature

The specific heats of reservoir rocks increase with increasing temperature, as indicated in Figures 17.11 and 17.12.

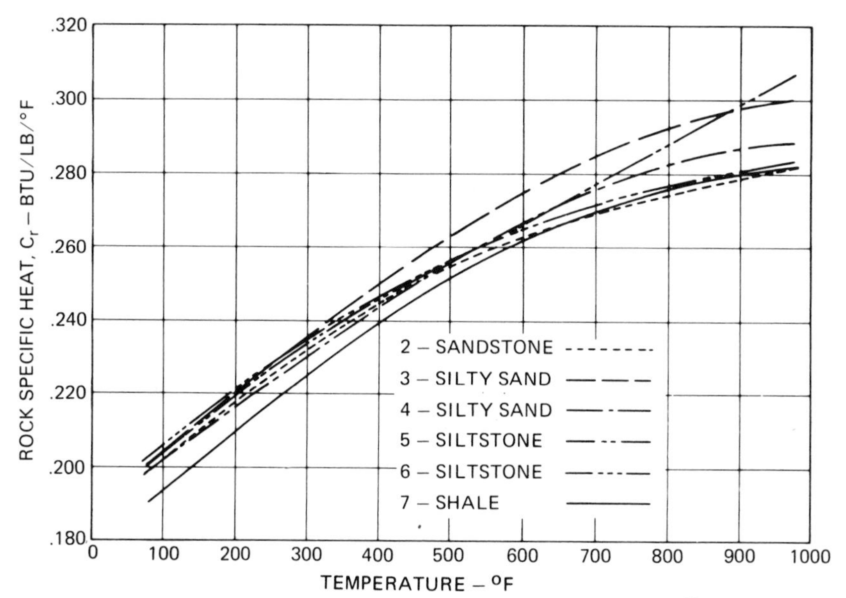

Figure 17.11. Experimental specific heats of dry reservoir rocks.[10] © 1958 SPE-AIME.

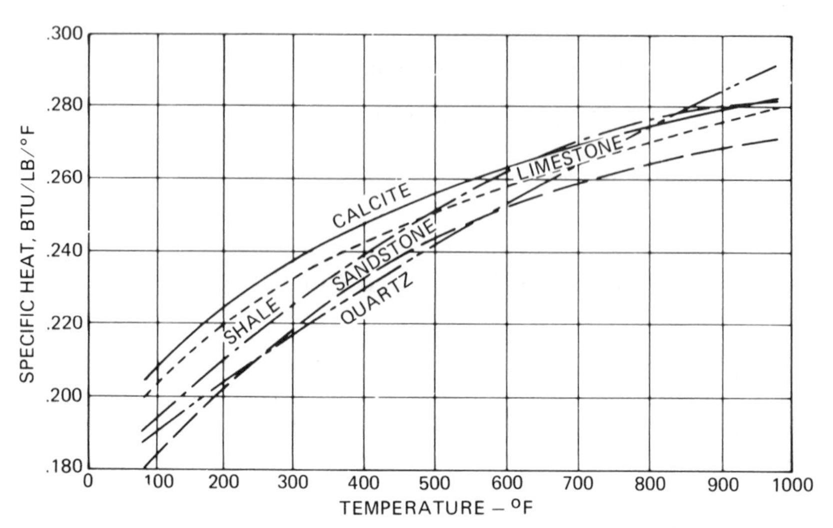

Figure 17.12. Experimental specific heats of various minerals[10] © 1958 SPE-AIME.

Volumetric Heat Capacities of Fluid-Saturated Rocks

Volumetric heat capacities of fluid-saturated rocks, $(\rho c)_{R+F}$, are calculated by

$$(\rho c)_{R+F} = \phi[S_g(\rho c)_g + S_w(\rho c)_w + S_o(\rho c)_o] + (1 - \phi)(\rho c)_r$$

where ϕ = rock porosity (fraction)
S = phase saturation (fraction)
ρc = product of density (lb/ft^3) and specific heat (Btu/lb-°F)

The subscripts g, w, o, and r refer to gas, water, oil, and reservoir rock, respectively.

Examples of calculated volumetric heat capacities of gas- and water-saturated rocks at 620°F are given in Table 17.3. Rock properties are tabulated in Table 17.4.

Thermal Conductivities and Thermal Diffusivities of Saturated Reservoir Rocks

Thermal conductivities and diffusivities $[K_h/(\rho c)_{R+F}]$ of the 10 samples of Table 17.4 are given in Table 17.5.

An equation developed by Krupiczka[23] gives the thermal conductivity of

Table 17.3 Calculated Volumetric Heat Capacities of Fluid-Saturated Rocks[a]

Sample	14.7 psia			500 psia		1500 psia		3000 psia	
	Dry	Methane	Water	Methane	Water	Methane	Water	Methane	Water
1. Sandstone	34.0	34.0	34.0	34.1	34.1	34.3	34.9	34.6	45.9
2. Sandstone	32.9	32.9	32.9	33.0	33.0	33.3	34.0	33.7	47.6
3. Silty sand	35.6	35.6	35.6	35.7	35.7	36.0	36.5	36.3	49.1
4. Silty sand	33.5	33.5	33.5	33.5	33.6	33.9	34.5	34.3	48.3
5. Siltstone	32.0	32.0	32.0	32.0	32.1	32.5	33.2	32.9	48.7
6. Siltstone	33.6	33.6	33.6	33.7	33.7	34.0	34.6	34.3	47.7
7. Shale	39.6	39.6	39.6	39.6	39.6	39.7	40.0	39.9	40.1
8. Limestone	35.4	35.4	35.4	35.5	35.5	35.7	36.2	36.0	46.8

[a]Heat capacity at 620°F and pressure indicated, Btu/ft^3-°F. From Ref. 10. © 1958 SPE-AIME.

Table 17.4 Description of Test Samples[a]

Sample	Description	Porosity	Principal Minerals		
			Quartz (%)	Clay Mineral	Other
1. Sandstone	Well consolidated, medium-coarse grain	0.196	80	Tr. Kaolinite	Tr. Pyrite
2. Sandstone	Poorly consolidated, medium-fine grain	0.273	40	Illite (?)	Feldspars
3. Silty sand	Poorly consolidated, poorly sorted	0.207	20	Kaolinite type	Feldspars
4. Silty sand	Medium hard, poorly sorted	0.225	20	Kaolinite type	Feldspars
5. Siltstone	Medium hard, broken	0.296	20	Kaolinite type	Feldspars
6. Siltstone	Hard	0.199	25	Illite	Feldspars
7. Shale	Hard, laminated	0.071	40	Illite–Kaolinite	—
8. Limestone	Granular, uniform texture	0.186	—	—	Calcium carbonate
9. Sand	Unconsolidated, fine-grained	0.38	100	—	—
10. Sand	Unconsolidated, coarse-grained	0.34	100	—	—

[a]From Ref. 10. © 1958 SPE-AIME.

Table 17.5 Calculated Thermal Diffusivities[a]

Sample	Bulk Density (lb/ft^3)		Heat Capacity (Btu/lb-°F)		Thermal Conductivity (Btu/hr-ft-°F)		Thermal Diffusivity (ft^2/hr)	
	Air	Water	Air	Water	Air	Water	Air	Water
1. Sandstone	130	142	0.183	0.252	0.507	1.592	0.0213	0.0445
2. Sandstone	90	115	0.200	0.374	0.285	1.050	0.0158	0.0244
(original)	109	126	0.200	0.308	(0.34)	(1.32)	(0.0156)	(0.0340)
3. Silty sand	90	115	0.202	0.376	0.285	(1.05)	0.0157	(0.0242)
(original)	119	132	0.202	0.288	(0.40)	(1.50)	(0.0167)	(0.0394)
4. Silty sand	85	112	0.200	0.393	0.261	1.110	0.0153	0.0249
(original)	116	130	0.200	0.286	(0.47)	(1.51)	(0.0202)	(0.0406)
5. Siltstone	96	118	0.202	0.351	0.338	1.036	0.0174	0.0250
(original)	105	123	0.202	0.318	(0.44)	(1.29)	(0.0207)	(0.0329)
6. Siltstone	120	132	0.204	0.276	0.396	(1.51)	0.0162	(0.0414)
7. Shale	145	149	0.192	0.213	0.603	0.975	0.0216	0.0307
8. Limestone	137	149	0.202	0.266	0.983	2.050	0.0355	0.0517
9. Sand (fine)	102	126	0.183	0.339	0.362	1.590	0.0194	0.0372
10. Sand (coarse)	109	130	0.183	0.315	0.322	1.775	0.0161	0.0433

[a]Values in parentheses estimated. From Ref. 10. © 1958 SPE-AIME.

the porous sand system, K_h, as a function of porosity and the thermal conductivity of the interstitial fluid.

$$K_h/K_{hf} = (K_{hs}/K_{hf})A + B \log K_{hs}/K_{hf}$$

where

K_h = effective thermal conductivity of the porous media, Btu/hr-ft-°F
K_{hf} = effective thermal conductivity of the fluid, Btu/hr-ft-°F
K_{hs} = effective thermal conductivity of the solid, Btu/hr-ft-°F
$A = 0.280–0.757 \log \phi$
ϕ = fractional porosity
$B = -0.057$

In testing this equation against 165 experimental data points, Somerton[24] reports that 76 percent of the calculated values agree within 30%. Comparison of this equation to California oil sand data is shown in Figure 17.13 using a value of $K_{hs} = 2.75$ Btu/hr-ft-°F.

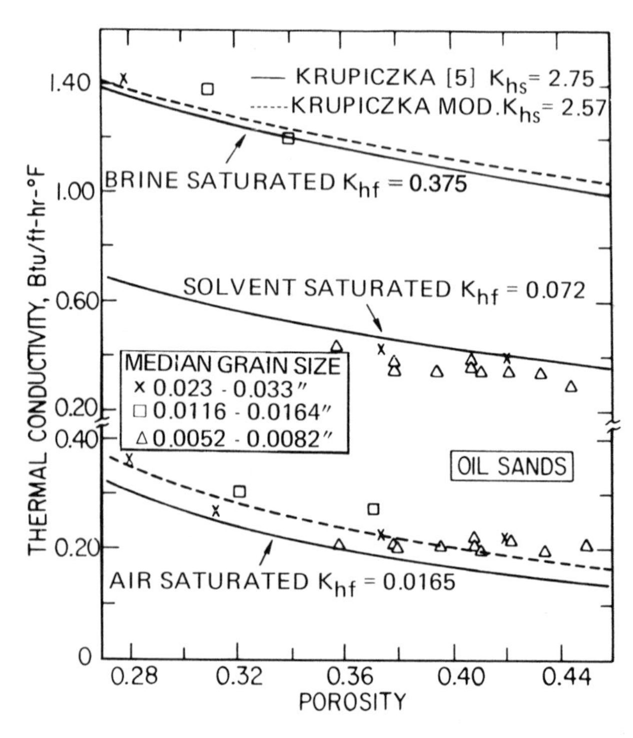

Figure 17.13. Thermal conductivity of oil sands [24] © 1974 SPE-AIME.

Somerton[24] reports a best fit of Kern River, California oil sand data shown in Figure 17.14 is obtained by the equation:

$$K_h = 0.735 - 1.30\phi + S_w^{0.5}$$

This equation may be used for oil sands having low quartz content and porosities between 0.28 and 0.37. For sands with higher quartz content, a correlation including the conductivity of the solid phase is needed and the above equation is modified to:

$$K_h = 0.735 - 1.30\phi + 0.39K_{hs}S_w^{0.5}$$

The solids conductivity can be estimated by

$$K_{hs} = 4.45Q + 1.65(1 - Q)$$

where Q is the fractional quartz content.

For sandstones saturated with water and gas, Tiknomirov determined the following empirical relationship for the thermal conductivity from a statistical analysis of 274 data points[11]:

$$K_h = 6.360[\exp(0.6(\rho + S_w))]T^{-0.55}$$

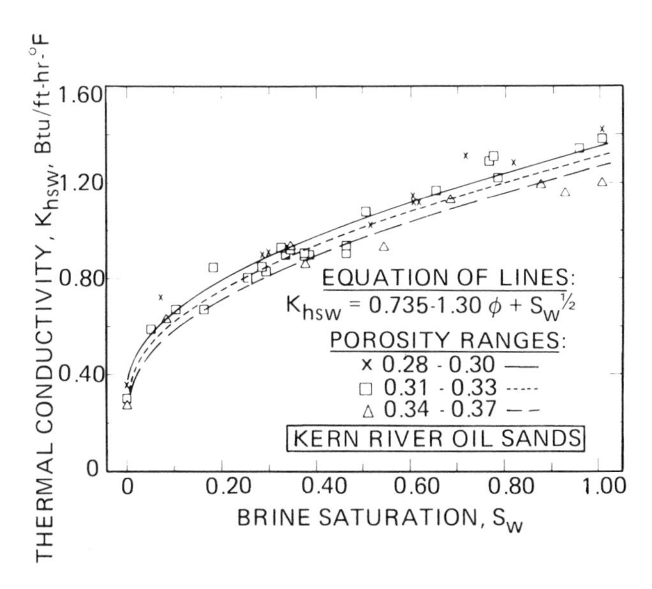

Figure 17.14. Thermal conductivity of Kern River oil sands [24]© 1974 SPE AIME.

where T = temperature (°K)

ρ = dry rock density (g/cm^3)

S_w = water saturation (fraction)

Results of this equation plotted in Figures 17.15 and 17.16 for porosities of 20 and 40% show a decrease in conductivity with increasing temperature and decreasing water saturation.

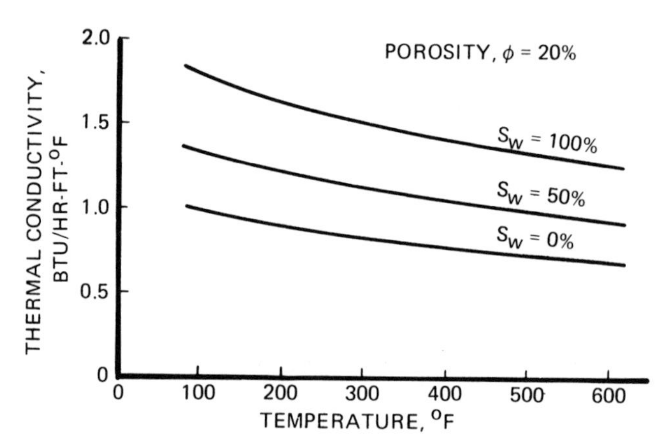

Figure 17.15. Prediction of thermal conductivity by Tiknomirov's equation for $\phi = 20\%$[11] © 1968 *Neftyanoe Khozaistov.*

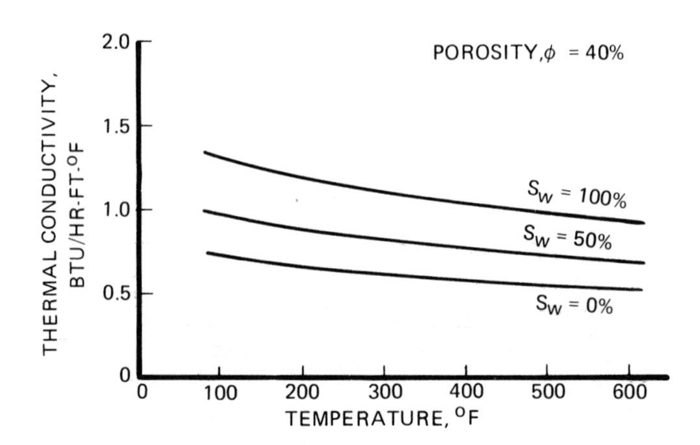

Figure 17.16. Prediction of thermal conductivity by Tiknomirov's equation for $\phi = 40\%$.[11] © 1968 *Neftyanoe Khozaistov.*

THERMAL PROPERTIES OF OTHER MATERIALS

Thermal Conductivities and Emissivities of Various Materials

Tables 17.6 and 17.7 give conductivities K_h and emissivities ϵ of various materials.

Thermal Properties of Insulating Materials

Thermal properties of conventional materials used for insulating pipe are given in Table 17.8.

Table 17.6 Thermal Conductivities of Various Materials[a]

	32°F (0°C)	212°F (100°C)	392°F (200°C)	572°F (300°C)	752°F (400°C)
Aluminum	117	119	124	133	144
Stainless steel[b]	—	9.4	10.2	10.9	11.7
Mild steel	—	26	26	25	23
Cement, dry	—		0.2 – 0.4		
Cement, wet	—		0.5 – 0.6		

[a]From Ref. 12. © 1954 McGraw-Hill.
[b]304, 316 Stainless Steel.

Table 17.7 Emissivities of Various Surfaces[a]

Iron	
Rusted	0.69
Dark grey	0.31
Steel	
Oxidized	0.74
Rough oxide layer	0.80
Rough plate	0.94–0.97
Stainless steel	
316	0.56
304	0.36–0.44
Aluminum, rough polish	0.18

[a]From Ref. 12. © 1954 McGraw-Hill.

Table 17.8 Thermal Conductivities of Insulating Materials at High Temperatures[a]

Material	Maximum Temperature (°F)	Mean Temperature (°F)									
		100	200	300	400	500	600	800	1000	1500	2000
Laminated asbestos felt (~40 laminations/in.)	700	0.033	0.037	0.040	0.044	0.048					
Laminated asbestos felt (~20 laminations/in.)	500	0.045	0.050	0.055	0.060	0.065					
Corrugated asbestos (4 plies/in)	300	0.050	0.058	0.069							
85% magnesia (13 lb/ft^3)	600	0.034	0.036	0.038	0.040						
Diatomaceous earth, asbestos and bonding material	1600	0.045	0.047	0.049	0.050	0.053	0.055	0.060	0.065		
Diatomaceous earth brick	1600	0.054	0.056	0.058	0.060	0.063	0.065	0.069	0.073		
Diatomaceous earth brick	2000	0.127	0.130	0.133	0.137	0.140	0.143	0.150	0.158	0.176	
Diatomaceous earth brick	2500	0.128	0.131	0.135	0.139	0.143	0.148	0.155	0.163	0.183	0.203
Diatomaceous earth powder (density, 18 lb/ft^3)	—	0.039	0.042	0.044	0.048	0.051	0.054	0.061	0.068		
Rock wool	—	0.030	0.034	0.039	0.044	0.050	0.057				

[a] From Ref. 13. [K = Btu/hr-ft-°F].

Table 17.9 Range of Physical Properties[a] of Sodium Silicate Foam

Density, lb/ft^3	5–15
Tensile strength, psi	30–80
Compressive strength, psi	30–100
Maximum service temperature, °F	1600
Thermal conductivity, Btu/ft-hr-°F	0.0167

[a]From Ref. 15.

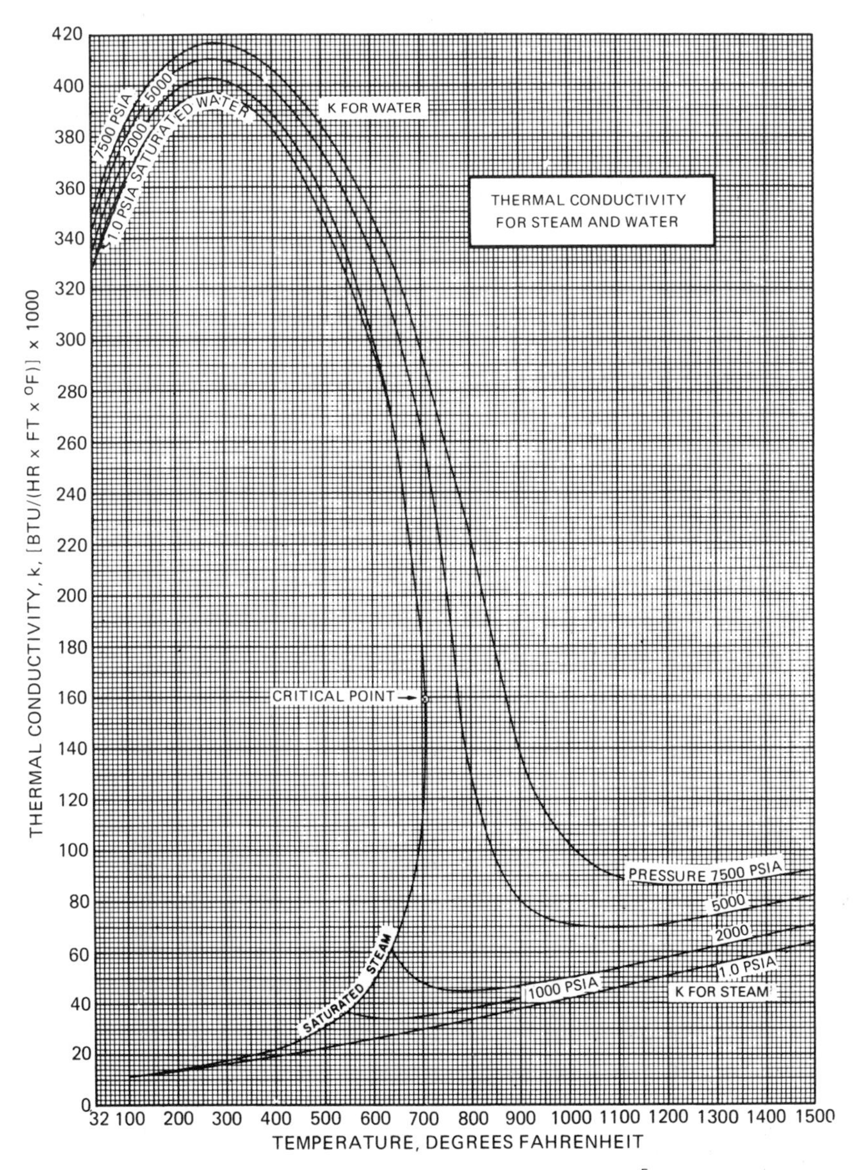

Figure 17.17. Thermal conductivity for steam and water.[5] © 1967 ASME.

Properties of Silicate Foam Wellbore Insulation

The properties of foamed sodium silicate formed during boil-off of an annulus loaded with sodium silicate solution when injecting steam down tubing vary, depending on the temperature of the injected steam and other factors.[14] Table 17.9 gives approximate properties of the insulation coating left on the tubing wall after boil-off. The thickness of the coating is normally $\frac{1}{4}-\frac{1}{2}$ in. thick.

Thermal properties of Steam, Water, and Air

Figure 17.17 gives the thermal conductivity of steam and water over a wide range of temperature and pressures. The data are useful in estimating the contribution of conduction in the annulus in wellbore heat loss calculations. After annulus boil-off when steam is injected, the annulus may be assumed to contain initially steam. Later, however, air will replace the steam in the annulus if it is vented to the atmosphere. Table 17.10 gives the physical properties of air at atmospheric pressure and various temperatures. The

Table 17.10 Some Physical Properties of Air[a]

Temperature (°F)	c_p (Btu/lb-°F)	μ (lb/ft-hr)	K (Btu/hr-ft-°F)	N_{Pr}	$Y = \frac{\rho^2 \beta g c_p}{\mu K}$, (1/ft^3-°F)
−300	0.2393	0.0153	0.00478	0.77	514×10^6
−200	0.2393	0.0242	0.00763	0.76	47.3×10^6
−100	0.2393	0.0322	0.0104	0.74	9.82×10^6
−50	0.2394	0.0360	0.0118	0.73	5.25×10^6
32	0.2396	0.0417	0.0140	0.72	2.21×10^6
100	0.2401	0.0460	0.0157	0.71	1.21×10^6
200	0.2412	0.0521	0.0181	0.69	0.570×10^6
300	0.2421	0.0578	0.0204	0.69	0.300×10^6
400	0.2446	0.0630	0.0225	0.68	0.174×10^6
500	0.2474	0.0679	0.0246	0.68	0.107×10^6
600	0.2504	0.0726	0.0266	0.68	6.98×10^4
700	0.2535	0.0770	0.0285	0.68	4.74×10^4
800	0.2566	0.0813	0.0303	0.69	3.34×10^4
900	0.2598	0.0854	0.0320	0.69	2.42×10^4
1000	0.2630	0.0892	0.0337	0.69	1.80×10^4
1200	0.2687	0.0967	0.0369	0.70	1.05×10^4

[a]From Ref. 6.

quantity $Y = \rho^2 \beta g c_p / \mu K$ is used in evaluating the effect of convection in the annulus. See Chapter 2 (Appendix).

VISCOSITY DATA FOR STEAM, WATER, VARIOUS GASES, AND PETROLEUM OILS

Steam–Water Viscosities

Figure 17.18, reproduced from the 1967 ASME steam tables,[5] gives the viscosity μ of steam and water at various temperatures and pressures. To convert from lbf-sec/ft^2 to centipoise, multiply by 4.788×10^4. Figure 17.19 gives the kinematic viscosity μ/ρ of steam and hot water in square feet per second.

Viscosities of Gases

Viscosities of gases at 1 atm may be estimated by the alignment chart (see Fig. 17.20 and Table 17.11).[16]

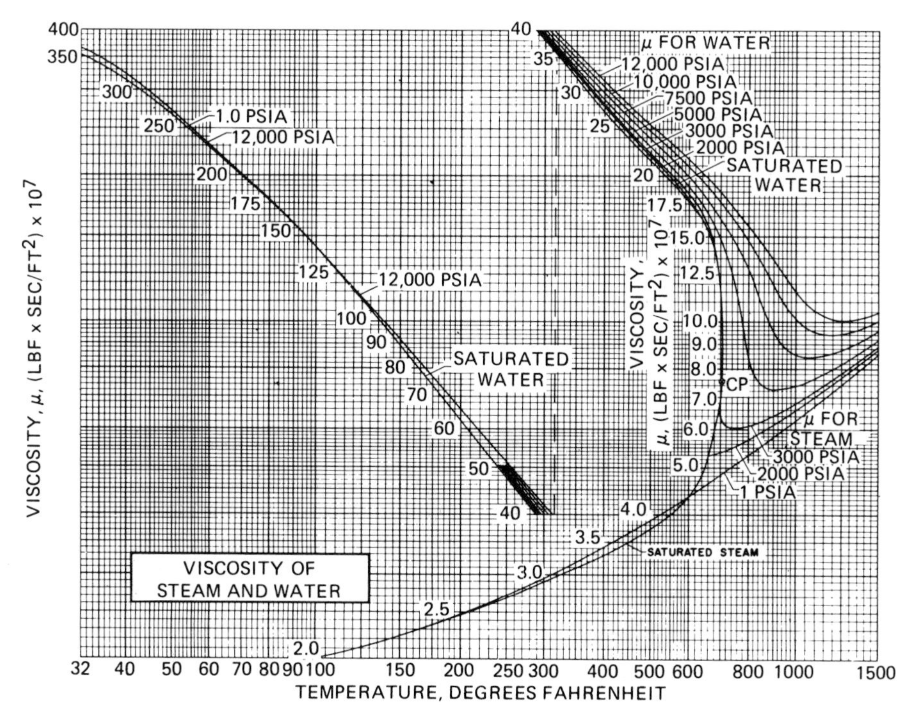

Figure 17.18. Viscosity of steam and water.[5] © 1967 ASME.

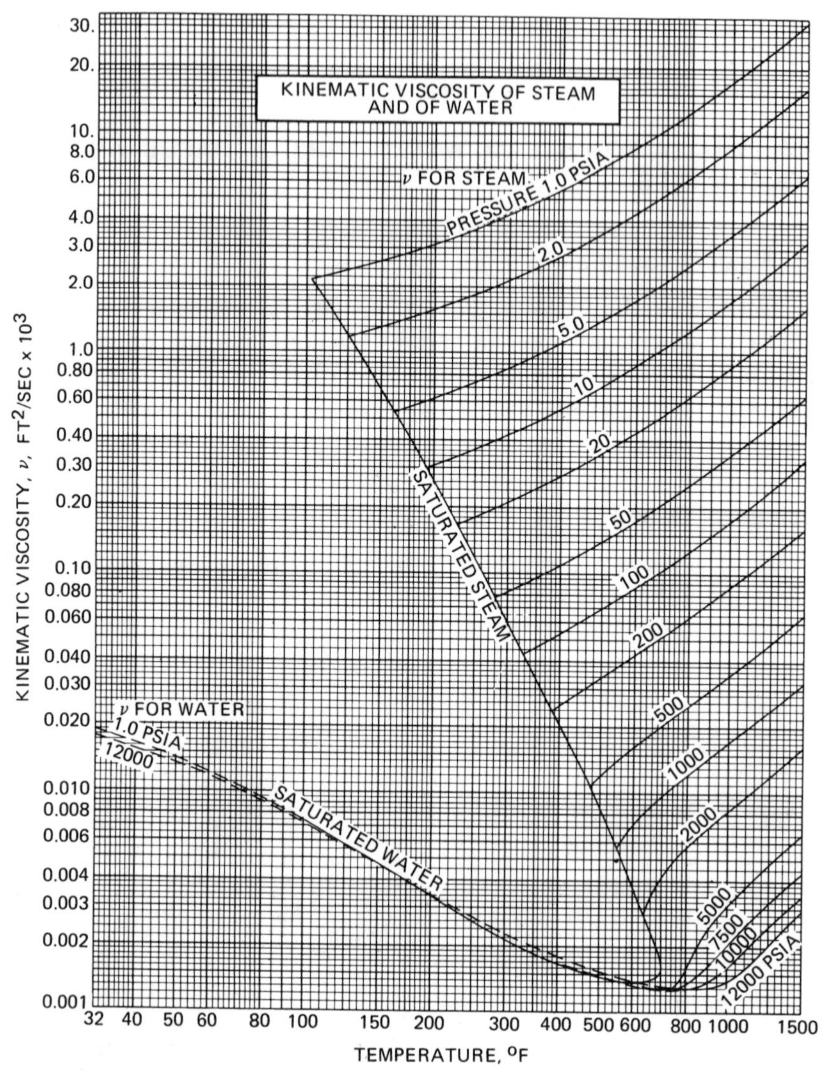

Figure 17.19. Kinematic viscosity of steam and water.[5] © 1967 ASME.

The effect of higher pressures on gas viscosities can be estimated using Figure 17.21.[17]

Viscosities of Petroleum Oils

The effect of temperature on kinematic viscosity of various crudes and distillates is given in Figure 17.22.[18]

Data for selected Canadian and Venezuelan crudes of potential thermal recovery interest are given in Figures 17.23–17.29.

Table 17.11 Coordinates for Viscosities of Gases for Use with Fig. 17.20[a]

Number	Gas	X	Y	Number	Gas	X	Y
1	Acetic acid	7.7	14.3	29	Freon-113	11.3	14.0
2	Acetone	8.9	13.0	30	Helium	10.9	20.5
3	Acetylene	9.8	14.9	31	Hexane	8.6	11.8
4	Air	11.0	20.0	32	Hydrogen	11.2	12.4
5	Ammonia	8.4	16.0	33	$3H_2 + 1N_2$	11.2	17.2
6	Argon	10.5	22.4	34	Hydrogen bromide	8.8	20.9
7	Benzene	8.5	13.2	35	Hydrogen chloride	8.8	18.7
8	Bromine	8.9	19.2	36	Hydrogen cyanide	9.8	14.9
9	Butene	9.2	13.7	37	Hydrogen iodide	9.0	21.3
10	Butylene	8.9	13.0	38	Hydrogen sulfide	8.6	18.0
11	Carbon dioxide	9.5	18.7	39	Iodine	9.0	18.4
12	Carbon disulfide	8.0	16.0	40	Mercury	5.3	22.9
13	Carbon monoxide	11.0	20.0	41	Methane	9.9	15.5
14	Chlorine	9.0	18.4	42	Methyl alcohol	8.5	15.6
15	Chloroform	8.9	15.7	43	Nitric oxide	10.9	20.5
16	Cyanogen	9.2	15.2	44	Nitrogen	10.6	20.0
17	Cyclohexane	9.2	12.0	45	Nitrosyl chloride	8.0	17.6
18	Ethane	9.1	14.5	46	Nitrous oxide	8.8	19.0
19	Ethyl acetate	8.5	13.2	47	Oxygen	11.0	21.3
20	Ethyl alcohol	9.2	14.2	48	Pentane	7.0	12.8
21	Ethyl chloride	8.5	15.6	49	Propane	9.7	12.9
22	Ethyl ether	8.9	13.0	50	Propyl alcohol	8.4	13.4
23	Ethylene	9.5	15.1	51	Propylene	9.0	13.8
24	Fluorine	7.3	23.8	52	Sulfur dioxide	9.6	17.0
25	Freon-11	10.6	15.1	53	Toluene	8.6	12.4
26	Freon-12	11.1	16.0	54	2,3,3-trimethylbutane	9.5	10.5
27	Freon-21	10.8	15.3	55	Water	8.0	16.0
28	Freon-22	10.1	17.0	56	Xenon	9.3	23.0

[a]From Ref. 16. © 1950 McGraw-Hill.

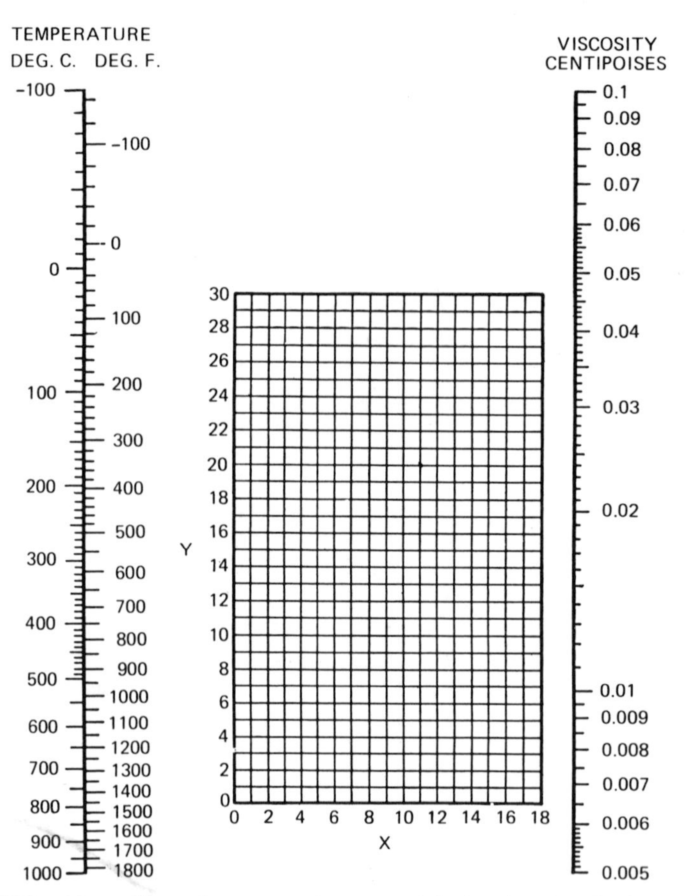

Figure 17.20. Viscosities of gases and vapors at 1 atm; for coordinates, see Table 17.11.[16] © 1950 McGraw-Hill.

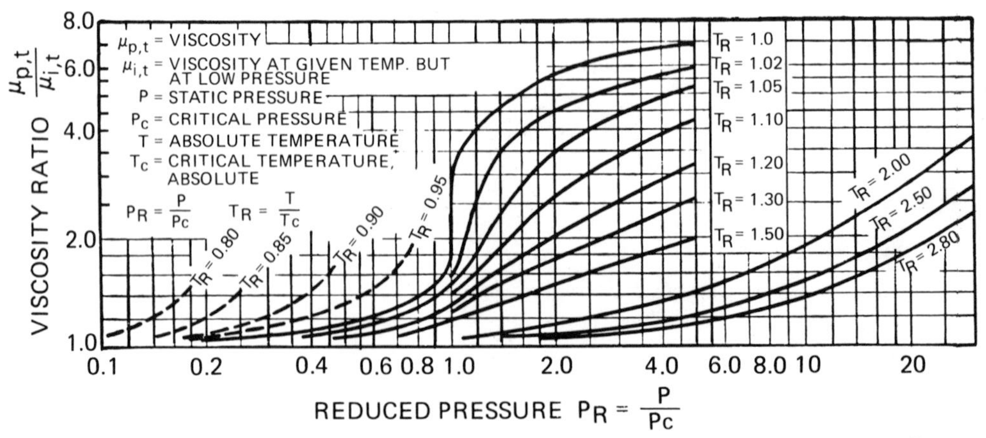

Figure 17.21. Effect of pressure on viscosities of gases at several temperatures.[17] Reprinted with permission from E. W. Comings and R. S. Egly, *Ind. Eng. Chem.*, **32,** 714–718. Copyright 1940 American Chemical Society.

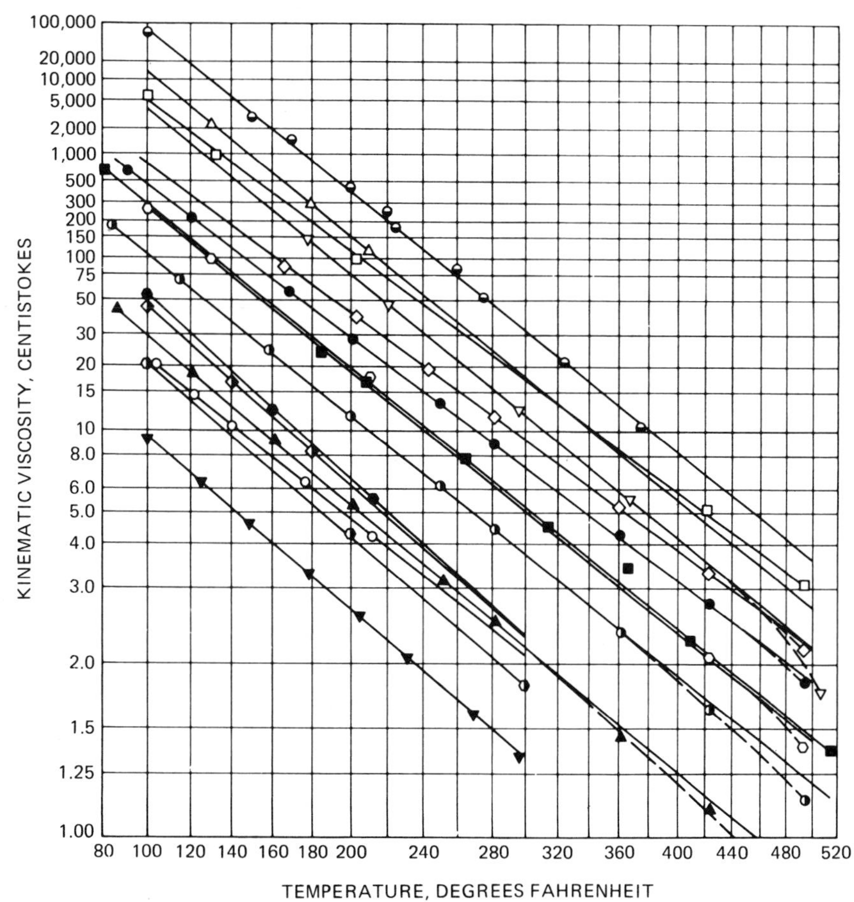

LEGEND

1. ◒ - CALIFORNIA CRUDE (9.9°API)
2. △ - COLOMBIAN CRUDE (10.9°API)
3. □ - MIDCONTINENT RESIDUUM (14.5°API)
4. ▽ - CALIFORNIA CRUDE (13.4°API)
5. ◇ - MIDCONTINENT CYLINDER STOCK (22.9°API)
6. ● - MIDCONTINENT HEAVY MOTOR OIL (23.5°API)
7. ■ - GULF COAST CRUDE (15.1°API)
8. ○ - NORTH LOUISIANA HEAVY CRUDE (18.3°API)
9. ◑ - MIDCONTINENT RED OIL (24.0°API)
10. ⬢ - SOUTH TEXAS CRUDE (17.8°API)
11. ◈ - GULF COAST CRUDE (20.5°API)
12. ▲ - MIDCONTINENT LIGHT PARAFFIN OIL (26.7°API)
13. ○ - PENNSYLVANIA VACUUM DISTILLATE (32.7°API)
14. ◐ - WYOMING CRUDE (21.4°API)
15. ▼ - MIDCONTINENT PRESSED DISTILLATE (29.2°API)

Figure 17.22. Viscosity versus temperature data for various petroleum oils.[18] © 1966 SPE-AIME.

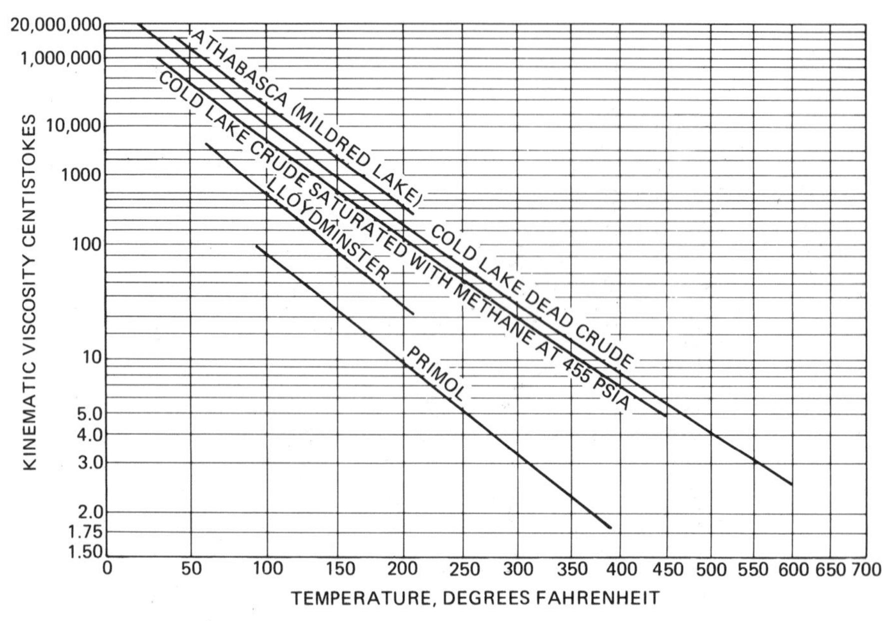

Figure 17.23. Viscosity of various Canadian heavy crude oils.[19]

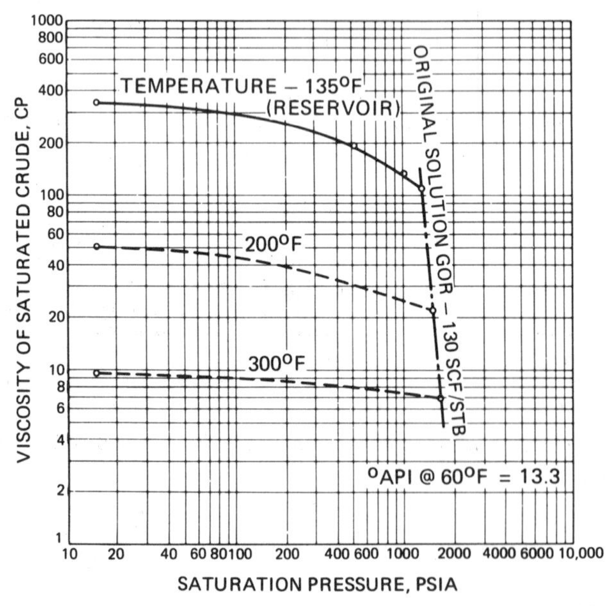

Figure 17.24. Viscosity of gas-saturated Pilon crude.[20]

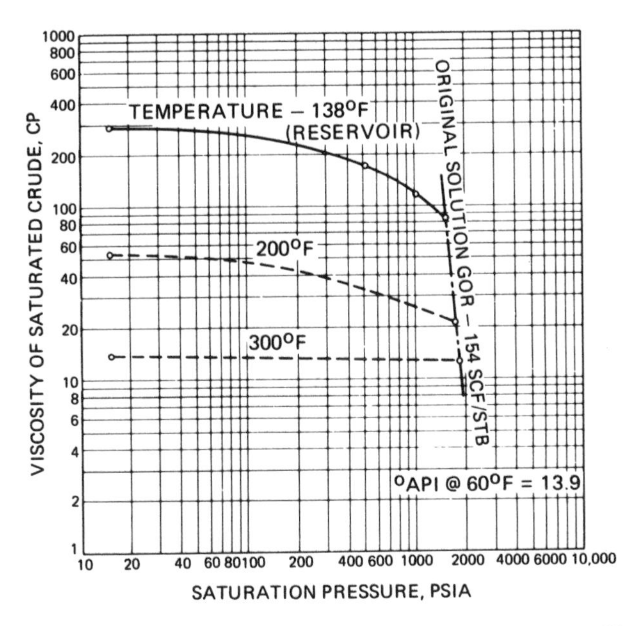

Figure 17.25. Viscosity of gas-saturated Jobo I crude.[20]

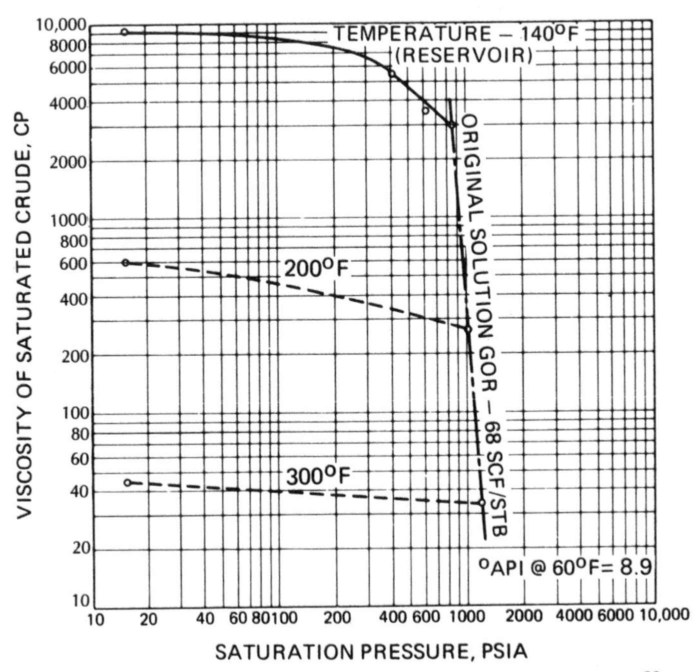

Figure 17.26. Viscosity of gas-saturated Jobo II crude.[20]

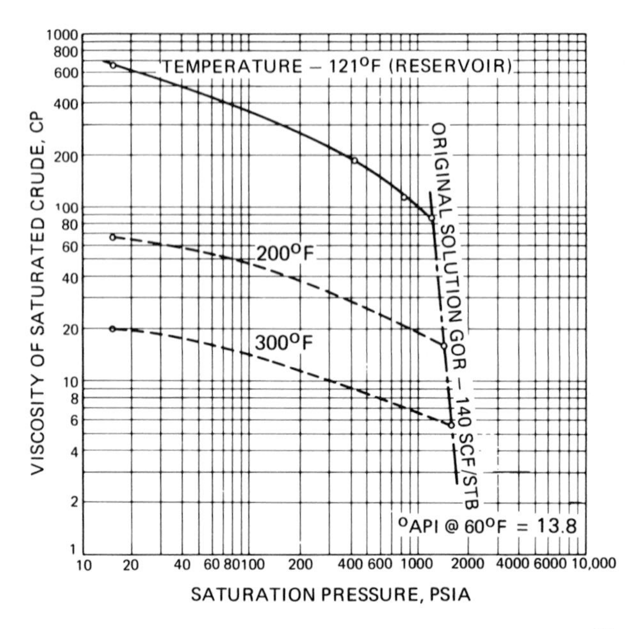

Figure 17.27. Viscosity of gas-saturated LL-4 crude.[20]

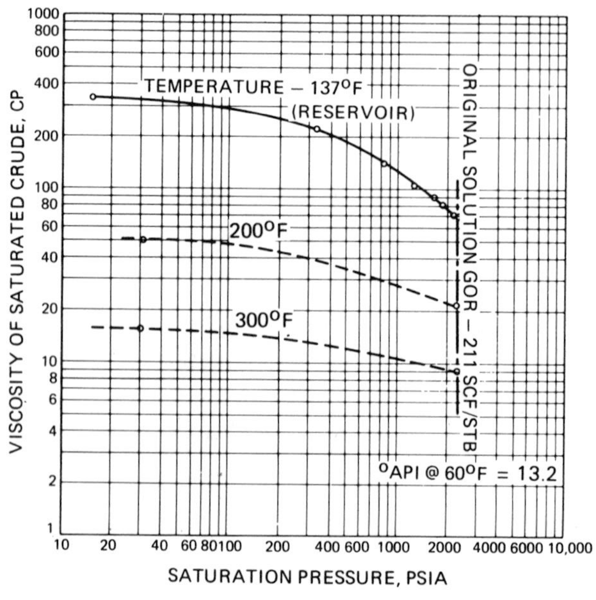

Figure 17.28. Viscosity of gas-saturated B-2 crude.[20]

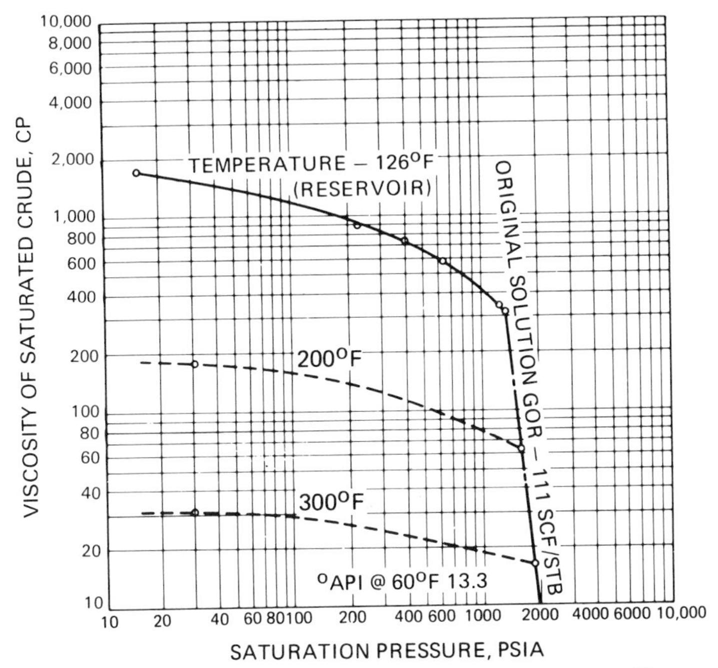

Figure 17.29. Viscosity of gas-saturated B-1 crude.[20]

HEATS OF COMBUSTION AND THERMAL CRACKING KINETIC DATA FOR PETROLEUM HYDROCARBONS

Heat of Combustion

The gross heat of combustion (higher heating value) for various oils as a function of API gravity and characterization factor (see p. 365) is given in Figure 17.30. The higher heating value is used if it can be assumed that water formed in the combustion reaction is condensed. The lower heating value is normally used in determining available energy to generate steam in a steam generator as these are normally designed so that flue gases exit above the dew point to avoid corrosion.

The heat of combustion of hydrocarbon gas as a function of molecular weight (29 times gas gravity) is given in Figure 17.31. The molecular weight of pure methane is 16.

Thermal Cracking Data for Petroleum Hydrocarbons

Table 17.12 gives kinetic data to estimate thermal cracking rates for several crudes. The percentage of conversion is calculated by:

$$\text{Percent conversion} = \frac{C_o - C}{C} \times 100$$

where

$$\ln \frac{C_o}{C} = kt$$

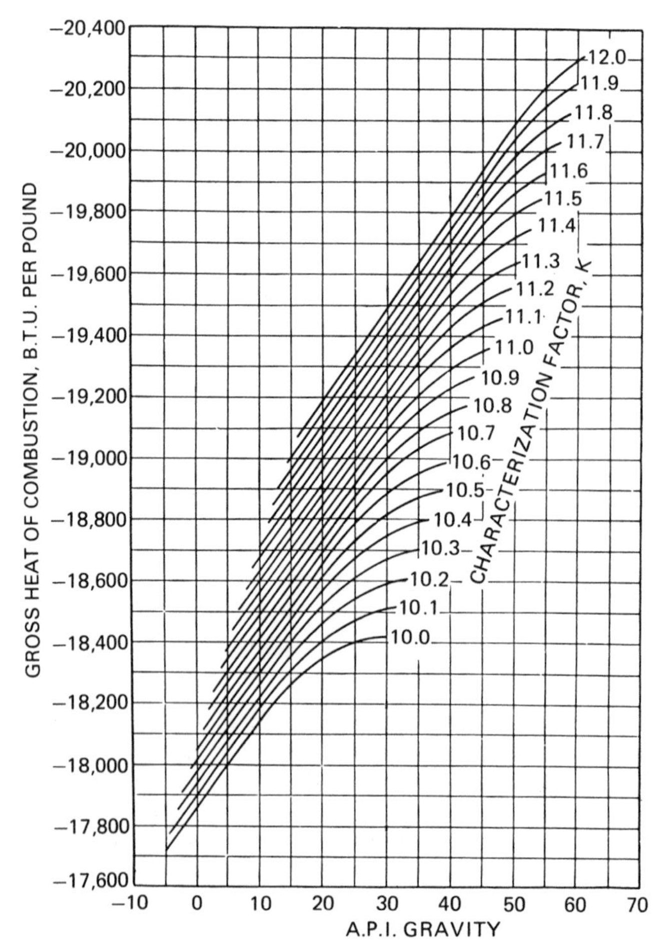

Figure 17.30. Gross heats of combustion of liquid petroleum hydrocarbons.[7] © 1949 McGraw-Hill.

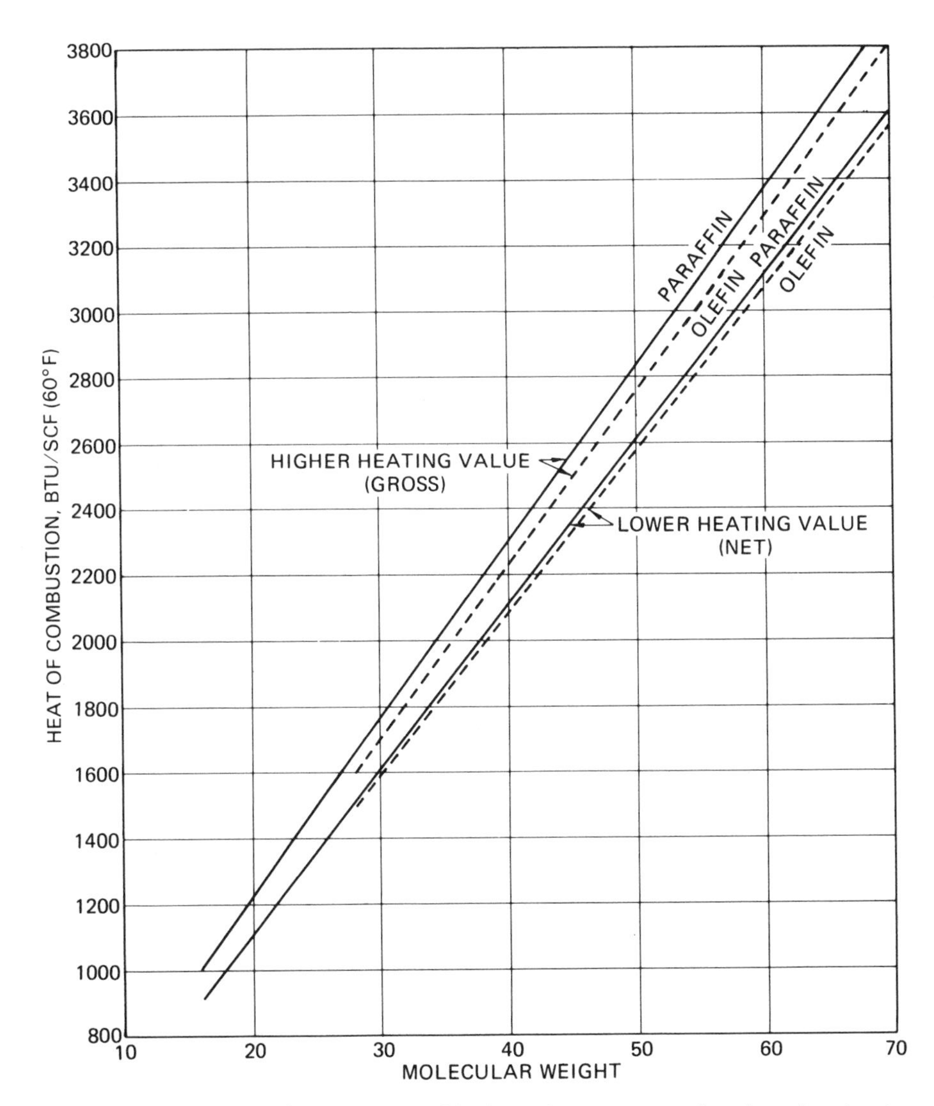

Figure 17.31. Heats of combustion of hydrocarbon gases as function of molecular weight.[21]

and C_o = weight fraction of oil present initially

C = weight fraction of oil remaining at time t

t = time (sec)

Also,

k = specific rate constant (sec^{-1}),

$= Ae^{E/RT}$

Table 17.12 Cracking Kinetic Properties of Several Crude Oils[a]

Oil	Gravity (°API)	Viscosity (cp)		Pour Point (°F)	Kinetic Parameters for Upgrading		Remarks
		150°F	210°F		Activation Energy E (kcal/mol)	$\log_{10} A$	
Athabasca	6.9	5242	457	—	49.0	11.53	Oil removed from tar sand by hot centrifuging
Crude B	15.2	190	43	—	54.0	12.92	
Crude C	15.5	105	28	15	49.0	12.13	
Crude D	16.8	84.5	23.1	−10	58.3	14.28	
Crude E	9.4	2202	219	75	43.6	9.86	
Crude F	24.5	13.9	5.8	90	62.2	15.46	Highly paraffinic crude
Crude G	12.4	483	98.6	—	47.5	11.07	

[a]From Ref. 22. © 1965 *J. Can. Pet. Tech.*, CIM.

Table 17.13 Typical Times and Temperatures Required To Upgrade Crude Oils[a]

Oil	Degree Upgraded (% reduction in distillation residuum)	Time Required				
		400°F	500°F	600°F	700°F	800°F
Athabasca bitumen	15%(μ_o: 200 cp, 150°F)	450 yr	1.9 yr	8.8 days	0.26 days	—
Crude B	15%(μ_o: 14 cp, 150°F)	3000 yr	8.3 yr	24 days	0.46 days	0.39 hr
	30% (pour point: 47°F)	—	53.8 yr	87 days	0.96 days	0.57 hr
Crude F	40% (pour point: 16°F)	—	76.9 yr	123 days	14 days	0.81 hr

[a]From Ref. 22. © 1965 *J. Can. Pet. Tech.*, CIM.

where A = rate coefficient (sec^{-1})
E = activation energy (Btu/lb-mol = 1800 kcal/mol)
T = absolute temperature (°R)
R = gas constant = 1.987 Btu/lb-mol-°R

Values for A and E are given in Table 17.12.

Table 17.13 gives example percentage of conversion (degree upgraded) results for three of the crudes. While the viscosity and pour points of many heavy crudes can be reduced by mild thermal cracking in the 550–650°F range (attainable temperatures during high-pressure steamfloods and in situ burning projects), this beneficial effect may be offset by the undesirable production of coke and unsaturates.[22]

REFERENCES

1. Bleakley, W. B. *Oil Gas J.* (January 11, 1965), p. 75; (February 15, 1965), p. 121.
2. Keenan, J. H., and Keyes, F. G. *Thermodynamic Properties of Steam*, 1st ed., Wiley, New York, 1936.
3. Nelsen, W. L. "Determining Densities · · · at High Temperature," *Oil Gas J.* (January 27, 1938), p. 184.
4. Gewers, C. Imperial Oil Ltd., unpublished data, December 12, 1965.
5. Meyer, C. A., McClintock, R. B., Silvestri, G. J., and Spencer, R. C., Jr., 1967 *ASME Steam Tables*, 2nd ed., ASME, New York, 1967.
6. Keyes, F. G. Technical Report 37, Project Squid, April 1, 1952.
7. Nelson, W. L. *Petroleum Refinery Engineering*, 3rd ed., McGraw-Hill, New York, 1949.
8. Holcomb, D. E. and Brown, G. G. "Thermodynamic Properties of Light Hydrocarbons," *Ind. Eng. Chem.*, **34,** 590 (1942).
9. Lewis, W. K., Radasch, A. M., and Lewis, H. C. *Industrial Stoichiometry*, 2nd ed., McGraw-Hill, New York, 1954, p. 15.
10. Somerton, W. H. "Some Thermal Characteristics of Porous Rocks," *Trans. AIME*, **213,** 375 (1958).
11. Tiknomirov, V. M. "Thermal Conductivity of Rock Samples and Its Relation to Liquid Saturation, Density and Temperature," *Neftyanoe Khozaistov*, **46**(4) 36 (April 1968).
12. McAdams, W. H. *Heat Transmission*, 3rd ed. McGraw-Hill, New York, 1954.
13. Marks, L. S. *Mechanical Engineers' Handbook*, McGraw-Hill, New York, 1951.
14. Penberthy, W. L., Bayless, J. H., and Edwards, H. R. "Silicate Foam Wellbore Insulation," *J. Pet. Tech.* (June 1974), pp. 583–588.
15. Baker, E. J., Jr. "New Applications for Sodium Silicate," Southwest Research Institute Paper presented at Twelfth National SAMPE Symposium, 1968.

16. Perry, J. H. *Chemical Engineer's Handbook*, 3rd ed., McGraw-Hill, New York, 1950.
17. Comings, E. W., and Egly, R. S. *Ind. Eng. Chem.*, **32,** 714–718 (1940).
18. Braden, W. B. "A Viscosity–Temperature Correlation at Atmospheric Pressure for Gas-Free Oils," *J. Pet. Tech.* (November 1966), p. 1487.
19. Gewers, C. Imperial Oil Ltd., unpublished data, December 16, 1968.
20. Orkiszewski, J. "Temperature Effects on the Viscosities of Saturated Venezuelan Crude Oils," Exxon Production Research Co., unpublished data, July 23, 1965.
21. Esso Research and Engineering Co., unpublished data, July 1950.
22. Henderson, J. H., and Weber, L. "Physical Upgrading of Heavy Crude Oils by the Application of Heat," *J. Can. Pet. Tech.* (October–December 1965), p. 206.
23. Krupiczka, R. "Analysis of Thermal Conductivity in Granular Materials" *Intl. Chem. Eng.* (1967) vol. 7, no. 1, p. 122.
24. Somerton, W.H., Keese, J.A., and Chu, S.L. "Thermal Behavior of Unconsolidated Oil Sands" *Soc. Pet. Eng. J.* (October 1974), p. 513.

INDEX